CHEMICAL SENSORS IN OCEANOGRAPHY

Ocean Science and Technology
A series of books in oceanology

Volume 1
Chemical Sensors in Oceanography
Edited by Mark S. Varney

This book is part of a series. The publisher will accept continuation orders which may be cancelled at any time and which provide for automatic billing and shipping of each title in the series upon publication. Please write for details.

CHEMICAL SENSORS IN OCEANOGRAPHY

Edited by

Mark S. Varney

School of Ocean and Earth Sciences
University of Southampton
UK

GORDON AND BREACH SCIENCE PUBLISHERS
Australia • Canada • France • Germany • India • Japan • Luxembourg
Malaysia • The Netherlands • Russia • Singapore • Switzerland

Amsteldijk 166
1st Floor
1079 LH Amsterdam
The Netherlands

British Library Cataloguing in Publication Data

A catalogue record for this book is available from the British Library.

ISBN 90-5699-255-4
ISSN 1561-5928

CONTENTS

PREFACE

Seawater is a complex liquid, and the sea is a hostile environment.

The study of chemical species at sea is a challenge – fraught with many practical difficulties. Chemical studies yield a wealth of fundamental information on biological, physical and geochemical processes in the ocean. Marine chemistry is challenging in three important aspects: the vast area of the oceans, their dynamic behaviour, and the low concentrations of species found in seawater. Improvements in the gaps in our present knowledge will come from collecting more measurements both vertically and horizontally, over wider geographical areas, and throughout the seasons.

Measurements of ocean chemistry are often made by lowering water-sampling vessels over the side of an oceanographic ship and bringing the water back to an on-board laboratory for analysis. For 'casts' of greater than 4000 m, this often means that the ship remains on station for periods of longer than 8 hours while the hydrographic wire is first unwound, the water samplers (or instruments) deployed, and the wire raised back up. Added to this, there is often a maximum number of sampling vessels that can be attached to such a wire, the problems of the measurement of chemical species at sea are immediately apparent. In other words: the sea is physically too large to properly measure it all. The number of (discrete) measurements on a typical oceanographic cruise, of up to 4 weeks in period, are necessarily limited. Most oceanographic stations are occupied only for a very limited period – usually less than 2 days at a time. Interpreting traditional oceanographic data assumes that the data is not affected significantly by fine structure in time or space. In the upper layers of the oceans, this assumption is invalid. In the future, marine scientists will want to study oceanographic processes *in situ*, in real time by adapting and using instruments underwater. Sensors for the measurement of chemical species have developed over the last few decades, but this 'industry' is still young. High resolution profiles containing high quality data allow marine scientists to identify small scale features and processes which cannot be detected by traditional sampling methods. Continuously recording, long-term *in situ* analysers will also be able to identify crucial events occuring on short timescales: blooms and primary production, nutrient cycling, ocean eddies, episodic upwelling, and vertical mixing.

'Chemical sensors' is a catch-all term, used to encompass both the sensors themselves, and the various sensing elements of an underwater instrument for the estimation of the concentration of various chemical species. A chemical sensor is experimentally designed to 'respond' to the quantity of a 'target' species (or class of species) diffusing passively to a detector. A chemical analyser 'moves' the seawater by mass transport through the instrument which then carries out the analysis. Both sensors and analysers face common problems when used to detect chemicals at low concentrations in seawater. In a marine sense, a chemical sensor can also be based on any characteristic of the target species that serves to distinguish it from the seawater matrix, including: its chemical reactivity, optical or electrical properties, or mass. Oceanographic chemical sensors therefore also encompass a field traditionally known as instrumental analysis. The popularity of the description 'chemical sensor' reflects the fact that the interface between the chemistry and the instrument (i.e. the sensor) is often more important than the rest of the instrument itself (the pressure housing, electronic design, power supply requirements, etc.). Although the 'cost' of a sensor can be quite low, the associated costs of manufacturing an analyser can make the whole instrument prohibitively expensive. There is a fine line between making simple systems more complex, and complex systems more simple. Critical judgements often have to trade-off long-term stability against speed of response, for instance, or accuracy versus long-term use. The installation of more options make the sensor more universal towards a greater range of analytes, or over a greater range of environmental conditions.

Ruzicka, a pioneer in the development of many analytical and electrochemical techniques, once quoted "...it may be argued that chemical analysers are too complex to comply with the accepted concept of a chemical sensor – whereas this is true for physical sensors, which are fully functional simple embodiments devoid of external mechanics. All chemical sensors designed so far have either remained structurally simple but could not fulfil the required performance criteria, or out of necessity they grew mechanically complex".

The objective of this book is to identify the novel areas where chemical sensors are being applied or developed, and to indicate their usefulness to marine science. There are five important aspects to any book dealing with chemical sensing: the basic sensors and their detecting principles, their advantages and disadvantages, the mathematics (or theory) of detection, the monitored parameters, and their application to the marine environment. It is also important to indicate how the instrumentation is changing, the importance (and problems of) calibration, together with signal processing routines, and software. I hope that the reader will appreciate that underwater instrumentation for the monitoring of chemical species cannot be simply reduced to an engineering problem. There is a need to understand the system as a whole in order to be able to effectively design and deploy instrumentation to realize the full potential of the results. Scientists who develop chemical sensors often combine talents acquired from many different disciplines: chemistry, oceanography, electronics, materials science, mechanical engineering and physics, to name just a few. This requirement has confounded progress in the development of chemical sensors for some appreciable time.

The sensors and sensing systems described herein portray the state of the art, in so far as chemical sensors have developed to date. The authors represent an excellent cross-section of leading scientists who have developed ideas to a level where instruments have been deployed at sea. Each chapter represents a chemical sensor (or sensing system) from fields as diverse as spectroscopy or electrochemistry, and lasers. There are a number of chemical sensor types that are not represented within this volume (for example: enzyme-immobilized biosensors, gas chromatographic instruments, and mass spectrometry). Some material cannot be included because of the industrial value of several current sensor programmes. Moreover, sensor technologies can be quickly surpassed by new findings and achievements in a relatively short time – since research proceeds apace in many academic and organisational laboratories. Many sensors have only demonstrated 'proof of concept' and have not yet developed beyond the laboratory bench. The sheer expense of developing instruments for oceanographic applications and a relatively small market demand contrive to limit progress in these areas.

Oceanographic chemical sensing is a new and expanding field and will see many more developments over the next few years. Very few oceanographic chemical sensors presently exist, but the present supply of chemically sensitive instruments is far outstripped by the demand to make these types of measurement, and as such represents an area still to be commercially exploited. We shall see future developments of these and other types of oceanographic instruments – it is only a matter of time. There are many more sensors yet to be designed and developed, we have only just seen the initial steps. The next generation of chemical sensors will be allied to improvements in electronics and signal processing techniques. The current level of technology will make this task relatively trivial.

ACKNOWLEDGEMENTS

Within this volume, I am delighted to be able to include many world-renowned experts in their respective fields. I am delighted at their response to the invitation to participate in this volume. I believe that an excellent cross section of current research in oceanographic chemical sensors is represented within these chapters, and I express my gratitude to them. I believe this to be a fundamental form of marine chemical research, which will serve the future of oceanographic progress, and I am delighted to be able to report it here. I am also indebted to my family, friends and colleagues who have helped contribute towards this edition, without whose assistance this volume could not have been prepared.

CONTRIBUTORS

Eric P. Achterberg	Department of Environmental Sciences, University of Plymouth, UK
M.M. Baehr	University of Montana, Chem. Pharm. Building, Missoula, USA
Stéphane Blain	UMR CNRS 6539, Institut Universitaire Européen de la Mer, Plouzané, France
Andrew R. Bowie	Department of Environmental Sciences, Plymouth Environmental Research Centre, University of Plymouth, UK and Plymouth Marine Laboratory, Centre for Coastal and Marine Sciences, UK
Charlotte Braungardt	Department of Environmental Sciences, University of Plymouth, UK
Robert F. Chen	Environmental, Coastal and Ocean Sciences (ECOS), University of Massachusetts-Boston, USA
Charles H. Clayson	Southampton Oceanography Centre, UK
Bill Davison	Institute of Environmental and Natural Sciences, Environmental Science Division, Lancaster University, UK
Michael D. DeGrandpre	Department of Chemistry, University of Montana, Missoula, USA
Tommy D. Dickey	Ocean Physics Laboratory, Department of Geography, Institute for Computational Earth Systems Science, University of California, Santa Barbara, USA
R. Fauzi	Plymouth Marine Laboratory, Centre for Coastal and Marine Sciences, UK
Greg A. Gerhardt	Departments of Psychiatry and Pharmacology, Rocky Mountain Centre for Sensor Technology, University of Colorado Health Science Centre, Denver, USA
Ronnie N. Glud	Marine Biological Laboratory, University of Copenhagen, Helsingor, Denmark
T.R. Hammar	Woods Hole Oceanographic Institution, USA
Gerhard Holst	Max-Planck Institut für Marine Mikrobiologie, Bremen, Germany
David J. Hydes	Southampton Oceanography Centre, UK
Hans W. Jannasch	Monterey Bay Aquarium Research Institute, Moss Landing, CA, USA

Kenneth Johnson	Moss Landing Marine Laboratories, CA, USA and Monterey Bay Aquarium Research Institute, Moss Landing, CA, USA
I. Klimant	Institute for Analytical Chemistry, Chemo and Bio-Sensors, University of Regensburg, Germany
Oliver Kohls	Institute for Analytical Chemistry, Chemo and Bio-Sensors, University of Regensburg, Germany
Michael Kühl	Institute for Analytical Chemistry, Chemo and Bio-Sensors, University of Regensburg, Germany
C. Mantour	Department of Environmental Sciences, Plymouth Environmental Research Centre, University of Plymouth, UK and Plymouth Marine Laboratory, Centre for Coastal and Marine Sciences, UK
Paul A. Moore	Laboratory for Sensory Ecology, Department of Biological Sciences, Bowling Green State University, USA
Mark B. Rawlinson	WS Ocean System Ltd, Alton, UK
Clare E. Reimers	Institute of Marine and Coastal Sciences, Rutgers University, New Brunswick, USA
Richard C. Sandford	Department of Environmental Sciences, Plymouth Environmental Research Centre, University of Plymouth, UK and Plymouth Marine Laboratory, Centre for Coastal and Marine Sciences, UK
John M. Tokar	NOAA Office of Oceanic and Atmospheric Research, Silver Spring, Maryland, USA
Mark S. Varney	Southampton Oceanography Centre, UK
David J. Whitworth	Department of Environmental Sciences, University of Plymouth, UK
Paul J. Worsfold	Department of Environmental Sciences, Plymouth Environmental Research Centre, University of Plymouth, UK
Paul N. Wright	Maritime Faculty, Southampton Institute, UK
Hao Zhang	Institute of Environmental and Natural Sciences, Environmental Science Division, Lancaster University, UK

1. INTRODUCTION

MARK VARNEY

University of Southampton, Southampton Oceanography Centre, European Way, Southampton S014 3ZH, UK

1.1 INTRODUCTION

The most interesting chemical species are those that interact with marine life, and it is universally those same compounds that vary most in concentration – in time and space. Nothing is ever found to be in equilibrium, and there is a complex interplay between biology, chemistry and physics. There is marine life to be found in all of the world's oceans, from the poles to the equator, and from the ocean surface to the bottom of the deepest abyssal plain. This biological community ultimately depends on the sun for its' supply of food and energy, and for this reason, most organisms tend to concentrate in the upper layers of the water column. The natural cycles of many elements are characterised by large temporal changes in concentration. As a consequence, measuring the fine vertical structure as a function of time is important in understanding the complex chemical, physical and biological interactions. The global movement and chemical forms of elements are closely linked to energy exchanges and water-driven processes.

Temperature variations, turbulence, physical mixing and tidal currents cause changes in concentration on horizontal scales of 100 m to 100 km. Meteorological conditions induce seasonal and inter-annual changes. Other climatic processes, such as the El Nino Southern Oscillation and the Monsoon in the Northern Indian Ocean (Arabian Sea), have a significant impact on land-sea-air interactions. Our understanding of the natural variability of atmosphere, ocean and land systems is very unclear, and is difficult to distinguish against the anthropogenically induced changes. Changes in concentration are often so fast that there is insufficient manpower, ships and time to be able to gather a clear, global view of all of the possible interactions at any one instant. In order to calculate accurate chemical budgets due care and attention must be paid to include all of the possible influences that may affect the sources, distributions and fates of chemical species. Experiments to measure the air-sea exchange rates, and fluxes within the water column and to the sediments, must be very carefully designed with proper reference to scale. There are a vast range of processes that occur in the ocean that influence the distribution of biological organisms, chemical species, and water masses. Each of these processes also have short-term and long-term variations.

Our understanding of the natural environment has changed considerably over the years. As we continue to explore the oceans, our ability to make better, high quality measurements has also improved. For instance, the concentration of many trace metals dissolved in seawater has appeared to decrease over the past 25 years (Figure 1.1). This is more a reflection of the lengths to which analysts have gone to improve their methodology and eliminate contamination than it has with actual changes in concentration. Trace metals are a good illustration of the inherent problems of chemical sensing as the measurements cannot be performed *in situ* – volumes of water have to be carefully sampled and brought back to a land-based laboratory for subsequent extraction and analysis. This entails a considerable number of analytical steps and therefore points at which contamination can ruin the measurement – as it obviously has in the past. The detection limits of most sensors and sensing systems are not nearly low enough to determine the majority of chemical species that occur naturally in seawater. Some metals, such as iron and manganese, have elevated concentrations in hydrothermal and estuarine plumes, but would not normally be detected in seawater where the concentrations are up to 10,000 times lower. The detection limits of trace metal sensors and other chemical systems need to be reduced by several orders of magnitude in order to be truely useful.

There is therefore a major requirement to develop a new generation of chemical sensors that can be deployed on ships of opportunity, underwater vehicles, seabed landers, and a wide variety of underwater platforms and installations to facilitate the collection of oceanographic data. Determining chemical species in seawater is a more exacting task than estimating physical parameters (such as:

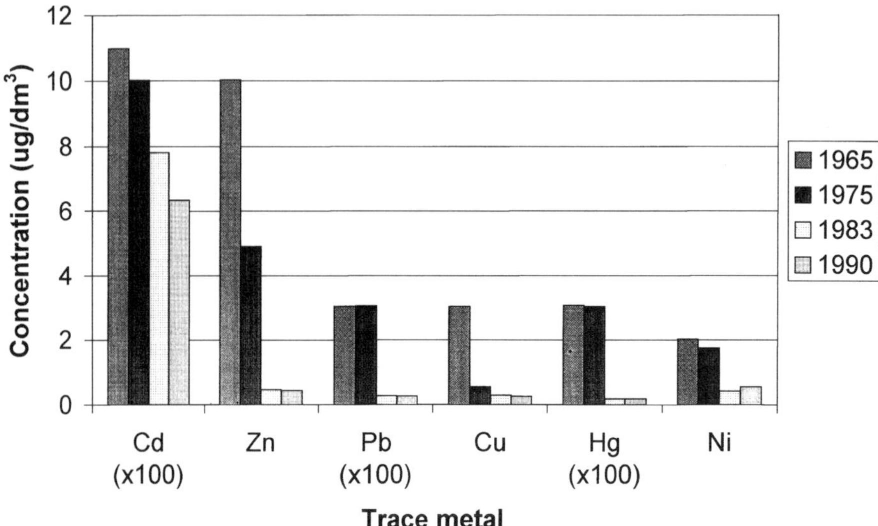

Figure 1.1 Plots of typical trace metal concentration levels taken from the literature over the past 25 years shows an apparent decrease in concentration, which is related more to better methods of sample handling and analysis than to oceanographic processes (adapted from Howard and Statham, 1993).

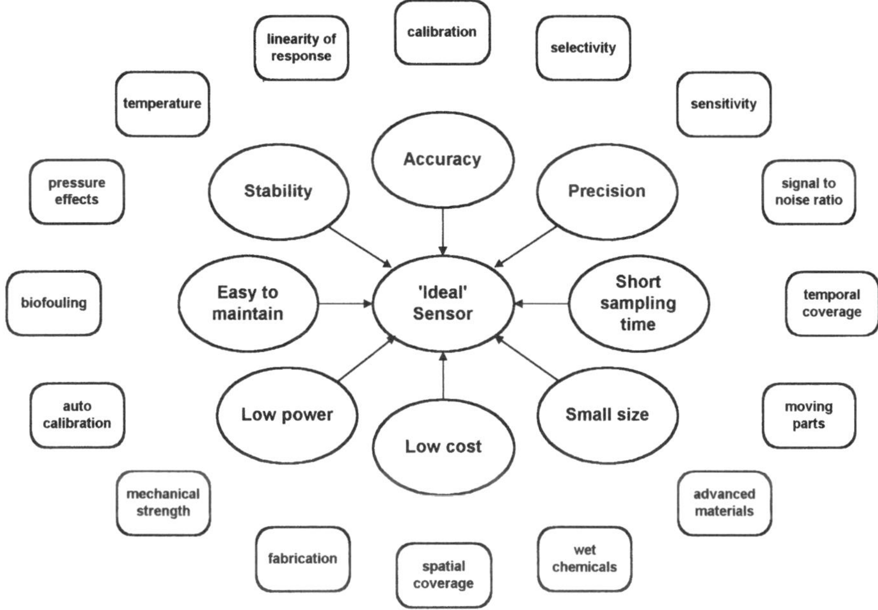

Figure 1.2 For successful adoption by the oceanographic community, a chemical sensor needs to meet a minimum "performance" or specification. However, the criteria used often differ according to the sensor itself, but an "ideal" sensor can usually be assigned certain key performance measures, and lesser associated criteria.

temperature, flow, or pressure) because in addition to requirements of accuracy, stability, speed of response and sensitivity, there is the need for selectivity. For example, the concentrations of cadmium and calcium are about 1×10^{-9} and 1×10^{-2} M, respectively. They are both implicated in biological utilisation by phytoplankton, and exhibit biolimiting behaviour. A sensor for cadmium must be extremely sensitive to measure the low nanomolar concentrations, but selective enough to discriminate against the 10^7 times higher concentration of calcium. Additional complications (Figure 1.2) arise through the need for long-term chemical stability, *in-situ* calibration, kinetics of the reactions under study, cross-sensitivity to other chemical species, biofilming and biofouling – which are discussed later in this chapter. Particular environmental problems may be expected with optical sensors (shadowing and biofouling), conductivity sensors (biofouling and mineral deposits), and gas sensing electrodes (ageing).

1.2 MEASUREMENT TECHNIQUES

Salinity sensors have traditionally been based on the measurement of conductance between two large surface area electrodes. A high frequency large

Plate 1.1 A standard "rosette" after deployment having water samples collected directly from the large volume GoFlo bottles for nutrient analysis in the shipboard laboratory. Also strapped across the bottom of the frame are a number of continuously recording sensors for salinity, temperature, pH, dissolved oxygen, fluorescence, turbidity and sound velocity. (Photograph courtesy of R. Scrivens, WS Ocean Ltd.). *See* Color Plate 1.

amplitude voltage (e.g. 800 Hz 2V p-p) passes an electrical signal through the seawater, without causing significant electrochemical reaction (i.e. oxidation-reduction) or polarisation at the electrode surface. The in-phase alternating current reponse is related to the salinity of the seawater through an empirical relationship which has undergone various revisions throughout the last few years. However, despite a number of instruments being manufactured, fundamentally the measurement of salinity has not changed over the past 10 to 15 years. Standard measurements of pH also continue. The bodies of pH electrodes are traditionally filled with a high concentration HCl solution, and the reference electrodes are filled with an electrolyte similar in composition to seawater. More modern designs have been based upon gel compositions which are less susceptible to external pressure changes, and changes in junction potential. There have been some changes in reference electrode design, such as the use of oil-filled capillary junctions and mixed-ion polymeric electrolytes. But, problems associated with the reference electrode determine the usefulness and reliability of pH measurements – saltwater composition buffers and calibration have recently been reviewed, for instance (Millero *et al.*, 1993). The fact is: there have been few novel or innovative sensors developed for the measurement of chemical species in seawater, and many oceanographers continue to use technology which has been tried and tested for use in the laboratory, but which is often impractical for use underwater.

The majority of sensor developments in oceanography have progressed in 3 main areas. These include:

- electrical phenomena (electrical conductivity, potentiometric, and amperometric methods);
- development of wet chemical methods (flow injection analysis);
- spectroscopic, optical fibre and waveguide-based sensors (absorption, transmission, fluorescence, chemiluminescence).

Many of these are outlined in later chapters. Acoustic, magnetic and pressure sensors are the domain of physical oceanographic sensors. Mass and thermal chemical sensors have seen very little development in oceanographic terms – perhaps for obvious reasons but somewhat puzzling because fibre-optics are temperature sensitive. Telecommunication engineers have traditionally avoided the problem by coating the outside of fibers with two layers: the first a silicone polymer (with a low thermal expansion coefficient), the second a tough protective polycarbonate jacket. There is intense interest in the use of quartz crystals, piezoelectric films, and Rayleigh and surface acoustic wave (SAW) devices in the biological, biochemical and clinical fields. However, their application to marine sciences remains low, particularly as they are predominantly gas phase devices which appear to be sensitive to environmental noise and temperature. The systems have potential if the appropriate engineering problems can be approached in a realistic way – but commercial systems are still a long way off. Films of polyvinylidene fluoride (PVDF) subjected to varying temperature and stretched along one or two axes with a simultaneous high DC field, can be made piezo- and pyroelectrically sensitive. Cooling the films (ranging in thickness from microns to several millimeters) leaves a residual polarisation which also allows an element of charge injection to occur. The piezoelectric effect has given rise to a number of sonar and speakerphone devices, and the pyroelectric effect has been utilised in imaging devices. In principle, biosensors can also be made from this material by immobilising an enzyme on one side of the PVDF film, and flowing seawater past the other side. A simple high impedance differential amplifier could be used to measure the signal between the two surfaces of the sensor. Although responses are non-linear, the output signal can be related to concentration. The key challenge to their successful implementation is designing sensors that respond selectively to a single target chemical or class of chemicals.

It is often said (in electrochemical terms) that an electrode can be made as active as fluorine, or as inert as argon. Electrochemical techniques are able to determine the "speciation" of various species. However, when the "species" are themselves dynamically distributed between different reactions or other species, the electrode may not be able to "sense" their true concentrations. Because of the various "timescales of measurement" electroanalytical techniques often do not

Table 1.1 Chemical sensors are often designed within very tight performance constraints. In recent years, most efforts have been focussed on the ability to meet certain accuracy criteria (e.g. the "WOCE standard" for CTD probes) or to provide a "suite" of measurments (e.g. *in situ* nutrients analyser for: phosphate, ammonium, nitrate, nitrite, pH, etc.) Few sensor systems have inherent sensitivities matching naturally-occuring concentrations. A figure often quoted for accuracy is 1% or less of the typically occuring range of concentrations for that analyte, and the precision an order of magnitude lower. This table lists typical requirements for a number of common sensors used for seawater measurements.

Parameter	*Typical Range*	*Accuracy*	*Resolution*
Temperature	−5 to 30 °C	0.005 °C	0.001 °C
pH	3.5 to 9.5 pH	0.01 pH	0.001 pH
Conductivity	0 to 65 mS cm^{-1}	0.002 mS cm^{-1}	0.001 mS cm^{-1}
Redox (Eh)	−400 to +600 mV	1 mV	0.1 mV
Chloride	1 to 14 pCI	0.01 pCI	0.001 pCI
Dissolved oxygen	0 to 500%	1%	0.1%
Chlorophyll (fluorescence)	0 to 100 mgC dm^{-3} (0 to 100 AUF)	0.1 AUF	0.01 AUF
Sulphide	1 to 14 pS	0.01 pS	0.001 pS

The response times for all sensors is generally needed to < 1 sec. Although there have been a number of automatic marine monitoring and observation systems (e.g. MerMaid, SeaWatch, and Marel) developed, they have tended to concentrate on systems design, power requirements, data telemetry and networking. Many of the sensors used on these stations do not completely fulfill the requirements for long-term deployment. Biofouling is perhaps the single most important factor that limits the confidence in data validity of autonomous measuring stations.

correspond with each other. Anodic stripping voltammetry (ASV) can strip metals out of solution, and away from organic complexes. Ion selective electrodes, on the other hand, directly measure an "equilibrium" concentration of the ion in solution, but do not have the sensitivity to operate at the ultra-low concentrations that many of these ions occur in seawater. Perhaps the main disadvantage of electrochemical techniques is their inherent lack of selectivity compared to other techniques. Electrochemical techniques have therefore become better adapted towards the measurement of "bulk" properties of seawater, such as conductivity. However, electrochemical sensors with a high degree of selectivity have been devised using a variety of different approaches. Microelectrodes, in particular, show real promise at being able to detect a number of different species at the same time (Brendell and Luther, 1995). In addition, electrodes, microelectrodes and microelectrode arrays have the potential to be used as biosensors (Figure 1.3), where enzymes can be immobilised on the electrode surface. Many of the commercially available biosensors use enzymes (preferably oxidases) as biological components and electrodes as transducers. Chemical or physical changes, evoked

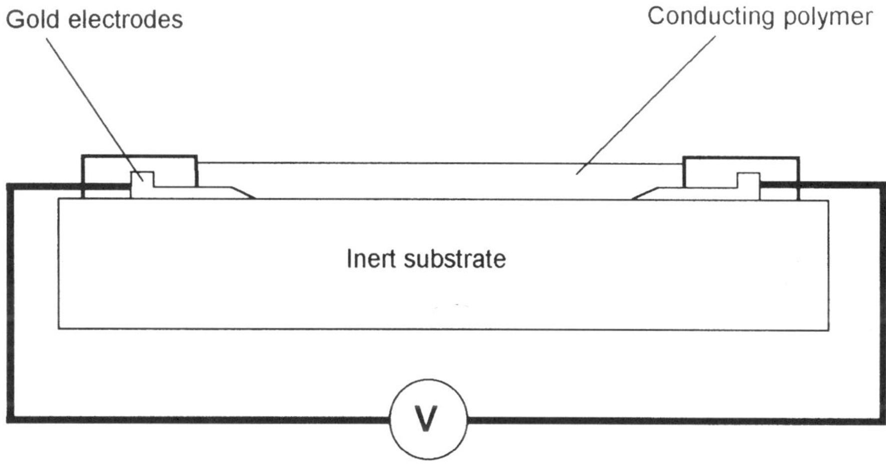

Figure 1.3 Conducting polymer electrode assembly. These devices are commonly manufactured using electrochemical or screen printing procedures, and lead to simple, cheap and disposable sensors. The conducting polymer commonly consists of 1-dimensional highly conjugated long-chain molecules. Typical "backbones" include the polymers: polyaniline, polypyrrole, and polythiophene. In a neutral state these materials are only semiconducting. The electronic conductivity is enhanced when dopants are introduced into the film by oxidation or reduction reactions, usually electrochemically. Relatively high levels of dopant can be used, typically up to 33%. The level and type of counterion used can greatly affect the resultant conductivity of the film. Nitrate sensors (Aylott *et al.*, 1997) based on immobilised nitrate reductase enzyme, for instance, respond quantitatively to seawater concentrations of nitrate ion. However, how the analyte molecules interact with the doped polymeric structure is complex and not yet completely understood.

by the recognition of the analyte by the biological component, are normally transformed into an electrical signal by the transducer. However, their use is presently confined to the clinical and biochemical sciences.

The most effective method to construct a long-term sensor for underwater use is to promote measurement principles that are not susceptible to interference or drift. Although electrochemical techniques are traditional and easy to adapt, very few are suitable for long-term use and do not possess long-term stability, resistance to high salt concentrations, mechanical ruggedness, or compatibility with various anti-fouling regimes. Spectroscopic detectors are easy to adapt for underwater use and new optical sensors have been developed by numerous research groups. There are a number of easily measured optical properties, such as transmission, absorption, polarization, luminescence, fluorescence and decay times. Transmission and absorption spectroscopy are well established techniques for underwater use, but remain unselective except to the highest concentration species. The use of infrared techniques in all aqueous environments is limited by the strong absorption of water. Attenuated internal reflection (ATR) infrared spectroscopy has yet to see a practical use in oceanography, although it is anticipated that a number of sensors could soon be developed for the *in situ* detection of chlorinated hydrocarbon compounds. However, the required detection limits (of the order of $\mu g\ dm^{-3}$) stretch the present systems to the edge of their performance – present prototype systems operate in the ppm range.

While a number of light-detecting systems have been developed over the past few years, light sources have generally suffered from a lack of stability. The determination of nitrate by the general absorbance of ultraviolet light is well documented, but because the background signal (due to dissolved organics) varies dramatically with wavelength, any drift in the lamp output may appear as an apparent (but erroneous) signal. Dual wavelength measurements have an inherent advantage as they differentially subtract out any instrumental change in response. For similar reasons, old techniques are being re-visited because of the increasing popularity of photodiode arrays (PDAs). PDAs and charged-coupled devices (CCDs) exhibit superior performance, particularly coupled with modern well-designed and sensitive electronics. There is great interest in fluorescence because of the ability to provide information on dissolved species at low concentrations without sample manipulation or preconcentration. Sensors based on fiber optics provide some interesting advantages over electrochemical and other sensors. Their simple, sturdy construction is well suited to oceanographic use. They are immune to electromagnetic or galvanic interferences, and require no reference electrodes. However, it is virtually impossible to prevent ageing processes in optical sensor elements under long-term operation in harsh environments. Signal drift compensation using sophisticated referencing techniques is technically possible, but adds considerably to the cost and complexity of the instrument. Fibre-optic Bragg Grating sensors have been in existence for several years. The technology has been typically used for strain sensing, but can easily be adapted for use in a thermistor chain (Figure 1.4). This is an excellent alternative to the traditional thermistor chain which has a bulky multiconductor cable and

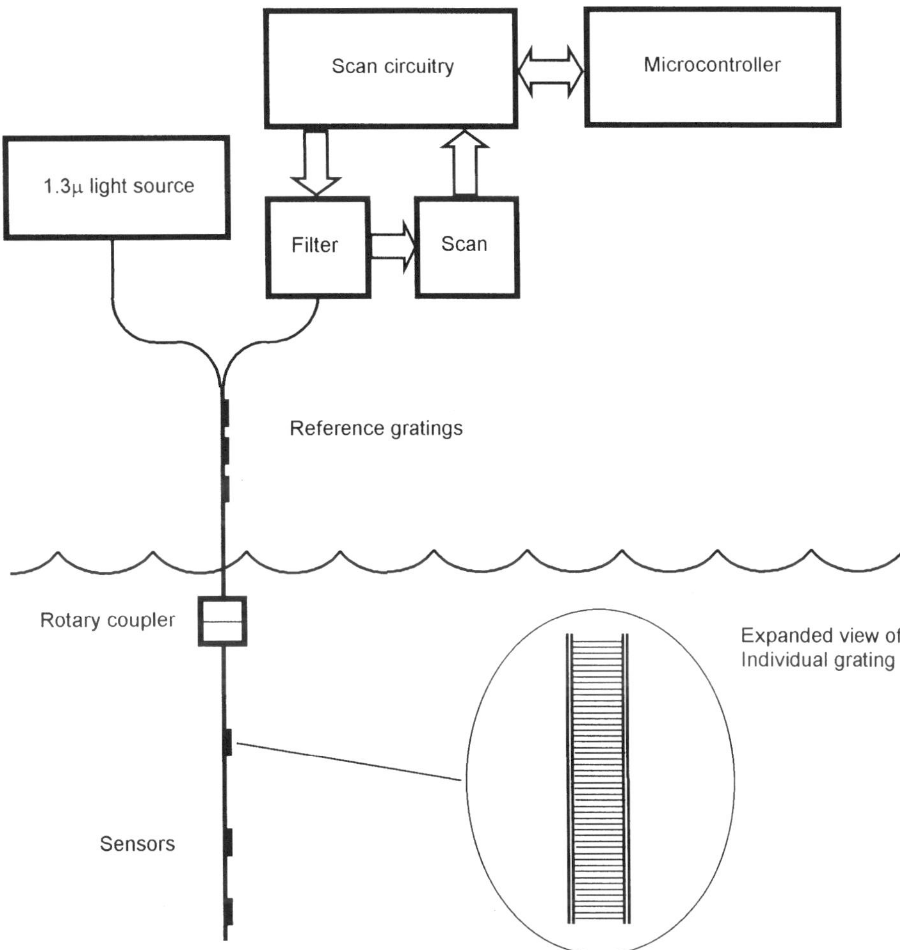

Figure 1.4 A schematic view of a fibre-optic based thermistor chain. The gratings are formed by exposing a length of fibre to a high power excimer laser through a phase mask: this creates a zone of alternating refractive index along the fibre core. These zones (each effectively several millimetres in length) act as interference filters reflecting light over a very narrow bandwidth. The resonant frequency is a function of the optical spacing of the refractive index maxima, which can be arranged to be a function of temperature or pressure. Individual "sensors" can be placed anywhere along the length of a fibre possessing separate non-overlapping wavelength ranges – allowing wavelength division multiplexing.

can be unreliable. Other methods of temperature profiling (undulators, towed fish, yo-yo's, etc) are not suited to long-term deployment.

The primary emphasis for the development of optical chemical sensors has been the formulation of solid phase indicators that interact with the analytes to produce an optically detectable signal. Much effort has gone into the search for porous polymers to serve as indicator substrates. Relatively few analytes can be detected directly by spectroscopy – it is necessary to add reagents to form a product that is optically active. Fibre-optic sensors for pH, dissolved oxygen and carbon dioxide have been developed. Many reactions have been devised that record the change in absorbance, transmission and fluorescence. Generally, these types of sensors incorporate fluorescent dyes, immobilised indicators or enzyme systems as the "sensing element" and are termed "optodes" or "optrodes". They have unique advantages: electromagnetic immunity, *in situ* use (at high pressures), very fast speed of response, and they can be easily miniaturised. However, they do possess some disadvantages: the "wet chemistry" can be difficult to reproduce, the sensing elements are prone to biofouling, it can be difficult to maintain a constant excitation energy output. Many of the interesting "chemistries" for which optrodes would be eminently useful occur in the blue end of the electromagnetic spectrum (often below 350 nm). Light sources for this light frequency are not easy to build into underwater instruments. The advent of blue lasers will "spur" the development of more chemical sensors that can utilise the extra information in this region of the visible and ultraviolet spectrum. Blue light emitting diodes (led's), and recently white led's, present a cheap and convenient energy source that cause a multitude of organic compounds to absorb energies towards the blue end of the visible electromagnetic spectrum (Figure 1.5). This new type of diode, invented at the Fraunhofer Institut fur Angewandte Festkorperphysik (IAF), can emit virtually any hue by varying the relative quantities of converter dyes or phosphors. The dye (or phosphor) is mixed in microgram quantities with the epoxy or acrylate resin directly in contact with the semiconductor chip. Typical brightnesses are 20 mcd at 10 mA – which is acceptable for many Oceanographic applications, and should encourage the development of many new sensor systems. However, a variable frequency single wavelength light source, capable of emission in part or all of the visible spectrum, currently remains elusive.

The detection of light over a range of wavelengths is less problematic. Holographic gratings are commonly available, for instance, and there are a considerable number of universal detector systems available. Interestingly, in the search for light systems that can emit in the blue and ultraviolet region of the spectrum there have been several detector developments that are specific to this region. Diamonds, "band gap" closely matches the energy of ultraviolet photons. Chemical vapour deposited (CVD) thin films of synthetic diamond are 5 orders of magnitude more sensitive to light between 180 and 220 nm, and are insensitive to all other wavelengths. There are also a number of electronically tunable filters, based mostly on Fabry-Perot cavities. These are essentially closely spaced two semi-transparent mirrors that vibrate at discrete frequencies. The oscillation of the walls interfers with the incident light beam, and transmits light that has an integral wavelength proportional to the distance between the mirrors. Originally developed for infrared applications, universal tunable filters extending into the visible spectrum are now becoming available. However, for very low wavelengths

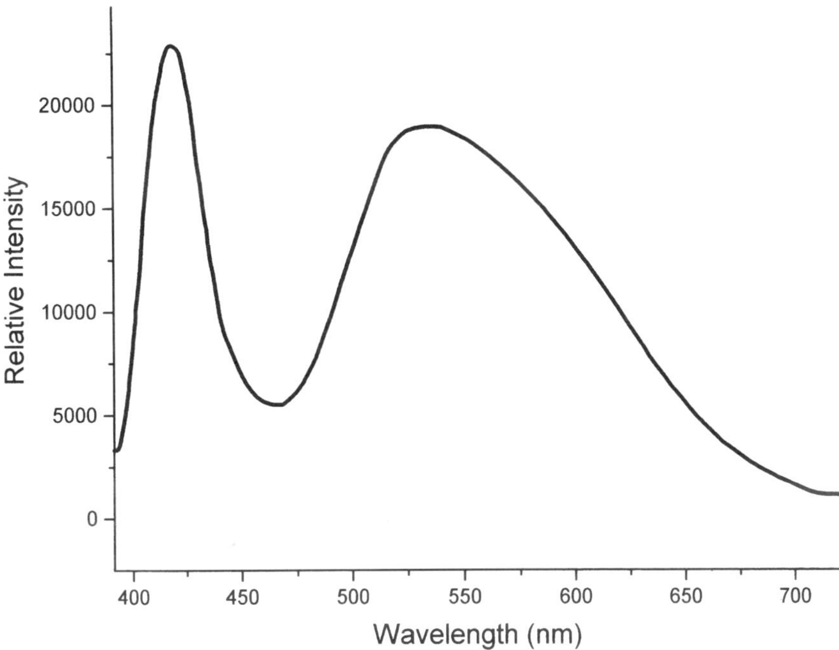

Figure 1.5 The emission spectrum of a Lucoled (Siemens™) LED that combines blue and yellow p-n junction LEDs to produce "white" light. The blue diode is based on gallium nitride to excite light from a Ce:YAG "converter" consisting of either organic dyes or inorganic phosphors in a yellow part of the spectrum to produce luminescence that involves a Stokes' shift towards longer wavelengths. The yellow then mixes with it's complementary colour, blue, to produce white. There is a constant on-going search for convenient (high intensity, low power) light sources that emit in the blue and ultraviolet end of the spectrum, either at discrete or over a range of wavelengths.

(e.g. < 500 nm) the internal distances involved are so small that the mirror faces virtually touch when the cavity vibrates and can be easily damaged. However, developments in microelectronics has produced fully integrated photodiode array systems that incorporate fixed holographic gratings for wavelength dispersion. Complete spectrophotometric detectors on a chip are now commercially available (Figure 1.6). Charge-coupled devices (CCDs) are also extremely sensitive over an extended wavelength range, and facilitate low level detection. Although the design of suitable excitation sources is still an issue, the problems of optical detection are fairly well resolved. Laser diode and fibre optic technology is developing all the time. Underwater spectrophotometers are becoming, at last, a reality.

Marine scientists have often compromised rapidity of measurement against precision and reproducibility of results, or the ease of use for the various techniques. The increased need for *in situ* measurements has increased the demand for better and more sophisticated instrumentation. It is apparent that as sensors

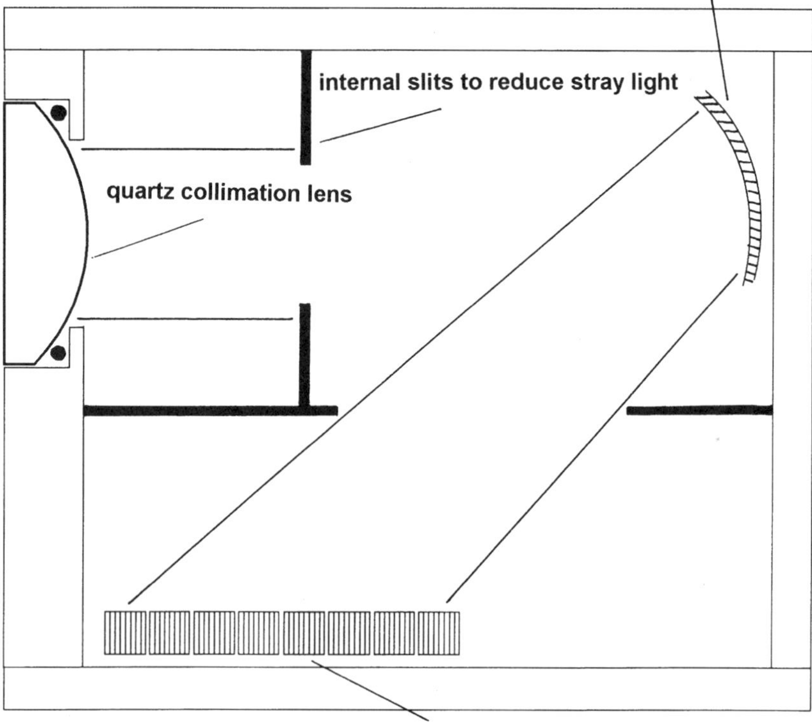

Figure 1.6 Fully integrated holographic grating and photodiode array systems form the basis of complete spectrophotometric detectors on a chip and are now commercially available from a number of different sources. The wavelength range is determined by the selection of the grating linewidths, and the resolution by the number of elements in the diode array, typically 512 or 1024. Baffles within the "chip" cavity reduce the amount of stray light. There are no mechanical moving parts, which makes this design ideal for extended underwater operation.

improve, there is a consequent demand for corroborative and associated measurements. As detection limits drop and environmental "noise" levels increase within data sets, it is extremely difficult to understand what the sensors actually respond to. All sensor systems need corroborative measurements to confirm that the signature is directly related to the measured parameter, and that it is not simply variation due to cross-sensitivity. High variability in datasets is common and needs explanation: when the sensor is remote and an observer is not present, it imposes great restrictions on the interpretation of the data and its reliability. It can be extremely difficult to understand the complexities involved in the design of underwater sensors, and data analysis.

Although there are a considerable number of "platforms" onto which chemical sensors can be deployed (moorings, CTD's, autonomous vehicles, benthic landers, etc.) the problems of routine operation, calibration, and autonomy means that we are heading towards the manufacture of chemical *analysers*, not sensors (see foreword). In a sense, the scientists involved are becoming "product engineers" possessing a mixed knowledge of: mechanical and electronic engineering, chemistry, software programming, and artificial intelligence. The sensors need to simple enough, but possess sufficient "smartness" to get the job done. The continued development of chemical sensors will always be a balance between simplicity and complexity.

Although chemical sensors have been developed for practically every area of marine science, the major requirement for sensors has probably been for the measurement of dissolved gases. The most important and valuable sensor known to man, the nose, contains around 30 million olfactory receptors, made up of 30 different types, which work in parallel to absorb a range of smell molecules. Most odorants are hydrophobic in nature, with molecular masses up to 300 Daltons, and usually contain a high proportion of polar functional groups within their structure. Imitating an olfactory cell is not easy: electronic equivalents, based on electrochemical cells have come close, but they are inherently limited to vapour detection and have not yet been developed for underwater detection. The majority of gas sensors involve the extraction of the gas molecules from seawater, in some fashion. It is often the "extraction" (a semi-permeable membrane, for instance) that defines the ability of a sensor to detect a molecule.

1.3 DISSOLVED GAS MEASUREMENT

Oxygen is one of the more important gases found in seawater. It is universal for life, and without it the seas would soon become sterile. The intimate involvement in photosynthesis and respiration has invoked more measurements of oxygen concentrations than for any other chemical parameter – except perhaps for total salt concentrations (salinity). It is important to emphasize the critical aspects of dissolved oxygen measurements, the physicochemical principles involved and the main factors that influence this measurement (i.e. temperature, pressure and salinity). Dissolved oxygen is, by no means, an easy measurement. There have been numerous methods and techniques developed for its measurement – each has its own complications. Long-term deployment, *in-situ* calibration, instrumental drift, and biofouling are just some of the factors that affect reliability. Although we are probably much more aware of the inherent problems of *in-situ* oxygen concentration measurements, present instruments are probably no better now than they were 10 to 15 years ago. Virtually every major oceanographic equipment manufacturer now has its own (or generic brand) oxygen sensor or sensing system (Plate 1.2). There is basically very little to choose between them, and they all offer equivalent performance. There have been very few advances in the technology – despite radical changes in the electronics and semiconductor

Plate 1.2 Many companies now retail "packages" that are capable of measuring a wide range of parameters. This instrument is capable of simultaneously measuring temperature, salinity (conductivity), dissolved oxygen, as well up to three other parameters. (Photograph courtesy of S. Tierey, YSI Ltd.). *See* Color Plate 2.

industries. Instrumental confidence in dissolved gas analysis is hard won, and easily lost.

Chemical sensors are typified by electrochemical sensors for dissolved oxygen. This, and the pH sensor, are probably the only two sensors that are routinely used for *in situ* work. The majority of electrochemical designs rely on the (passive or forced) irreversible reduction of oxygen at platinum or gold electrodes. Because of the variable nature of seawater composition, the electrodes are isolated behind a semi-permeable membrane in an electrolyte of controlled composition. The commercial development of pulsed oxygen electrodes has generally led to an improvement in the long-term stability and reliability over earlier galvanic and potentiostatic designs. Such sensors are now increasingly used to monitor oxygen concentrations over periods of weeks to months (Wallace and Wirrick, 1992). However, calibration is a difficult issue on such moored systems, and is usually accomplished only at the beginning and end of the deployment interval. Any instrumental drift has to be estimated by alternative methods.

Carbon dioxide, CO_2, is another important gas in the atmosphere and is directly implicated in controlling the worlds' atmospheric temperature and therefore climate through an enhancement of the "greenhouse effect". Vast quantities of the gas are transported into and out of the ocean surface according to oceanic air-sea exchange, dissolution and biological uptake (Watson, 1993). Water masses contain considerably different quantities of the gas. Various physical, chemical and biological processes produce significant fluxes of carbon dioxide (and many other gases) between the sea and the atmosphere. The atmospheric, and hence oceanic, concentration of carbon dioxide is increasing from the burning of fossil fuels. Present estimates of the oceanic uptake of anthropogenic carbon are approximately 2.0 gigatons of carbon per year, roughly one-third of the global fossil fuel emissions into the atmosphere each year. The ocean has a natural ability to buffer the increasing levels through dissolution of the gas, especially at high latitudes. Model predictions suggest that the ocean will continue to play a major role in taking up excess atmospheric CO_2 over the next few centuries. The projections are based almost entirely on simulations with the physical circulation and biogeochemical fields of the ocean fixed at current observed states. World Ocean currents and circulation patterns account for some of the observed large scale variations in the amounts of dissolved carbon dioxide. Primary productivity in the surface waters also "fixes" carbon obtained from the atmosphere, and is the basis for virtually all marine food chains. However, there are very divergent opinions as to the relative importance of physical and biological controls on the carbon dioxide chemistry of the upper water column. There is much work to be done relating the timescales of addition and removal processes to those established by physical mixing. A key task is determining which processes are central to the storage of carbon in the ocean and how they might respond to atmospheric changes. The exchange of carbon dioxide between the atmosphere and the oceans is one of the most important questions that can be addressed by marine chemists today.

There are three main controls on the total alkalinity and carbon dioxide concentrations in seawater, these are: (1) the air sea exchange of carbon dioxide gas, (2) the photosynthetic activity with consequent formation of decomposition of body tissue, and (3) the formation or dissolution of calcium carbonate. Phytoplankton photosynthesis can cause large departures from *in-situ* near-surface equilibrium concentrations of carbon dioxide and oxygen (with respect to their atmospheric partial pressures). A number of important chemical and biological phenomena show a strong diurnal tendency. Temperature differences between day and night create convection cells in the upper water column, resulting in a considerable diel change in the mean depth of the mixed layer. This mechanism may influence the nutrient supply in the euphotic zone, and particle dynamics in the mixed layer. Particle dynamics are related to many photochemical reactions. Phytoplankton photosynthesis, respiration and migration are governed by the solar inputs at the surface. Primary productivity is highly variable, being dependent on: season, depth, light, nutrient availability and phytoplankton species. The productivity may be estimated indirectly by measurement of chlorophyll fluorescence, and then by comparison with satellite colour images. There is an urgent need to correlate CO_2 uptake from the atmosphere with oceanic heat flux, to estimate the fluxes of carbon from surface to deeper waters, and to study temporal and seasonal changes in CO_2. So, although the variations in the concentrations of individual gas species are interesting and of undoubted importance, it is also important to be able to simultaneously measure other chemical, biological and physical parameters. Although biological productivity must have a significant role, it is particularly difficult to be able to assess the global significance of carbon fluxes without the measurement of: water temperature, photosynthetic activity, total CO_2 concentrations, alkalinity, air-sea gas exchange rates, physical mixing of water masses, and so on.

Many gases participate in a wide range of chemical and biological processes in the water column and in the atmosphere (Aylott *et al.*, 1997). Atmospheric methane (CH_4) is an integral component of the greenhouse effect, second only to CO_2 as an anthropogenic source. Methane's overall contribution to global warming is large because it is estimated to be twenty-one times more effective at trapping heat in the atmosphere than CO_2 over a 100-year time horizon (IPCC, 1996). Over the last two centuries, the concentration of methane in the atmosphere has more than doubled, due largely to increasing emissions from anthropogenic sources (landfill, agricultural activities, fossil fuel combustion, coal mining, the production and processing of natural gas and oil, and wastewater treatment). Landfills are the largest single anthropogenic source of methane emissions in the United States (USEPA figures), and whilst all future releases of volatile hydrocarbon gases are regulated, we have no appreciation of how they interact with the oceans. Methane is thought to be generated in the upper oxic water layers by bacterial "microclimates" which maintain their own anaerobic environment by forming an organic protective coating around colonies of their own organisms. The limited available data suggests that the oceans absorb about as much methane as they produce. New scientific findings based upon a number of measurements taken in several locations around the globe show that open oceans apparently absorb methane and methyl bromide, while coastal areas and upwellings produce these gases. Organosulphur and organohalogens also have their own respective distributions and fates with respect to air-sea interactions and involvement in biological processes (Davison *et al.*, 1996). These gases are commonly measured by headspace analysis and gas chromatography (Malin *et al.*, 1993). For most of the trace gases studied so far, usually analysed immediately after sampling using ship-board dedicated facilities, the observed variations in concentrations and vertical fluxes represent a non-steady state arising from rapid fluxes. However, all are involved in extremely complex biogeochemical cycles, and are implicated in the global greenhouse effect. Unfortunately, in virtually every instance, we know very little about the mechanisms and rates of reactions involved. The ability to measure CO_2 (and all other biologically important gases) would make substantial contributions to oceanographic research.

The equatorial Pacific is the largest oceanic source of carbon dioxide to the atmosphere and has been proposed to be a major site of organic carbon export to the deep sea (Murray *et al.*, 1994). Study of the chemistry and biology of this area from 170° to 95 °W suggests that the variability of remote winds in the western Pacific and tropical instability waves are the major factors controlling chemical and biological variability. The reason is that most of the biological production is based on recycled nutrients; only a proportion of the nutrients transported to the surface by upwelling are taken up by photosynthesis. Biological cycling within the euphotic zone is efficient, and the export of carbon fixed by photosynthesis is small. The fluxes of carbon dioxide to the atmosphere and particulate organic carbon to the deep sea were about 0.3 gigatons per year, and the production of dissolved organic carbon was about three times as large. The data establish El Niño events as the main source of interannual variability.

Another important parameter is the measurement of the micronutrients: nitrate, phosphate and silicate. Traditional methods rely on the addition of reagents which react with the native species to form a coloured dye that can be measured colorimetrically – often in a flow injection analyser. Nutrient systems are therefore closer to chemical analysers than they are to sensors (Plate 1.3). Often the requirement for calibration and standardisation presents enormous logistical and mechanical problems to their routine deployment. These type of systems are mechanically complex, not particularly compact, and involve complicated hard- and firm-ware for them to operate correctly. Chapter 2 describes a non-mechanical flow injection analyser by Hans Jannasch which overcomes many of the inherent problems of chemical analysers, based on osmotic pumps. Limited kinetic formation of azo-dyes, temperature effects on flow rates, biofouling, and temperature-dependant kinetically limited reactions are just a few examples of the types of processes that still plague continuous or segmented flow injection analysers.

1.4 Eh (REDOX) MAPPING

Sensor development has not been restricted to dissolved species or gases in seawater. It has long been known that sediments are a dynamically active zone, and exhibit significant differences in chemical parameters within very short disances. Redox potentials, or Eh, are a valuable indication of the oxidative or reductive power of the sediments, and is primarily responsible for the trapping, or remobilisation, of species delivered to the seafloor such as managanese or iron. An inert electrode, usually platinum, will respond to the oxidation or reduction potential within the sediment. In fauna and aerobic bacteria within the surficial

Plate 1.3 A flow-injection analyser attached to a standard CTD frame prior to a hydrographic cast. The FIA instrumentation is attached on the left hand side of the frame, with the plastic reagent bags below the pressure housing containing both optical and electronic components. The inlet manifold is already submerged. Conductivity, temperature, depth, oxygen (CTDO) sensors and two Niskin water bottles are mounted in the middle of the rosette (for calibration purposes), and the battery housing is on the right. This equipment has a battery endurance of 8 hours and is used up to 6000 m depth. (Photograph courtesy of S. Blain). *See* Color Plate 3.

sediments utilises dissolved oxygen, leading to an increasingly negative Eh potential. Below an Eh of 0 mV (when all oxygen has been used up) anaerobic baceria become dominant, principally sulphate-reducers. Increasing concentrations of S^{2-}, HS^- and H_2S reflect the progressive destruction of SO_4^{2-}, an alternative oxidant. Sulphides are significant reductants, and sediment potentials often reach as low as – 250 mV. The distance down to where there is significant sulphide concentration (the "sulphide zone") depends on the availability of organic carbon. This can vary widely, from depths of centimeters in coastal areas, bays and estuaries, to several meters on continental slopes where carbon inputs are generally very low. The "seepage" of petroleum hydrocarbons along fault lines enhance the amount of organic carbon being metabolised by the bacteria (Spies and Davis, 1979). Consequently, where there is a local seepage of oil in deepwater sediments, the enhance microbial activity will cause the sulphide zone to occur nearer the surface (Figure 1.7). The discovery of seeps has a major influence on the exploration and drilling activity of oil companies.

Ocean floor exploration is an excellent technique for finding oil and gas. Discovery of large-scale seepage has led to the establishment of many of the

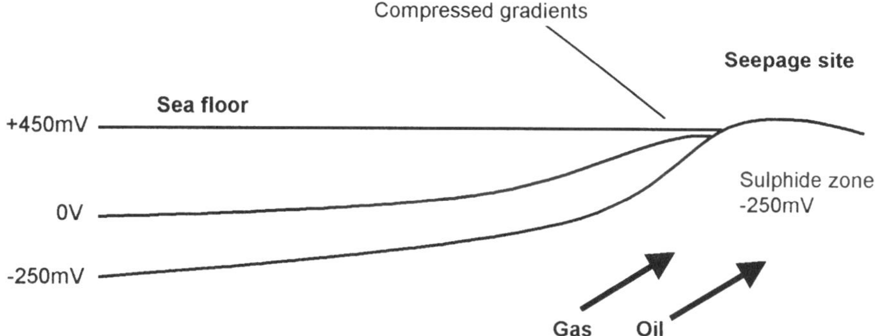

Figure 1.7 Schematic model of the approach of oil and gases towards the seafloor surface giving rise to decreased redox potentials. The presence of oxygen increases the redox potential (towards +600 mV). Oxygen is used up by sulphate-reducing bacteria, and the hydrocarbons fuel the reduction, pulling the redox potential to values of −250 mV. The electrochemical "gradient" is an indication of the proximity of the sulphide zone and therefore how likely a seepage may be in the vicinity.

world's most important petroleum production fields, including those in the United States, Iraq, Iran, Kuwait (Burgan), Trinidad, East Borneo, and North Sumatra. Presently, over 30 major seepages are known to exist in the deep waters of the Gulf of Mexico. Unfortunately, though, hydrocarbon seepages covering several square kilometers of the sea floor usually remain undetected. Land-based "surface geochemistry," capable of detecting part-per-million traces of seepage petroleum, has become a significant industry. Oil exploration on the seafloor, however, has not reached the same level of maturity.

Since sea floor seepages are not subject to the same degree of erosive forces (groundwater movement, soil formation, and the growth of vegetation), they are usually much more vast and long-lasting than those originating on land. The industry needs new technology to tap into the enormous resources of deep sea petroleum. At present, only three techniques are available for finding ocean floor oil seepage. Coring, which involves recovering small cylindrical samples of sea floor sediment, was the first method used, but it is ponderous and costly. Side-scanning sonar detects sea floor topography only suggestive of seepage, and seismic surveys reveal anomalies that may be caused by gas-charged sediment. These are also expensive and are not definitive tests.

A novel sensor development (Thompson *et al.*, 1997) uses a series of redox sensors to rapidly assess the presence or absence of petroleum hydrocarbons via the measurement of Eh, and is a valuable complement to seismic or acoustic methods and a faster alternative to coring. Taken at fixed intervals along parallel grid lines, these readings reliably indicate where further detailed examinations should be conducted. A major strength of this system is that decisions can be made to explore the area in greater detail or to look elsewhere, while the survey ship is still on site.

1.5 PROBLEMS OF DATA STORAGE AND COMMUNICATIONS

Development of sensors for underwater use has not resulted in a clear trend towards miniaturisation, perhaps for obvious reasons. There have been, on the other hand, clear trends to increase sensitivity and speed of response – some aspects of which are covered within this volume. However, the primary considerations for most sensors and sensor systems are: size, weight and power consumption. Another trend in instrument development is to integrate an ever increasing number of measurements into a single instrument "package".

A major problem is keeping seawater away from the instrumentation and the electronics. It is said that putting an instrument several miles underwater is technically more difficult than sending a space craft into outer space. The pressure at 4 miles depth (the average depth of the Atlantic) is 10000 pounds per square inch – more than the test pressure of most commercial gas cylinders! Instruments have to be designed to operate under some of the harshest conditions on the surface of this planet and the ocean can be the most hostile and unforgiving place on earth. Usually this is accomplished by keeping all the sensitive electronics within a pressure housing, pressure-rated to at least 1.5 times the working (or operational) depth of the instrument. This is expensive and often involves complicated wiring arrangements utilising multiconnector bulkhead connectors. These, the armoured cable, and other fixtures used to hold the assembly totally out-weighs the cost of the instrument inside. A solution successfully used in the Southampton Oceanography Centre has been to simply encase the electronics in resin. The adhesion of the resin to the leads and other connections prevents seawater intrusion, and there is no pressure housing. Certain electronic components cannot be used because they would be compressed as the device was lowered into the sea (relays, for example). This technique also means that the instrument is regarded as "disposable" because it is impossible to retrieve or repair the electronics once encapsulated. However, the absense of a pressure housing produces considerable savings.

Sensitive measurements often require the measurement of extremely low currents. Currents in the microamp range (10^{-6} amps) are not uncommon, and nano- and femtoamps have been known. Such currents are normally converted to measureable voltages which are buffered and/or multiplexed before conversion into a digital data stream. Several excellent data converters, based on sample-and-hold technology allow these low curents to be digitally sampled without any complex analogue electronics, often being fed directly into a microcontroller or memory device. Such devices have wide application throughout all sensor technologies.

Data sampling rates as high as 20 Hz can present potential problems if, for example, several sensors are simultaneously recorded together with a date/time stamp. However, data storage is relatively easy, and many excellent and versatile devices exist. 20-bit data sampling, averaging a 10 Hz frequency and recording results every 60 seconds requires an approximate storage capacity of 2 Mb for 1 months work. This is not a great demand for modern computer equipment, but does represent a considerable design consideration.

Hydrographic cabling oftens presents a significant inductance to the transmission of signals along its length: a high grade wire might typically have a resistance of 20–30 ohm/km and a capacitance of 100–150 pF/m. RS232 transmission protocol is not good enough to be transmitted more than 50 m – so some form of frequency multiplexing is often employed to superimpose the signals on the armoured (single conductor) power cable. Frequency shift key (FSK) format is a commonly employed method. However, the maximum data transfer rate is often limited, for instance the maximum baud rate transmitted along a 3000 m cable is 19200 bps, and for that along a 6000 m cable is 9600 bps.

Data transmission across a multi-sensor instrument, such as a moored instrument buoy or chain, requires that the data is communicated around the system in much like an InterNet protocol. Several inexpensive RS232 transducers have evolved, principally for building system management markets, which have been successfully adapted for marine systems (LonWorx is one such example). There is an oceanographic autonomous vehicle (auv) which has been designed to "pop" up to the surface, transmits its data to a special VHF relay facility which is then viewed in real-time on the web. "Virtual instruments" are those that can be controlled or interrogated over the InterNet and can similarly distribute their data and results. Virtual instrumentation is still in its infancy, but it will inevitably become an important aspect of future electronic design. Interactive web sites and/or virtual instruments are emerging as an especially valuable resource for observers and modellers alike.

When sensor systems are beyond line of sight or the reach of radio transmission, the overhead of communicating the data in real time can be considerable. Data streams from remote instruments can place a considerable burden on satellite communications systems, for example, which have extremely limited data transmission bandwidths. The time to recover a message relayed from an Argos data relay carried on board low earth orbiting (NOAA series) satellites is dependent upon the length of time the message is allowed to transmit and the latitude. A typical length of time to recover a relayed message is about 8 hours for a 500-byte message when received from a polar location, and 12 hours for a 250-byte message from a mid-latitude location. Longer messages can be accommodated but require longer transmission times. There are two satellites about 850 km (527 miles) above the Earth. Both circle the Earth from pole to pole 14 times a day, orbiting approximately every 101 minutes, scanning a swath over 5,000 km (3,000 miles) wide.

1.6 PROBLEMS OF SENSOR DESIGN

The majority of problems concerning the development of an underwater instrument concern some very practical problems. Oceanographic sensors for deployment on databuoys must be small, low in power consumption, rugged, stable, sensitive and precise enough for periods of typically 30 days or more. For deployment on autonomous underwater vehicles (auv's) or towed platforms at speeds of up to 4 m s^{-1} the sensors also need to have fast response times (<10 s) and compensate for the effects of flow, temperature, and pressure changes. If a sensor is towed through the water at (say) 10 knots, and is only capable of taking measurements every 10 minutes, then the finest spatial resolution that can be observed will be just over 300 metres across. The offered resolution needs to be appropriate for the expected spatial coverage. If measurements cannot be made continuously, then the sampling time needs to be related to the speed through the water (or water past the sensor). The discrete nature of the measurements limits the number of samples that can be taken *in-situ* or per unit distance. Ideally, sampling time should be as small as possible, commensurate with the expected changes in concentration per unit distance. Electronic design is an area that requires special attention and detail. Ground or earth loop problems can be problematic, and initially difficult to identify. Battery operation is essential – which in itself often alleviates earth loop problems. However, galvanic interactions with the instrument housing, the hydrographic wire, or the carousel can plague readings. Severe meteorological conditions can toss a surface buoy around with phenomenal force, and for long continuous periods of time. The boards, cables, wiring and all interconnections have to be selected and secured appropriately. Corrosion of the materials used in the instruments can also be a serious consideration – especially when an instrument deployment interval of several months is being considered.

Oddly, size has never introduced a restraint on marine sensor design. Small size, low power requirements and low production costs do not compare with the cost of going to sea. Underwater housings have always been (relatively) large – there is ample room to include as much electronics (and wiring) as necessary. Probably for this reason alone, there has been no tendency towards microfabrication. Because of the individual nature of chemical sensor instruments, they largely use traditional printed circuit techniques, or even wire-wrap – which is especially prone to corrosion caused by salt. Thick film sensors have seen limited use but are seriously inhibited by the impure composition of the (gold, silver, platinum and carbon) inks used to screen print the sensor electrodes. The presense of small quantities of contaminant metals often causes the response to differ from that of the pure metal electrode. The Canadian Navy, however, has investigated the use of disposable thick film gold electrodes for the rapid measurement of dissolved oxygen sensors remarkably like an XBT (expendable bathythermograph).

Datalogging and direct computer interfacing eases the data management burden. The demands of many modern sensor systems, which often employ a number of sensor "suites", requires computing power that is beyond the capacity of most simple microcontrollers. There are many commercial PC systems (using 486 processors, and up) which have been minaturised, and which are very capable of performing the complex arithmetic and data assimilation. Artificial neural network analysis is a new tool that could be valuable for examining the spectral content (or "fingerprint") of optical sensors – especially scanning spectroscopic methods (such as fluorescence) which involve time-dependant 3D data. Large numbers of spectra can be generated quickly, and it is the "pattern" of peaks and valleys that identify the component. However, peak searching and matching is a laborious task for which existing software is not entirely suitable.

1.7 PROBLEMS OF CALIBRATION

Calibration of underwater sensors and systems is difficult, and there are many attendant complications. Calibration is essential to the absolute accuracy of a sensor, and often to its long-term deployment. "Drift" can often be ascribed to variations in temperature, but may also be caused by deterioration of response, degradation of the sensor chemistry, and a multitude of other problems. The stability of many reagents cannot be guaranteed for the length of time many systems need to be kept in the water. The thiosulphate solutions used in the standard Winkler titration for the determination of dissolved oxygen, for instance, begin to degrade after only a few days. The majority of flow injection analysers are now designed with a valve inlet that can select either the sample stream, a blank or a standard solution of known concentration. Newer systems are now being developed that can perform an operational *in-situ* serial dilution of a standard solution, based on stepper motor driven injection syringes.

Calibration is a difficult issue, and should be included as a feature early on in the design and prototyping of sensors. It can be difficult to maintain calibration, and there is often a trade-off between sensitivity and response rate. Many underwater instruments now contain self-calibration procedures, both to determine consistency in the reaction chemistries, and to maintain performance of the electronic circuitry. For reasons which are often more obvious to electrical engineers, calibration routines are usually combined with fault detection and fault diagnosis – which is a continuous real-time requirement. Once a fault is determined (such as a calibration measurement going beyond a set point, or low- or non-response from a sensor) a warning should be issued and appropriate action taken (reconfigure instrument into "safe" state, power-down, rerun calibration standards, etc). However, "trends" in the data may also reflect long-term variations in instrumental behaviour or stability. Deviations between expected and observed behaviours may suggest or predict an upcoming fault.

Calibration of electrochemical sensors present some very interesting challenges, especially for the dissolved gases. It is not usually possible to manufacture "standard solutions" containing a known concentration of a dissolved gas in seawater, that can be stored for use in a flow injection system, or diluted easily for a series of standard solutions. Oxygen sensors can easily be calibrated at zero gas concentration, but other concentrations are difficult to manufacture and keep stable. If the sensor can be temporarily exposed to the atmosphere, this may serve as the basis of a single-point calibration. However, this approach has associated technical difficulties. It is possible to relate the measured current during the electrolysis of water to the concentration of gas evolved at the cathode, and use this as the basis of the calibration of a dissolved oxygen sensor.

Commercial bench-top systems (AMT Analysenmezztechnik GmBH) presently use water for the on-line production of pure oxygen standards, but there has not, as yet, been any further development of seawater calibrant systems.

Calibration *in situ* is often not absolutely necessary if a water sampler is used to simultaneously collect discrete samples which can provide subsequent calibration or reference points for the detector response. Large volume water sampler bottles (Plates 1.5 and 1.6), which can be triggered at predetermined times or locations, are ideal – but still require operator intervention. Calibration remains an area that has yet to be coherently addressed for the majority of sensors. Designing a system to "pump" or introduce calibration fluids or reagents to a sensor, effectively converts a sensor into a chemical analyser (see preface for an

Plate 1.4 A series of six platinum redox electrodes (50 cm apart) are contained within a 3 m steel tube. The instrument is lowered to the seafloor and its own weight pushes it into the surface sediments, making measurements of redox potential at discrete intervals. The instrument is then lifted a few meters above the sea floor and "towed" to the next survey site. (Photograph courtesy of "Petrosurveys inc., Dallas" and Sea Technology, Compass Publications). *See* Color Plate 4.

Plate 1.5 A view of a recently developed multi-port water sampler. A stator in the center of the apparatus rotates around each of the ports on the outside and pumps a filtered water supply into any number of bags connected by tubes to the nipples. On retrieval of the instrument, the tubes are then tied or knotted and then individually removed for subsequent analysis. (Photograph courtesy of R. Scrivens, WS Ocean Ltd.). *See* Color Plate 5.

explanation between the terms sensor and analyser). This introduces additional mechanical complexity which often makes sensors difficult to operate in the ocean, and yet this is the very direction that the majority of recent developments have taken. Sensors need to be simple, robust and make appropriate measurements, with the minimum of complexity.

Plate 1.6 The multi-port water sampler located within an undulating towed fish prior to deployment. In this scenario, the water sampler would be programmed to take samples at timed intervals or commanded from the surface via an umbilical. (Photograph courtesy of R. Scrivens, WS Ocean Ltd.). *See* Color Plate 6.

1.8 BIOFILMING AND BIOFOULING

Marine biofouling is the (undesirable) biological growth of organisms on underwater surfaces (Plates 1.7 and 1.8). Seaweed (algae), sponges, anemones, barnacles, limpets, tubeworms, and mussels are some of the most common visible forms of biofouling. The calcareous encrustations left by these animals can be permanent and long-lasting – and of great financial concern to the marine sector in general. The effects of biofouling is also of great concern to the sensors industry – there is no immunity and the long-term reliability of certain sensors is therefore questionable. From practical experience of year-round operation of instruments at sea, the response of transmissometers can alter dramatically after continuous immersion of only 3 to 4 days. The inability to control biofouling has serious implications for the deployment of underwater instrument packages for long-term environmental monitoring, particularly those which have sensor surfaces exposed directly to seawater (such as electrochemical and spectroscopic sensors).

It is widely accepted that biofouling involves at least three main stages: biofilming, bacterial/microbial growth, and colonisation (see box) (Characklis and Escher, 1988). Adsorption of dissolved organic substances onto immersed

(a)

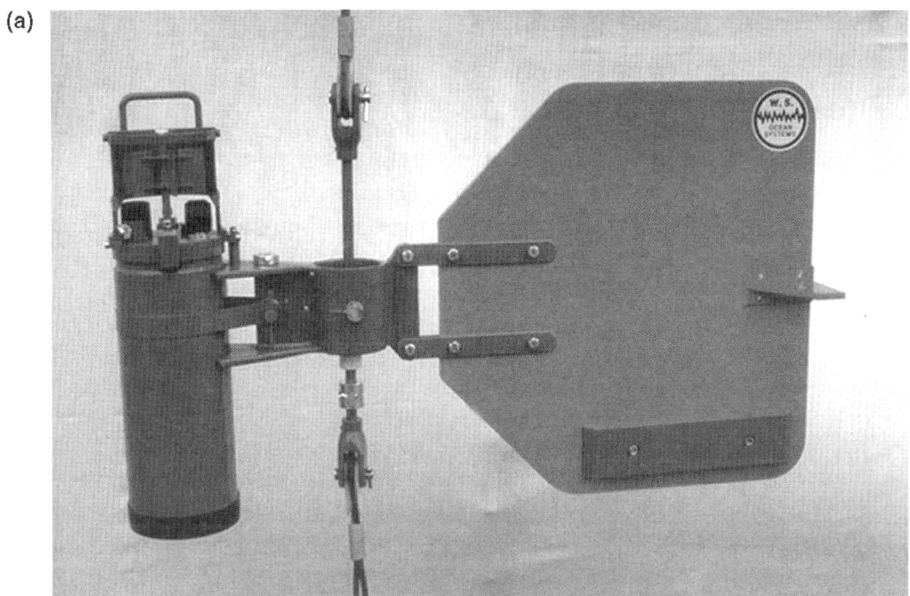

(b)

Plate 1.7 (a) and (b) Before and after shots of an Anderaa current meter deployed in the tropical Atlantic after only 2 months. Shellfish growth all over the outside of the instrument stopped the rotor from turning within 6 weeks of the experiment. *See* Color Plate 7.

(a)

(b)

Plate 1.8 (a) This was the underwater section of a moored buoy system to measure nutrients in estuarine and coastal zones. The frame and instruments have been covered in algal growth (seaweed) after only 6 weeks on station. (b) The disconnected individual instruments shows that the cables, connectors and sensor surfaces are totally shrouded in algal growth. The left-hand housing was a sample inlet system, the middle instrument was used for optical measurements, and the housing on the right contained electronic components used to control the equipment. (Photographs courtesy of P. Wright, Southampton Institute). *See* Color Plate 8.

surfaces constitutes a very early stage in biofouling, and has a direct effect on the sequence of events leading ultimately to macrofouling. A "conditioning" (sub-nanometer) macromolecular layer, generally glycoprotein in nature (Baier, 1980; Chamberlain, 1992), allows subsequent colonisation by bacteria and other microbes (Charackis and Cooksey, 1983; Marshall, 1985). The presence of only a small quantity of surface-active material is sufficient for successful attachment. The second stage of biofouling is slower, and environmentally more variable. It is evident that "films" adsorb virtually the moment that a surface is immersed in seawater, and continue growing (often at an exponential rate) for considerable periods afterwards. The surfaces of materials immersed in seawater may be altered by the deliberate addition of "primary films" to enhance or delay the onset of biofouling (ZoBell and Allen, 1935). Adsorption of specific compounds (or anti-fouling agents) onto a surface is sufficient to prevent the simultaneous colonisation of microorganisms and thus change the suitability of that surface for subsequent biological settlement. This has long been recognised, for instance, by marine biologists who exposed a number of model surfaces to cell-free extracts of adult barnacles (*Balanus amphitrite*, for example) to enhance the settling of the larvae (Crisp *et al.*, 1985).

Box: Stages involved in biofouling

Filming, or conditioning, stage	*Bacterial/microbial growth*	*Colonisation stage*
Physico-chemical adsorption of hydrophobic organic compounds. Simple- and multi-layer molecular films attach or adhere themselves to surface. Typical timescales range from seconds to hours, and can be significantly affected by surface characteristics, and physico-chemical properties (temperature, water movement, ionic concentration, etc.)	Bacterial and microbial activity within the film causes the release of extracellular exudates. These "biofilms" exhibit a polysaccharide/proteinaceous nature, often referred to as a "slime", and can stabilise within hours. This biofilm is believed to determine subsequent events. This stage can also initiate or accelerate biodegradation and/or biocorrosion.	The presence of the exudates and the bacterial/microbial colonies traps further particles and organisms. These are likely to include algal spores, marine fungi and protozoa – some of which may involve chemo-sensory stimuli. The further attachment of higher organisms, especially the free-swimming larval forms of barnacles, limpets, bivalves, etc, occurs over much longer time scales (weeks to months). Many of these organisms respond to chemical cues. Barnacles, for example, are attracted directly to their own species, and can only reproduce if the adults are within a few millimeters of each other.

There are a variety of methods known to prevent or minimise biofouling: ultrasonics, ultraviolet light, mechanical wiping, sparging with compressed air, surface-soluble films, poisonous coatings, and *in situ* generation of chlorine. There is no one single effective or satisfactory technique. The use of ultrasonics, *uv* radiation and compressed air are power-hungry. The use of wipers is limited to flat surfaces, and prone to mechanical complexities and problems. Protective coats and paints may not be entirely compatible with electrode systems and/or membranes. In short, biofouling is a complex process that requires more fundamental understanding of the problem before better solutions can be sought. Very little research has been carried out into the causation (and even less into the control of) biofouling.

Adsorbed organic compounds form rapidly on a variety of surfaces, and can be observed by a number of techniques. Biofilming has a crescendo, cumulative effect on many chemical sensors – especially upon optical windows and electrodes. The adsorption of dissolved organic compounds may alter surface properties in ways that would affect subsequent attachment and colonisation. The attraction of an organism for a surface may be influenced by the changes in hydrophobic/hydrophilic changes, the surface charge and charge density, and the chemical functional groups available for reaction. Baier *et al.* (1968) have reviewed the surface chemistry and physics of biological adhesion emphasising wetting phenomena. Marshall *et al.* (1971) emphasise the importance of electric charge and dipole interactions in attracting bacteria toward surfaces, at an early stage in film formation. Certain bacteria produce compounds that are attractants to larvae (although they are not an essential prerequisite). It is possible through careful observation of the early stages of biofilm formation that subsequent stages of macrofouling can be retarded, if not eliminated.

Describing the second stage of biofouling as a "film" is somewhat misleading because there is a complex 3-dimensional association of bacteria, diatoms and polymeric molecules of widely differing properties and character (Plate 1.9). Diatom biofilms typically grow up to 500 μm in thickness with up to 3×10^5 cells cm^{-2}. Biofilms can be observed by a number of different methods, including spectroscopy and electrochemistry. Both methods are simple, direct and non-destructive, and are therefore invaluable for the investigation of the growth and characterisation of the films as well as the performance of the substrates.

Biofouling is a highly dynamic process. Water flow, seawater properties, surface characteristics and a host of other environmental factors all influence the onset and duration of biofouling. Fouling is more intense in the tropics, for instance, where the breeding season is often uninterrupted, and growth is rapid in the warm waters. The present control measures (mostly paints containing organotin or organometal compounds) may reduce the macrofouling aspect, but do not prevent the initial biofilming. Some slow-release paints containing herbicides have been tested with some success, but they all contaminate seawater to some extent – sometimes eliciting irreversible toxic effects. Once initiated, filming can subsequently lead to localised or general biocorrosion of the surface.

It is not yet fully understood why certain organisms do not foul or do not become fouled themselves, and there is some evidence that there is no clear succession driving biofouling (Whetstone, 1992). Clear chemical cues enhance the colonisation of surfaces. Effort is being directed towards the investigation of the basic stages of colonisation, including: the role of calcium (which is thought to inhibit signal transduction within the organism)(Clare, 1996b), pheromone

Plate 1.9 Scanning electron micrograph of the surface of an antifouling paint fouled by a marine diatom biofilm. Large diatoms are Achnanthes; small diatoms are Amphora spp. Diatoms and bacteria (not visible at this low magnification) produce extracellular polymeric substances (EPS) which forms a matrix, binding the cells together into a compact biofilm. The scale bar is 100 μm. (Photograph courtesy of M. Callow, University of Birmingham).

transmittor chemicals (glycoprotein in nature), and synthetic furan and lactone-type compounds (Figure 1.8). Over 90 natural product antifouling compounds have so far been identified (Clare, 1996a). There would appear to be certain relationships between chemical structure and activity which can be used in the search for synthetic antifouling compounds, or paints. There is a massive search for synthetic analogues to emulate the natural antifoulants produced by certain key marine organisms such as echinoderms, sponges and corals. The epidermis of such organisms is composed of a negatively charged, hydrophilic proteoglycan (McKenzie and Grigolava, 1996) thought to mimic a non-stick coating. The formulation of natural and synthetic chemical antifoulants has produced a variety of complex and exotic molecules, some of which show great promise (e.g. polysiloxanes) in minimising the effects of colonisation. Such complex polymeric films also exhibit a "self-healing" ability when small scars are made across a surface, which may reduce the later effects of biocorrosion on the underlying structure/material. Few research programmes incorporate screening programmes to select new toxic compounds because of the current legislative and environmental safety aspects. The two fundamental research directions presently followed entail

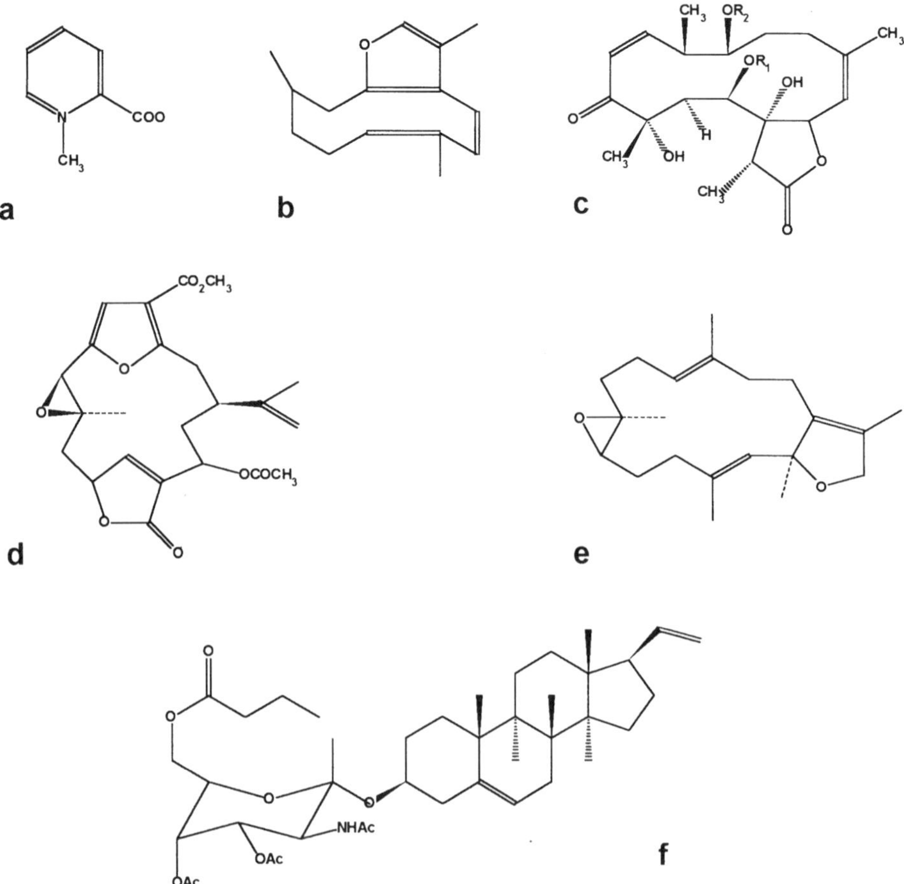

Figure 1.8 A few examples of natural compounds containing furan and lactone-type rings, that exhibit antifouling properties. Structure **a** is a homarine (Tergett *et al.*, 1983); structure **b** a furangermanacrene (Izac *et al.*, 1982); structure **c** is a generic terpenoid structure termed a renillafoulin (Keifer *et al.*, 1987); **d** is a pukalide (Rittschof *et al.*, 1985); **e** is 14-hydroxycembra-1,3,7,11-tetraene (Coll *et al.*, 1987) and **f** is a aminogalactose saponin (Bandurraga and Fenical, 1985). All compounds have been extracted from octocoral species.

(i) formulation of "release" coatings ("self polishing copolymers") that interfer with the ability of an organism to adher to a surface, thus rendering it easier to clean, and (ii) a natural product approach, where natural marine products are incorporated into a coating to repel marine organisms, without killing them.

Spectroscopy and electrochemistry are simple, direct and non-destructive methods used to study the growth and characteristics of molecular films. Fibre optic sensors have been used to indicate the development of fouling layers. A small diameter tip (1 mm) can be easily integrated into any surface to be

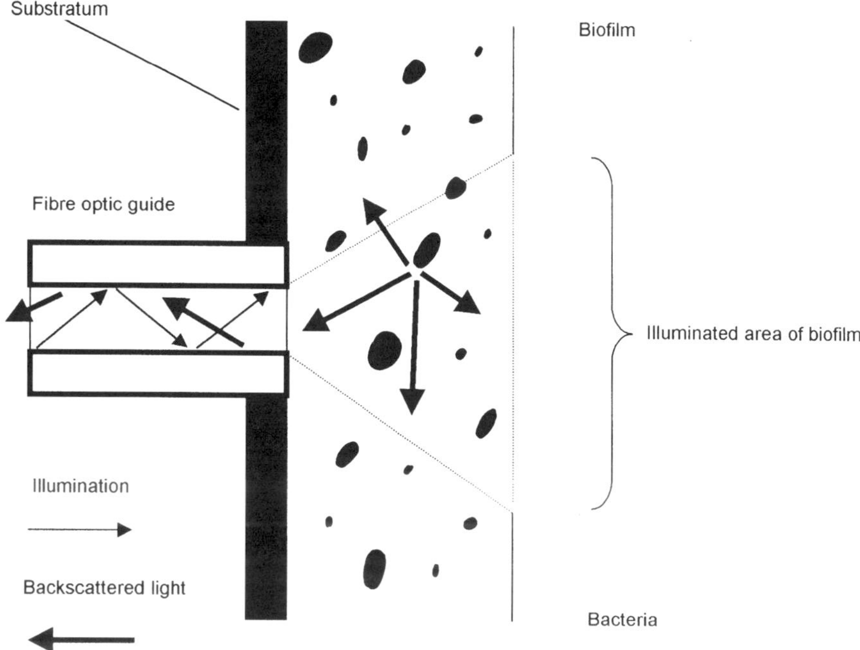

Figure 1.9 Fibre optic sensors have been used to indicate the development of fouling layers. The tip illuminates and reflects light within a well-defined area adjacent to the surface. The light beam is randomly scattered by various objects (bacteria, diatoms, etc.) within the film and the intensity is approximately proportional to concentration of scattering particles (different particle sizes will reflect different amounts of light). A small diameter tip (1 mm) can be easily integrated into any surface to be monitored. The light signal usually needs to be incorporated into some form of lock-in amplification to reduce effects of ambient light interference.

monitored. The tip illuminates and reflects light within a well-defined area adjacent to the surface (Figure 1.9). A simple and expensive arrangement, based upon an LED emitting light at 580 nm modulated at 150 Hz, demonstrates the cumulative growth of organic compounds at glass surfaces (Figure 1.10). Although this simple optical experiment cannot accurately measure the actual film thickness (because the initial deposits mask and possibly quench the reflected light from subsequent layers) it demonstrates that biofilming occurs as soon as a surface is immersed in seawater, and that logarithmic accumulation of organic films occurs over intervals of hours to days. Further experiments transmitting infrared beams through various optically transparent iR windows (also known as "coupons") exposed to recirculated seawater reveals the chemical nature of biofilm development (Figure 1.11). Although the instrument is sensitive to water absorption, and compensating the signal for ageing effects is complicated, good quality and reproducible iR spectra can be obtained over periods of several days.

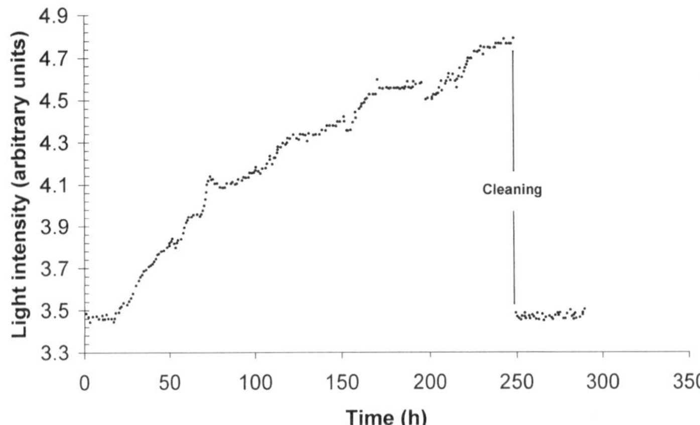

Figure 1.10 A simple and expensive arrangement, based upon an LED emitting light at 580 nm modulated at 150 Hz, demonstrates the cumulative growth of organic compounds at a glass surface. The medium had an artificially high amount of nutrients to illustrate the exponential growth, but the cleaning of the light cell immediately restored the initial signal. The biofilm develops on both the light source and the detector windows (1 cm diameters), so the signal is effectively doubled.

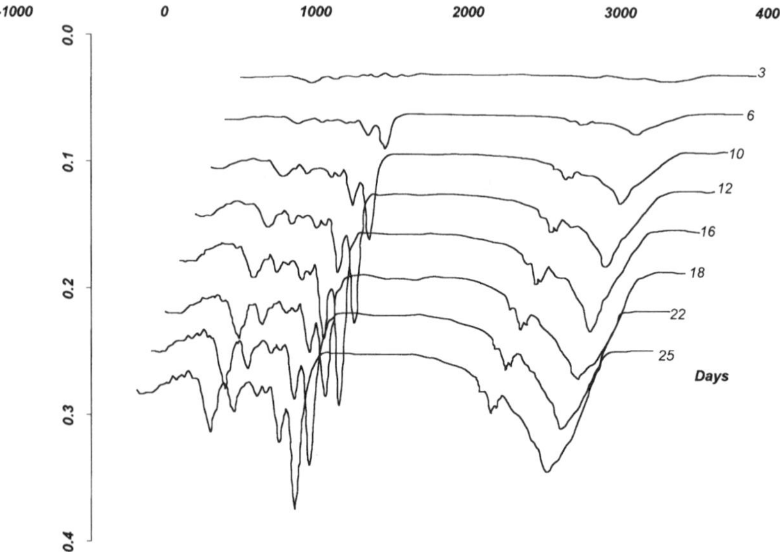

Figure 1.11 Biofilming occurs as soon as a surface is immersed in seawater, and logarithmic accumulation of organic films occurs over intervals of hours to days. Experiments transmitting infrared beams through various optically transparent iR windows (also known as "coupons") exposed to recirculated seawater reveals the chemical nature of biofilm development.

The increase in the protein absorption and sugar (polysaccharide) absorption bands (1500–1700 and 900–1100 wavenumber regions) indicates an approximate rate of deposition of biofilm on the iR window. Tryptophan fluorescence (in the ultraviolet) can also be used as a "marker" for the growth of biological material on various surfaces.

AC impedance and electrochemical noise measurements, so-called "transient" techniques, can be applied to the direct analysis of surface-active substances in seawater (Formaro and Trasati, 1968). Electrochemical methods are generally more sensitive than spectroscopy. All adsorbed organic compounds of biological origin lower the surface energy, and induce an overall negative electrical charge. An electrochemical response is not due to a single isolated event, such as a single electron transfer process. A charge transfer occurs at the end of a succession of coupled phenomen, perhaps in parallel with each other. These include: (i) the faradaic reaction (usually a reaction of interest such as the reduction of dissolved oxygen), (ii) the reaction of adsorbed material, and (iii) non-faradaic charging of the double layer. The "double layer" capacitance is depressed markedly depending on the chemical nature of the adsorbing molecules, as well as on the physical nature of the surface. It is easy to demonstrate the rapid formation of films on platinum electrodes. The electrochemical method, AC tensammetry, indicates a fast (almost exponential) rate of film deposition over timescales of minutes to hours (Figure 1.12) but unfortunately gives little information about the chemical character of the film.

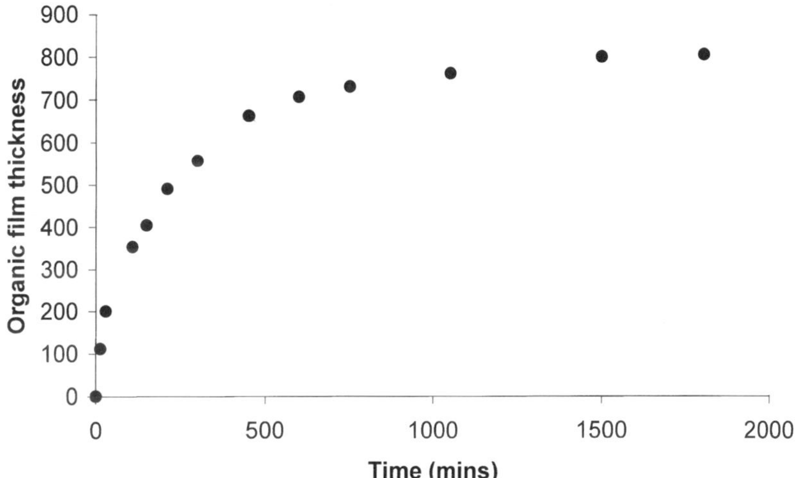

Figure 1.12 The electrochemical method, AC tensammetry, indicates a fast (almost exponential) rate of film deposition over timescales of minutes to hours but unfortunately gives little information about the chemical character of the film. Molecular films form very rapidly on fresh platinum surfaces during the first few minutes of exposure, growing at an appreciable rate for a period of hours, and continuing to grow thereafter. For reference, 24 hours is equivalent to 1440 minutes.

Biofouling is an undesireable phenomenon and we are only part way towards effective countermeasures in its control. Nickel/chromium electroplating is especially effective at protecting exposed steel surfaces from the effects of corrosion, and anti-fouling paints protect metallic and plastic surfaces from fouling, but sensitive chemical sensor elements (optical lenses, electrodes, membranes, and so on) are particularly vulnerable and cannot easily be protected without significantly altering their response. Synthetic chemicals formulated to emulate natural compounds that exhibit anti-fouling properties have the same disadvantage. There has been some success with immobilising anti-fouling substances into optically transparent materials, such as quaternary ammonium compounds into polymethacrylate (Parr *et al.*, 1997). Electrochemical methods probably hold the greatest potential for the control of biofilming because the techniques used to measure the extent of film growth can also be used to regenerate, and condition, the electrode surface (Figure 1.13). One method is to make transparent sensor surfaces electroactive by depositing (sputtering) a very thin SnO_2 layer. A "train" of voltage signals can then be applied (a form of mild electrolysis) to activate or passivate the surface, accordingly. Alternatively, an electrolysis voltage can be applied to two electrodes (glassy carbon, for instance, is stable in seawater) within the general vicinity of the sensing surface. The gaseous

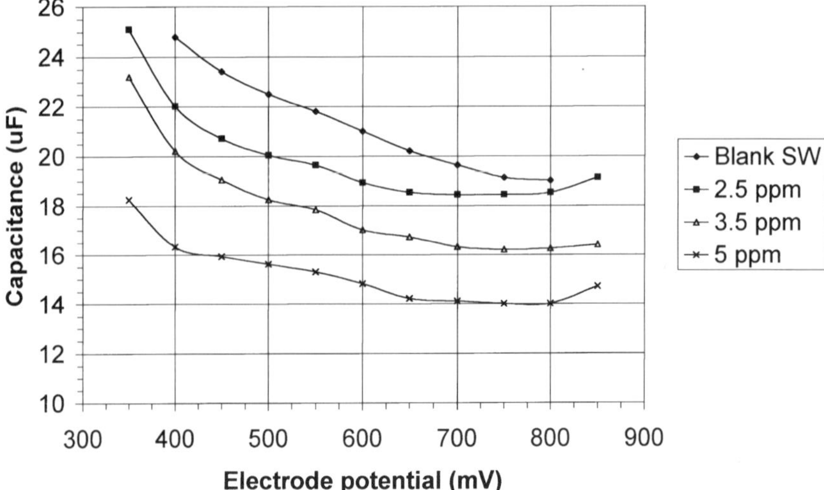

Figure 1.13 Electrochemical impedance is a function of both electrode potential and organic matter concentration. Here, several seawater samples of different organic carbon concentration were taken from around the coastline of the South of England and the electrode capacitance (impedance) estimated. Dissolved organic carbon (DOC) concentrations were subsequently analysed by a wet chemical oxidation method. The electrode is cleaned *in situ* inbetween the individual measurements by applying a voltage "train" of pulses to alternately oxidize and reduce the electrode surface. This removes the film and regenerates the electrode prior to the next measurement.

by-products of electrolysis are often sufficient to sterilise the seawater, and prevent macrofouling – this has been used successfully on a larger scale within harbours and marinas for the prevention of algal growth on over-wintered boats and yachts.

Biofilming and biofouling is not a big problem for sensor systems onboard towed vehicles (towed fish, undulators, for example). The movement of water across the sensing surfaces, and the frequent deployment and retrieval, mean the opportunity for colonisation is much reduced. Static, free-floating and moored systems present greater consideration to the long-term effects of biofouling, and spectroscopic lenses, optics and fibres will be much more vulnerable. Effective countermeasures against biofouling depend on the monitoring of biofilm formation (Flemming *et al.*, 1996). However, chemical sensors are rather specialised and require an alternative technology. It is likely therefore, that the ultimate development of chemical sensors in marine environments may be retarded by the lack of suitable protection to minimise or prevent fouling. It is undoubtedly an area that would benefit from further research, and will require a multidisciplinary approach.

Investigations of the surfaces of certain marine organisms have, however, often found a notable lack of fouling. Some biological systems are therefore employing very successful antifouling mechanisms. Biological mechanisms of antifouling are of intense interest both because they are manifestly superior in effect and persistence over human technologies and because they may be "environmentally friendly".

The challenge, therefore, to marine biologists and materials scientists, is to discover how these biological antifouling processes operate and transfer this technology to the engineering of antifouling surfaces.

There is a considerable body of research into biological aspects of fouling. Much is concerned with actual fouling organisms with the aim of determining the factors that promote and inhibit settling of fouling organisms on man-made structures. Research into how marine organisms prevent or ameliorate fouling on themselves is a much smaller and less developed field. Interest has mainly focused on the release of natural biocides. Information on potential biocides is extensive but unsatisfactory because there is usually no evidence that these compounds are actually released nor that antifouling is their primary role. Other studies have looked at the surface of the organisms, in particular the physicochemical properties of the surface and its rate of turnover. Copious production of mucus secretions is a commonly suggested antifouling mechanism in both marine plants and animals. There is some evidence that this indeed may be its function. Work on the actual physico-chemical nature of natural antifouling surfaces is rare.

At present, while existing research has highlighted some promising lines of enquiry, there does not appear to be sufficient knowledge to produce a complete physical analogue of a natural antifouling mechanism. This is largely due to the fact that previous studies rarely considered more than one component of an antifouling mechanism and there is little information concerning the relative importance of different components to the overall mechanism. Artificial antifouling mechanisms are rarely completely effective against the whole range of fouling organisms. This is particularly true of bacteria whose growth is often observed to be a prerequisite for macrofouling to occur. Success of natural antifouling processes is likely to be due to the use of more than one mechanism. Three possible mechanisms can be broadly defined. These are:

- Having a surface that is of appropriate surface energy and charge (too slippery) for fouling organisms to adhere to;
- Replacing the surface rapidly so that any adhering organisms are lost;
- Having a surface too toxic or unfavourable for fouling organisms to settle, metamorphose, grow or reproduce.

1.9 IMPORTANCE OF CHEMICAL SENSORS TO OTHER SCIENTIFIC DISCIPLINES

Data assimilation is an objective method for combining or comparing observed data with modelled results. For example, if one had an irregularly spaced grid of measurements, linear interpolation would generate a complete "map" of an area. If additional information is known about the region (such as circulation, or heat gain/loss) it would be possible to enhance the map to include both the observations and associated dynamical processes at work. A common problem with physical, biological and geochemical models is the consideration of errors and uncertainties in the original data. Although data sets are undoubtedly growing, there is still a scarcity of information available to supply the demand for modelling the ocean. The lack of data limits the ability to develop and assimilate computer models and is directly linked to the cost and effort required to get man, instruments and equipment to sea.

Much has happened in the few years prior to the publication of this book. There have been a number of international multidisciplinary oceanographic research programs aimed to collect oceanographic data sets to enable computer modelling to provide an unprecedented view of the global ocean. Ocean general circulation models (OCGMs) There are large and diverse data sets being collected, only some of which are chemical in nature. The scientific community is currently organising a global plan to study natural and human-induced phenomena which both affect and mirror global change, to see how the Earth works as a system. Central was a NASA contribution known as "Mission to Planet Earth" which concentrates on intensively mapping the thermal properties of the entire Earth's surface. The World Climate Research Program, sponsored by the International Council of Research Unions (ICSU) and the World Meteorological Organisation (WMO) will focus on the physical aspects of climate, clouds and radiation, ocean circulation, and air-sea inter-relationships. ICSU has also established the International Geosphere-Biosphere Program to explore the interactive chemical and biological processes regulating the total Earth system. The total effort integrates various short-term, simultaneous research missions, and provides the first coordinated measurements between the atmosphere, oceans, solid earth, and hydrologic and biogeochemical cycles. A number of satellite and Earth probes are now in constant use supporting this effort. Data collected by the Nimbus-7 Coastal Zone Color Scanner (CZCS) in orbit between 1978 and 1986 were used in planning the North Atlantic Bloom Experiment, the Equatorial Pacific Process Study, and the Arabian Sea Expedition. NASA recently launched (October 1997) the Sea-viewing WIde Field of View sensor (SeaWIFS) that provides broad synoptic ocean colour imagery to support field work. The temporal and spatial variations in surface ocean productivity that these images reveal are dramatic.

Many national research councils and funding bodies have announced sensor-related initiatives to supply the demand for various sensors in all scientific fields. Smaller sub programmes promote sensor development, and increase industrial interaction – mostly funded through international funding agencies, such as: the European Communities Marine Science and Technology (MAST), joint global ocean fluxes (JGOFS), and world ocean circulation experiment (WOCE), climate variability and prediction research programme (CLIVAR), global ocean ecosystem dynamics (GLOBEC). We need to improve our knowledge of the current state of the ocean and our ability to predict future variations on monthly, seasonal, annual and decadal time scales. EuroGOOS (the European component of the Global Ocean Observation Systems, GOOS, programme) has identified key chemical parameters in its "FerryBox" Project (a project designed to examine how ferries could be used to gather routine oceanographic data). These include: photosynthetically available radiation (PAR), temperature, conductivity, turbidity, transmissivity, fluorescence, phosphite/phosphate, nitrite/nitrate/ammonia, silicate, pH. Other additional instruments, beyond this "standard suite" of measurements, can be deployed for specific mission requirements.

It would appear, however, that the majority of funding agencies, and the work of the various research establishments and authoritories is still "capacity building". Implementation of a number of sensor systems onto buoys will find uses in a number of environmental monitoring and surveillance coastal seas programmes. These will, in the future, provide a basis for an operational and information system for the management of regional seas and will have long-term political and economic implications. A long-term goal is to transfer the data and information into useful products and services required by a wide range of users, including governments, industry, science and individuals. Chemical sensors are only one component amongst many others essential to this goal, including: data collection, analysis and modelling, telecommunication, and continued product development.

1.10 MODELLING OF DATA FROM MEASUREMENT TECHNIQUES

Modelling of physical and chemical data is relatively "mature", and a number of acceptable computer programs exist. Atmosphere-ocean general circulation models play a central role in studies of natural variability and of anthropogenic climate change. However, there is insufficient knowledge of their behaviour. Currently, almost all (coupled) models drift away from the observed climate unless a large adjustment is made between the heat flux leaving the ocean and entering the atmosphere: this is an indication of problems in understanding the complex patterns and processes occuring in the environment. Biological dynamics are even less well defined. There is an apparent lack of data for the corresponding biogeochemical models – especially to predict the response of the oceanic carbon cycle to changes in climate, for instance. Initial efforts have crudely represented a number of fundamental processes such as: primary and secondary production, grazing, particle transport, remineralisation and the cycling of dissolved and particulate organic matter. Such models have so far been restricted to local or regional scales. The use of ocean colour data collected by satellite will benefit the development of this generation of models. Topics that are poorly understood but will have immense interest include: nitrogen fixation, nitrogen versus phosphorus limitation, trace-element limitation, non-Redfield ratios of carbon to nitrogen, active biological transport, vertical migration, planktonic size structures, eddy pumping, and the production/consumption of labile and semi-labile dissolved organic carbon.

1.11 FUTURE NEEDS AND REQUIREMENTS FOR GLOBAL COVERAGE OF CHEMICAL PARAMETERS

The need for chemical sensors in marine monitoring is growing. Basic marine water quality monitoring in developing nations is clearly necessary. The opportunity for market development is particular attractive as there are viable needs and applications in developed nations as well. The current emphasis on longer term automatic monitoring is to provide data on the status and recover of rivers, estuaries and in-shore regions as environmental programs move beyond point-source discharge abatement, and focus more on non-point source control, environmental control, and integrated coastal zone management. This approach requires water quality measurements throughout marine ecosystems, not just at selected discrete points. This also entails an integrated scheme of meauring both basic water quality parameters (dissolved oxygen, nutrients, chlorophyll, etc) as well as specific anthropogenic and industrial pollutants (pathogens, pesticides, and radionuclides). Different geographic regions have different priorities (Table 1.2). North America has instigated a considerable number of aggressive programs, including the U.S. Environmental Monitoring and Assessment Program (EMAP), National Oceanic and Atmospheric Administrations (NOAA) National Status and Trends Program, and Canada's Ecological Monitoring and Assessment Program. Western Europe has sponsored a number of inter-government programs to combat pollution in the Mediterranean, Baltic and The North Sea. Marine environmental monitoring programs are emerging in the Red/Black Seas, Eastern Europe, and Russia, but the level of monitoring is more basic.

Table 1.2 Relative priorities for contaminants among marine areas.

Regional Sea	*Caribbean*	*North Sea*	*Baltic Sea*	*Mediterranean*	*Red Sea*	*The Gulf*	*Asian Seas*	*Black Sea*
Algal toxins	High	Medium	Medium	Medium	Low	Low	High	Medium
Radionuclides	Insignificant	Medium	Low	Low	Insignificant	Insignificant	Insignificant	High
Herbicides & Pesticides	High	Medium	High	Medium	Insignificant	Insignificant	High	High
Litter & plastics	Medium	Low	Medium	High	Medium	Low	Medium	High
Human pathogens	High	High	Medium	High	Medium	High	High	High
Nutrients	High	High	High	High	Low	High	High	High
Dissolved oxygen	Insignificant	Insignificant	High	Medium	Insignificant	Insignificant	Medium	High
Synthetic organics	Low	Medium	Medium	Medium	Low	Low	Low	High
Petroleum (oil)	High	Medium	Low	Medium	High	High	High	High
PAHs	Low	Medium	Medium	Medium	High	High	Medium	Medium
Suspended particulates	High	Low	Medium	Low	Low	High	High	Medium
Trace metals	Low	Medium	Low	Low	Low	Low	Low	Medium
Phytoplankton pigments	Medium	High	High	Medium	Low	Insignificant	High	High
Pharmaceuticals	Insignificant	Medium	Low	Low	Insignificant	Low	Low	Low

(From: Health of the Ocean Panel of the Joint Scientific and Technical Committee for GOOS, "A strategic plan for the assessment and prediction of the Health of the Ocean: A module of the Global Ocean Observing System" U.S. Interagency GOOS project Office, NOS/NOAA, Silver Spring, Maryland 1996.)

CARICOMP (Caribbean Coastal Marine Productivity Program) coordinates the marine water quality monitoring activities in 25 laboratories in 19 countries. Instrumentation is often of the simplest variety: hand-held (thermistor, refractometer, Sechi disk, etc) and cost-effective. Technology requirements in Southeast Asia, where population growth is rapid, are modest by Western standards.

Coordinated, international directions for emerging marine water quality monitoring and instrumentation is provided under the auspices of GOOS (Global Ocean Observing Systems), particularly through the Health of the Ocean (HOTO) module. A key component is the identification of the priority contaminants and pollutants, and the likely timescales and magnitude of impact. The second stage is the establishment of a set of reliable, easily applicable biological distress indices which will define the extent and impact of water quality degradation. The HOTO panel has developed a "Framework for the Preparation of Regional Blueprints" which allow each region to set management goals, priority measurements and observations, sampling strategy and data products.

The technology for measuring many of the key chemical and biological parameters has advanced steadily over the past few years. Although these parameters have often involved intensive sampling procedures and lengthy laboratory analysis, the technology is gradually shifting towards automated, *in situ* instrumentation, sometimes with direct data transmission to ship or shore (Figure 1.14). This has prompted notable advances in platform development. We already have a number of surface, floating, moored and bottom-mounted platforms capable of transmitting data in real-time. In addition, AUV and ROV systems will be easily adapted to support marine monitoring exercises. The United Kingdom research community has developed AutoSub: an autonomous 5 m full-ocean-depth rated submersible, with 5 to 20 days endurance. AutoSub is used for transect work across ocean regions, deep-sea canyons, under-ice measurements, and shelf-edge transects. It has been designed in such a way to facilitate various payloads. Similar packages may be placed on ferries that follow precisely determined and predictable routes many times each year. Already, trials of prototype systems are underway in the Baltic, on the Norwegian coast and in The English Channel (termed "ferryboxes"). EuroGOOS has a fundamental interest in evaluating instrument packages to provide data on a routine basis to numerical models (Bosman *et al.*, 1998). However, the further development of chemical sampling and analysis technology still presents something of a considerable problem.

There is a need to explore the various seawater chemistries that can be easily measured, or exploited, to provide a broader range of chemical sensors. This book illustrates the general approaches currently taken to measure nutrients, essential elements, carbon dioxide, and other analytes. These research programs will undoubtedly continue to develop and improve their design and performance. *In situ* detection of carbon dioxide, for instance, is a major area that is fundamental to understanding the role of the ocean in controlling the Earth's climate. There are still too many unknowns about the sources and sinks of carbon dioxide to make accurate or reliable predictions about the future. The intimate

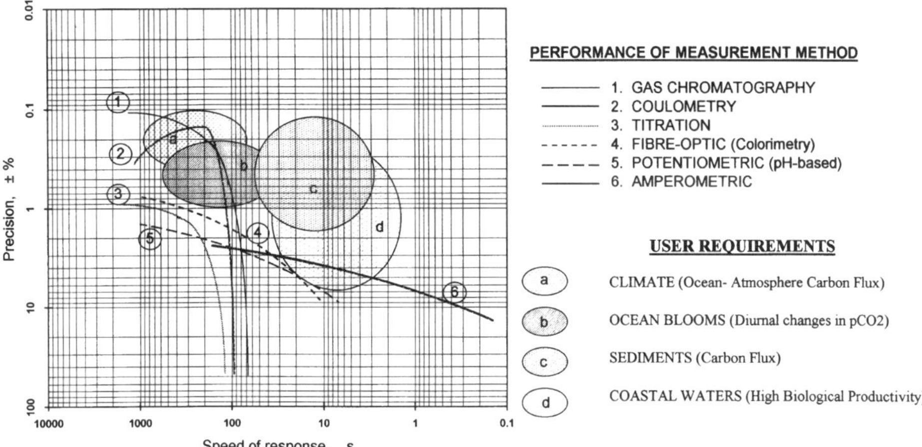

Figure 1.14 Comparison of various methods and precision versus speed of response for the determination of pCO_2 in seawater. Various users' requirements (the broad circular areas, (a) to (d) need either very high precision data taken over long time intervals (a), or high quantities of data with lesser precision (d). The various lines indicate the range of precision over which selected techniques presently measure carbon dioxide. As indicated, there are few techniques having both the necessary precision and speed of response to be able to fulfil the mission requirements of an instrument for carbon dioxide. Either gas chromatography or coulometry, for example, could be used for climate and oceanic productivity studies. Spectroscopic and electrochemical techniques, on the other hand, are inherently faster techniques but lack the required precision. Lines are drawn from available data quoted in recent scientific publications (Walt *et al.*, 1993; Goyet and Snover, 1993 and Robertson *et al.*, 1993).

association of carbon dioxide with fossil fuel burning and the greenhouse effect can only be properly interpreted with reliable and consistent *in situ* time-series measurements at a series of geographical locations and depths. Only continuous monitoring will allow us to understand the complex biogeochemical cycles that occur in the ocean, and identify anthropogenic influences. We still need to quantify trace gas exchange rates and mechanisms, physical mixing of water masses, nutrient and pollutant dispersion, biological and physical influences on the water cycle, and the adaptability of organisms and communities to environmental change. We have to advance and test our understanding of small- and large-scale biogeochemical and ecosystem processes, and their behavior under environmental perturbation.

To address the challenges of future research demands, scientists must have access to a wide range of observational data, covering the human, physico-chemical and biological components of the Earth system. For the natural science and marine applications, specialised research facilities, and research platforms are needed. These include: satellites, airborne sensors, mobile, *in situ* and underwater measurements, and sea- and land-based experimental and monitoring facilities. Because of the high capital and operational costs of some of these items, their future support is somewhat uncertain and may ultimately depend on Governmental strategic support. However, all data-gathering operations (and associated data interpretation and modelling efforts) are dependent on complementary laboratory facilities. Considerable effort is needed to gather and consolidate the considerable volumes of data produced, and to utilise that information effectively.

The *in-situ* measurement of chemical species is of global importance but there is not enough manpower, shiptime or equipment to be able to measure all of the important properties of seawater throughout the ocean. Many nations only have one or a small fleet of research ships. These ships currently provide the only means to obtain chemical, biological and physical data over the full vertical depth range of the ocean: they are also essential for the calibration and validation of remotely-sensed data. There is therefore a future requirement to instrument the majority of the water masses throughout the oceans. The sheer volume and surface area of the oceans makes this a monumental task. The fact is: very few chemical parameters within the ocean have yet been measured. There have been some innovative and interesting attempts to remedy this situation. The United States Real-time Environmental Information Network and Analysis System (REINAS), for instance, is an InterNet meteorological data gathering database. REINAS collects meteorological and oceanographic information, and generates on-line real-time or historical plots of user-selected parameters. There are numerous land stations and oceanographic data buoys around Monterey Bay, California USA transmiting meteorological data (surface temperature, wind speeds/direction, humidity, radiance, etc) and ocean data (CTD, transmittance, fluorescence and ocean surface current data) that have already revealed an incredible level of detail. Such data is of immense value to single-user and collaborative research groups. There is no doubt that other sites will be developed to collect and integrate meteorological and oceanographic data. On a 10 to 20 year timescale, the use of autonomous long-range submersibles may reduce the need for research ships and research cruises, and could provide the basis for a sophisticated ocean observation system. There is a need to balance their (relatively) high development costs against their potential large observational advantage. Investments in automated technology could involve improvements to buoy- and sub-surface systems, and the equipping of the existing research ships with better sensor packages – all resulting in clear scientific benefits.

By examining past developments it is possible to envisage systems which could meet most of the requirements of today: unfortunately data requirements will not stand still and we need to look to the future for the next generation of instruments. There are many clear and immediate benefits of real-time and continuous data on chemical analytes that complement satellite images of sea surface temperature and colour. A comparison of continuous horizontal profiles of a number of variables would be of great importance in assessing the ability of satellites to provide quantitative data on biological processes. High density, high resolution remotely-sensed information would prove invaluable for chemically monitoring the aquatic environment, for developing circulation and climate models, predicting distributions of various chemical species, and dispersal of pollutants. Synthesis of data and modelling efforts are underway in many laboratories worldwide. The Hawaian Ocean Time series (HOT) has currently undertaken about 100 cruises to a site termed Aloha (22°45'N 158 °W) and assembled the largest existing data set of biologically important parameters for ecological modelling. The results have already yielded unexpected results challenging previous theories of biogeochemical cycles in the oligotrophic North Pacific (http://hahana.soest.hawaii.edu). This data set reveals a rich spectrum of variability in a system that was thought (or expected) to be mostly featureless. It also highlights the lack of accurate conceptual frameworks on which to model the data. It would seem for the time being that as more is learnt, less is understood.

The future is uncertain, and there are many research problems remaining to be solved. Advances in scientific understanding mean that there is (and must be) increasing emphasis on the use of sensors and observations in analysis and modelling. Programs of research and development must be closely integrated in order to make the best use of the instruments and data, and to fully justify the high costs involved. The search for suitable and improved sensors must continue.

References

Aylott, J.W., Richardson, D.J. and Russell, D.A. (1997). Optical biosensing of nitrate ions using a sol-gel immobilized nitrate reductase, *Analyst*, 122(1), 77–80.

Baier, R.E. (1980). Substrata influences on the adhesion of microorganisms and their resultant new surface properties in "Adsorption of Microorganisms to Surfaces" (Bitton, G. and Marshall, K. eds) pp. 59–104, New York: John Wiley.

Bandurraga, M.M. and Fenical, W. (1985). Isolation of temuricins. Evidence of a chemical adaption against fouling in the marine octocoral *Muricea fruticosa* (Gorgonacea), *Tetrahedron*, 41, 1057–1065.

Bosman, J., Flemming, N.C., Holden, N. and Taylor, K. (1998). The EuroGOOS Marine Technology Survey, EuroGOOS publication No. 4, Southampton Oceanography Centre, Southampton ISBN 0-904175-29-4.

Brendell, P.J. and Luther, G.W., III. (1995). Development of a gold amalgam voltammetric microelectrode for the determination of dissolved Fe, Mn, O_2 and S(-II) in pore waters of marine and freshwater sediments, *Environmental Science and Technology*, 29, 751–761.

Chamberlain, A.H.L. (1992). The role of adsorbed layers in bacterial adhesion in "Biofilms – Science and Technology" (Melo, L.F., Bott, T.R., Fletcher, M., Capseville, B. eds) pp. 59–68. NATO ASI series E: Applied Sciences, Volume 223. Dordrecht, Netherlands: Kluwer Academic Publishers.

Characklis, W.G. and Cooksey, K.E. (1983). Biofilms and microbial fouling, *Adv. Appl. Microbiology*, 29, 93–138.

Characlis, W.G. and Escher, A.R. (1988). Microbial fouling: initial events in "Marine Biodegradation" (Thomson, M.F., Sarojini, R. and Nagabhushanam, R. eds) Balkema, A.A., Rotterdam, pp. 249–260.

Clare, A.S. (1996). Marine natural product antifoulants: status and potential, *Biofouling*, 9, 211–229.

Clare, A.S. (1996). Signal transduction in barnacle settlement: calcium revisited, *Biofouling*, 10, 141–159.

Coll, J.C., Price, I.R., Konig, G.M. and Bowden, B.F. (1987). Algal overgrowth of alcyonacean soft corals, *Mar Biol.*, 96, 129–135.

Crisp, D.J., Walker, G., Young, G.A. and Yule, A.B. (1985). Adhesion and substrate choice in mussels and barnacles, *J. Colloid Interfacial Sci.*, 104, 40–50.

Davison, B., ODowd, C., Hewitt, C.N., Smith, M.H., Harrison, R.M., Peel, D.A., Wolf, E., Mulvaney, R., Schwikowski, M. and Baltensperger (1996). Dimethyl sulfide and its oxidation products in the atmosphere of the Atlantic and Southern Oceans, *Atmospheric Environment*, 30(10), 1895–1906.

Formaro and Trasati (1968). Capacitance measurements on platinum electrodes for the estimation of organic impurities in water, *Anal. Chem.*, 40, 1060–1067.

Hendey, N.I. (1951). Littoral diatoms of Chichester Harbour with special reference to fouling, *J. Roy. Microscopical Soc.*, 71, 1–86.

Howard, A.G. and Statham, P.J. (1993). Inorganic trace analysis, John Wiley & Sons, Chichester.

Izac, R.R., Bandurraga, M.M., Wasylyk, J.M., Dunn, F.W. and Fenical, W. (1982). Germacrene derivatives from diverse marine soft-corals (Octocorallia), *Tetrahedron*, 38, 301–304.

Keifer, P.A., Rinehart, K.L. and Hooper, I.R. (1986). Renillafoulins, antifouling diterpenes from the sea pansy, *Renilla reniformis* (Octocorallia), *J. Org. Chem.*, 51, 4450–4454.

Malin, G., Turner, S., Liss, P., Holligan, P. and Harbour, D. (1993). Dimethylsulfide and Dimethylsulphoniopropionate in the Northeast Atlantic during the summer Coccolithophore Bloom, *Deep-Sea Research Part I-Oceanographic Research Papers*, Vol.40(7), 1487-1508.

Marshall, K. (1985). Mechanisms of bacterial adhesion at solid-water interfaces in "Bacterial Adhesion" (Savage, D.C. and Fletcher, M. eds) pp. 133–161. New York: Plenum Press, USA.

McKenzie, J.D. and Grigolava, I.V. (1996). The echinoderm surface and its role in preventing macrofouling, *Biofouling*, 10, 261–272.

Millero, F.J., Zhang, J.-Z., Fiol, S., Sotolongo, S., Roy, R.N., Lee, K. and Kane, S. (1993). The use of buffers to measure the pH of seawater, *Mar. Chem.*, 44, 143–152.

Murray, J.W., Barber, R.T., Roman, M.R., Bacon, M.P. and Feely, R.A. (1994). Physical and biological controls on carbon cycling in the equatorial Pacific: US JGOFS EqPac process study, *Science*, (266), 58–65.

Parr, A.C.S., Smith, M.J., Beveridge, C.M., Kerr, A., Cowling, M.J. and Hodgkiess, T. (1997). Optical properties of a fouling-resistant surface (PHEMA/benzalkonium chloride) after exposure to a marine environment, *Water Res.*, (submitted for publication).

Rittschof, D., Hooper, I.R., Branscomb, E.S. and Costlow, J.D. (1985). Inhibition of barnacle settlement and behaviour by natural products from whip corals, *Leptogorgia virgulata* (Lamarck, 1815), *J. Chem. Ecol.*, 11, 551–563.

Spies, R.B. and Davis, P.H. (1979). The infaunal benthos of a natural oil seep in the Santa Barbara Channel, *Marine Biology*, 750, 227–237.

Targett, N.M., Bishop, S.S., McConnell, O.J. and Yoder, J.A. (1983). Antifouling agents against the benthic marine diatom, *Navicula salinicola*. Homarine from the gorgonians *Leptogorgia virgulata* and *L. setacea* and analogs, *J. Chem. Ecol.*, 9, 817–829.

Wallace, D.W.R. and Wirrick, C.D. (1992). Large air-sea gas fluxes associated with breaking waves, *Nature*, 356, 694–696.
Watson, A. (1993). Air-sea gas exchange and carbon dioxide, in "The global carbon cycle" (Heiman, M. ed.) pp. 397–411, Berlin: Springer-Verlag.
ZoBell, C.E. and Allen, E.C.J. (1935). *J. Bacteriol.*, 29, 239.

Color Plate 1 A standard "rosette" after deployment having water samples collected directly from the large volume GoFlo bottles for nutrient analysis in the shipboard laboratory. Also strapped across the bottom of the frame are a number of continuously recording sensors for salinity, temperature, pH, dissolved oxygen, fluorescence, turbidity and sound velocity. (Photograph courtesy of R. Scrivens, WS Ocean Ltd.). *See* Page 4.

Color Plate 2 Many companies now retail "packages" that are capable of measuring a wide range of parameters. This instrument is capable of simultaneously measuring temperature, salinity (conductivity), dissolved oxygen, as well up to three other parameters. (Photograph courtesy of S. Tierey, YSI Ltd.). *See* Page 14.

Color Plate 3 A flow-injection analyser attached to a standard CTD frame prior to a hydrographic cast. The FIA instrumentation is attached on the left hand side of the frame, with the plastic reagent bags below the pressure housing containing both optical and electronic components. The inlet manifold is already submerged. Conductivity, temperature, depth, oxygen (CTDO) sensors and two Niskin water bottles are mounted in the middle of the rosette (for calibration purposes), and the battery housing is on the right. This equipment has a battery endurance of 8 hours and is used up to 6000 m depth. (Photograph courtesy of S. Blain). *See* Page 18.

Color Plate 4 A series of six platinum redox electrodes (50 cm apart) are contained within a 3 m steel tube. The instrument is lowered to the seafloor and its own weight pushes it into the surface sediments, making measurements of redox potential at discrete intervals. The instrument is then lifted a few meters above the sea floor and "towed" to the next survey site. (Photograph courtesy of "Petrosurveys inc., Dallas" and Sea Technology, Compass Publications). *See* Page 24.

Color Plate 5 A view of a recently developed multi-port water sampler. A stator in the center of the apparatus rotates around each of the ports on the outside and pumps a filtered water supply into any number of bags connected by tubes to the nipples. On retrieval of the instrument, the tubes are then tied or knotted and then individually removed for subsequent analysis. (Photograph courtesy of R. Scrivens, WS Ocean Ltd.). *See* Page 25.

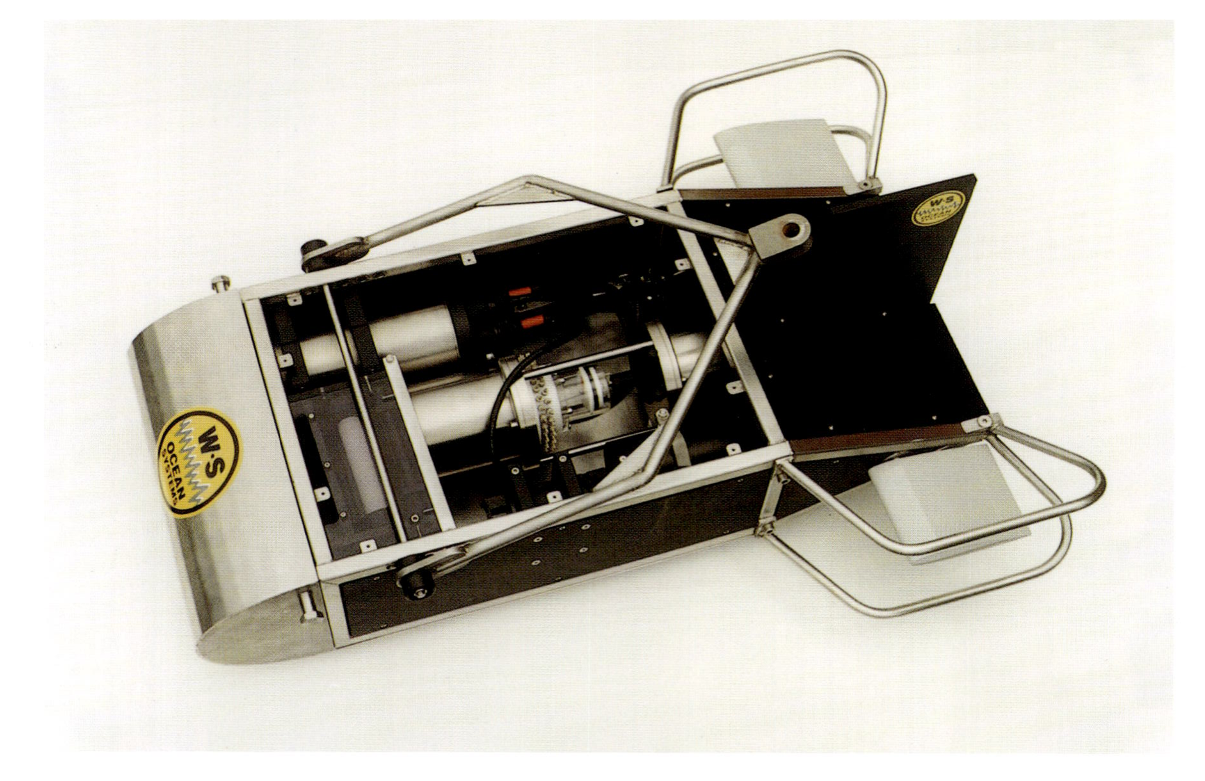

Color Plate 6 The multi-port water sampler located within an undulating towed fish prior to deployment. In this scenario, the water sampler would be programmed to take samples at timed intervals or commanded from the surface via an umbilical. (Photograph courtesy of R. Scrivens, WS Ocean Ltd.). *See* Page 26.

(a)

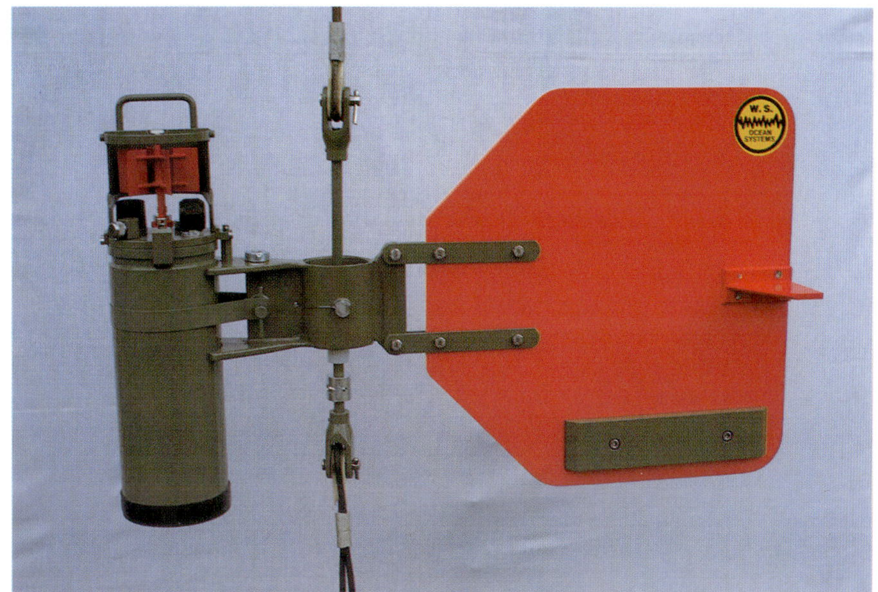

(b)

Color Plate 7 (a) and (b) Before and after shots of an Anderaa current meter deployed in the tropical Atlantic after only 2 months. Shellfish growth all over the outside of the instrument stopped the rotor from turning within 6 weeks of the experiment. *See* Page 27.

(a)

(b)

Color Plate 8 (a) This was the underwater section of a moored buoy system to measure nutrients in estuarine and coastal zones. The frame and instruments have been covered in algal growth (seaweed) after only 6 weeks on station. (b) The disconnected individual instruments shows that the cables, connectors and sensor surfaces are totally shrouded in algal growth. The left-hand housing was a sample inlet system, the middle instrument was used for optical measurements, and the housing on the right contained electronic components used to control the equipment. (Photographs courtesy of P. Wright, Southampton Institute). *See* Page 28.

Color Plate 9 Pumping technology: peristaltic pump. The picture shows the modified peristaltic pump for *in situ* measurement with the FIA analyser (Floch *et al.*, 1998). The rotor and the rollers are equipped with the PVC tubings. The pressure plates (white and black articulate plastic parts) are released. Other details of the *in situ* analyser are also visible on this picture. On the bottom the white cylindrical box contains the thermostating device. On the top two rotary valves mounted on the oil filled housing. One of the valves is equipped with teflon and peek tubings. *See* Page 51.

Color Plate 10 Photographs (A) microprofiling and (B) benthic chamber instruments. Sensors are mounted directly on the bottom end cap of the aluminum housing of the microprofiler and to the lid of the benthic chamber. *See* Page 250.

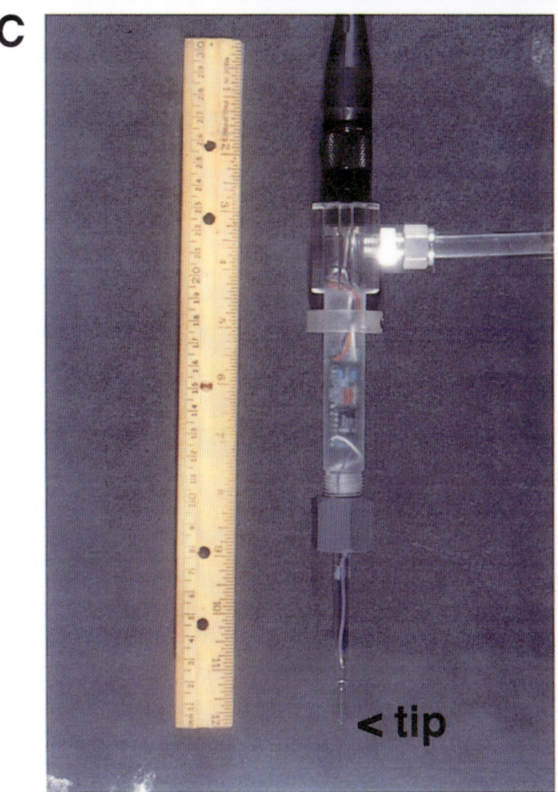

Color Plate 11 (A) Line-drawing of a holder for mounting microsensors on deep-sea instrument packages. This design is screwed directly to the pressure housing and pictured in (B). The design pictured in (C) allows for a cabled connection to an underwater electrical connector/penetrator. Both these designs have been mated to *in situ* microprofiler instruments by C. Reimers. The photo in (B) was taken in 1990 at 4100 meters on the continental rise off California and shows a resistivity sensor and needle-style O_2 and pH microelectrodes in use. Pressure has collapsed the compensating bulbs on the holders. The photo in (C) shows a Clark style O_2 microelectrode assembled with an external pre-amplifier within the holder. This design has been attached to an underwater motorized manipulator operated apart from the main Microprofiler electronics housing on a Remote Operated Vehicle. *See* Page 265.

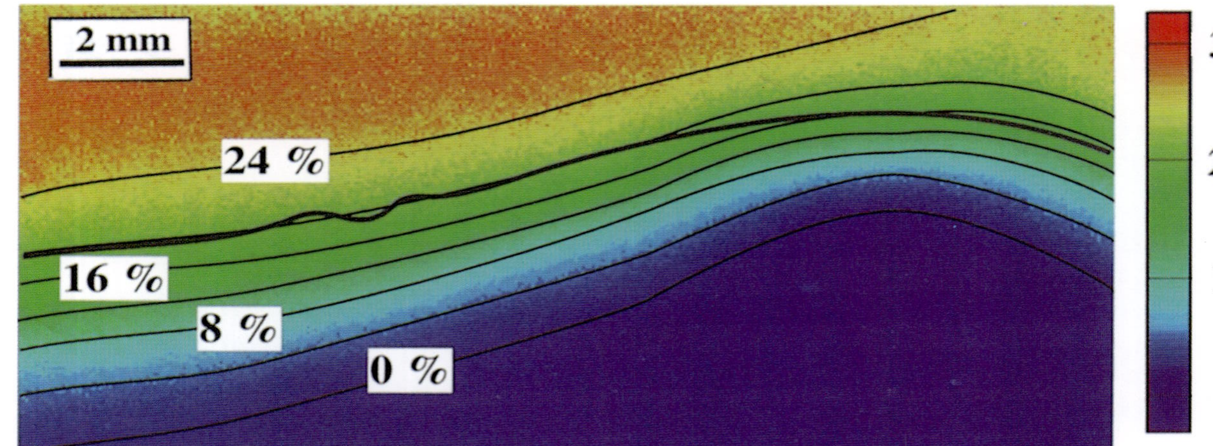

Color Plate 12 The O_2 distribution at the benthic interface of a tidal flat sediment overlain by water equilibrated with a gas mixture of 24 % O_2. Isolines of 0, 4, 8, 12, 16, 20 and 24% O_2 are indicated together with the position of the sediment surface (thick line). The O_2 partial pressure is expressed on a linear scale bar with 256 colors (scale bar shown) for visualization purposes. The O_2 penetration and the calculated diffusive O_2 uptake vary from 1.8–3.5 mm and 10–25 mmol m^{-2} d^{-1}, respectively. Flow direction from left to right (from Glud *et al.*, 1996a). *See* Page 276.

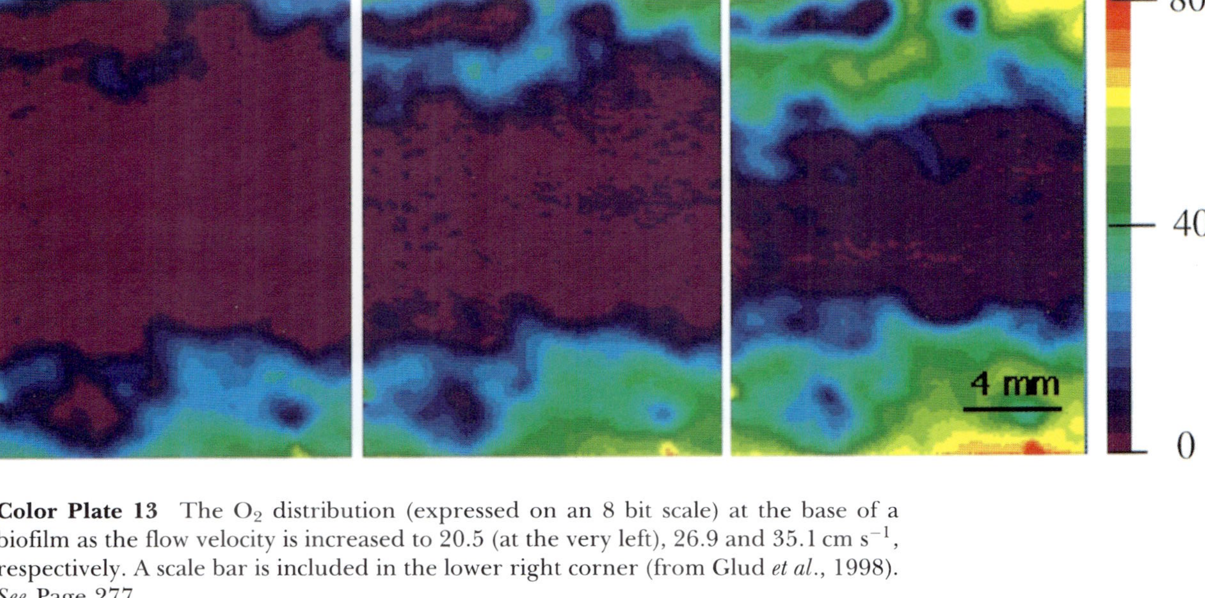

Color Plate 13 The O_2 distribution (expressed on an 8 bit scale) at the base of a biofilm as the flow velocity is increased to 20.5 (at the very left), 26.9 and 35.1 cm s^{-1}, respectively. A scale bar is included in the lower right corner (from Glud *et al.*, 1998). *See* Page 277.

2. *IN SITU* CHEMICAL ANALYSERS WITH COLORIMETRIC DETECTION

STÉPHANE BLAIN[a], HANS W. JANNASCH[b] and KENNETH JOHNSON[c]

[a]*UMR CNRS 6539, Institut Universitaire Européen de la Mer, Place Nicolas Copernic, 29280 Plouzané, France,*
[b]*Monterey Bay Aquarium Research Institute, PO Box 628, 7700 Sandholdt Road, Bldg.A, Moss Landing, CA 95039, USA and*
[c]*Moss landing Marine Laboratories, PO BOX 450, Moss Landing, CA 95039, USA (Also at Monterey Bay Aquarium Research Institute)*

2.1 HISTORICAL INTRODUCTION AND DEFINITION

One of the basic concepts of analytical chemistry is the transformation by a chemical reaction of the chemical species to be measured to a product easier to detect. When the product is coloured due to absorbing light in the visible region of the electromagnetic spectrum, these methods are called colorimetric, photometric or spectrophometric analytical methods. In the history of marine analytical chemistry the first colorimetric methods appeared in the early twentieth century. Atkins (1923a; 1923b) published such methods for phosphate and silicate determination. A nitrate colorimetric determination was also reported by Harvey (1926). Supplanting the previous time-consuming gravimetric or titrimetric methods, they considerably improved the determination of numerous dissolved chemical species in sea water. The development of the colorimetric methods allowed the first nutrient determination to be carried at sea during the Discovery expedition in 1925–1927 (Deacon, 1933). Chemical analysers resulted from the logical automation of these manual and tedious wet chemical assays. This work was pioneered by Brewer and Riley (1965). Driven by the increase of the number of samples to be analysed, as well by the requirement for higher quality measurements, this evolution was followed by concomitant improvements in technologies (mechanic and electronic) and the occurrence of commercial instruments (e.g. Technicon® autoanalyser).

As discussed in the foreword, the main feature which defines a chemical analyser is the active transport of the sample (sea water) through a manifold where a chemical reaction takes place between the analyte (a dissolved chemical species) and reagents. This point differentiates the chemical analyser from the chemical sensor where a passive transport (e.g. diffusion) occurs. After the chemical reaction, the coloured product flows through a detector. The light absorbance measured in a flowcell is converted to concentration by calibration with the analysis of known standard solutions having a similar matrix.

In-situ (from Latin: in its original place) means that all the steps of the analysis (sampling, chemical reaction, calibration and detection) are carried out under the sea surface. In comparison with laboratory or shipboard analysis, *in situ* analysis implies important variations of environmental parameters. Pressure varies from atmospheric pressure to hundreds of bars when deep profiles are performed whereas temperature ranges from –2 °C up to 30 °C depending on the location of the studied area and/or of the depth. These difficulties add to the complexity created by the sea water matrix. Variations in the matrix (e.g. salinity gradient in an estuary) can also greatly complicate the optimisation of the reaction chemistries.

This chapter describes an *in situ* chemical analyser as defined above. It focuses on the technology of transport (pumping systems), chemistry and technology for colorimetric detection. Examples of deployment at sea clearly demonstrate the benefit and the limits of such instruments.

2.2 TECHNOLOGY

2.2.1 Pumps

2.2.1.1 Peristaltic pumps

Most of the commercial automated analysers (segmented flow analyser or flow injection analyser) dedicated to laboratory chemical measurements use peristaltic pumps to propel the fluids. These pumps can deliver fluids at very precise flow rates using calibrated PVC tubing placed between a rotor equipped with rollers and a pressure plate (Plate 2.1). A motor (e.g. stepper-motor) actuates the rotor moving the rollers across the tubing (Figure 2.1a). The rotation rate and direction are set electronically. Some of the main advantages of this technology are: no contamination of the fluid by the pump, no contamination of the pump by the fluids, positive displacement, insensitive to pump head orientation. The use of such a pump *in situ* requires it to be electrically isolated from sea water. Usually a pressure resistant housing (e.g. aluminum or titanium) contains the electronic modules whereas the motor can be placed in an oil-filled container equipped with a rubber diaphragm insuring equal pressures inside and outside the housing (Figure 2.1a). Regarding the pump head and the associated tubing, two solutions have been explored. Johnson *et al.* (1986a) and Gamo *et al.* (1994) placed the head pump in the same pressure compensated container as the motor. With such a design, no corrosion occurs but the tubing (tygon) deteriorate rapidly and are difficult to service. On the other hand Daniel *et al.* (1995a) placed the head of the pump in sea water. This design facilitates the changing of the tubing and adjustment of the flow rates. Replacing some parts of the commercial pump (Gilson) with home-made pieces avoids rapid corrosion of the metallic parts. However, all systems based on peristaltic pumps suffer from deterioration of the tubing which precludes long time deployment.

Plate 2.1 Pumping technology: peristaltic pump. The picture shows the modified peristaltic pump for *in situ* measurement with the FIA analyser (Floch *et al.*, 1998). The rotor and the rollers are equipped with the PVC tubings. The pressure plates (white and black articulate plastic parts) are released. Other details of the *in situ* analyser are also visible on this picture. On the bottom the white cylindrical box contains the thermostating device. On the top two rotary valves mounted on the oil filled housing. One of the valves is equipped with teflon and peek tubings. *See* Color Plate 9.

2.2.1.2 Piston pumps

Taylor *et al.* (1993) described a positive displacement pump based on a principle widely used in high performance liquid chromatography (HPLC) (Figure 2.1b). The pump head (volume 10 μL) is operated at 30 to 60 strokes per minute by a rotating "swatch" plate resulting in a truncated pulsed flow. A combination of four pumps provides constant flow through the detector and has been successfully used *in situ* up to 2000 m depth. Other applications of piston pump are provided by the coastal chemical analyser described in chapter 4 (D. Hydes).

2.2.1.3 Osmotic pumps

The production of osmotic pumps was initially described by Theeuwes and Yum (1976) for biomedical applications. They were first applied to *in situ* analysis by Jannasch *et al.* (1994), where osmotic pumps were used to both pull in seawater and add reagents without the need for any electrical power. The driving force within an osmotic pump is caused by the osmotic pressure created across a rigid semi-permeable membrane by solutions of differing salinity. Osmosis results in a net diffusion, and thus flow, of water through the membrane from the fresher to the more saline side. The osmotic pressure is maintained constant throughout a pump's life by keeping a brine solution saturated with an excess salt (NaCl), and

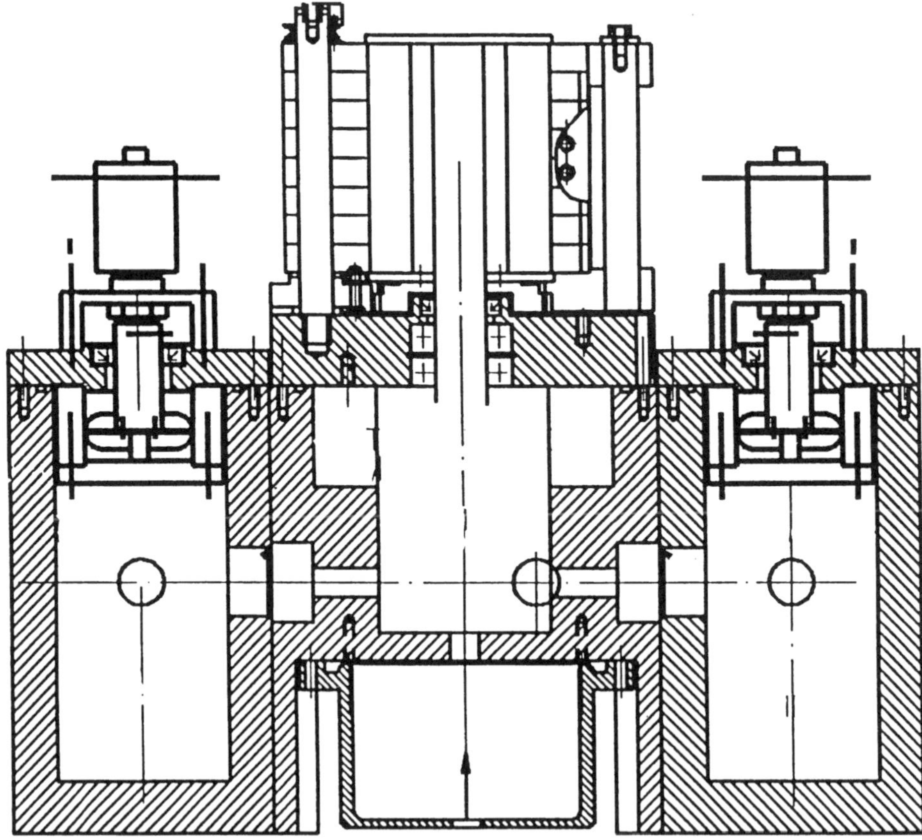

Figure 2.1a The technical drawing (courtesy B. Leide DITI IFREMER) shows the oil-filled container. The stepper motors are placed within the container filled with oil (white zones). The container is equipped with a rubber diaphragm (vertical arrow at the bottom) insuring pressure balance inside and outside the housing. The rotating seal are made waterproof with joints located around the rotating axis of the motors (pump and valves). They are represented by the small arrows on each sides of the axis.

maintaining zero salinity (or seawater) on the opposing side of the membrane. In practice, flow rates of osmotic pumps are predominantly dependent on temperature. Reagent pumps incorporate an impermeable flexible membrane to separate the reagent from the brine, (Figure 2.1c). Typical flow rates that can be obtained by these pumps are 1 to 40 $\mu L\ h^{-1}$ at 20 °C. Jannasch *et al.* (1994) showed that spurious signals may be due to temperature fluctuations, particularly when relatively rapid ($>1\,°C\ h^{-1}$). The temperature dependence is mainly caused by two effects; the change of the diffusion coefficient of water and the difference of the thermal expansion coefficient between the pump housing and the contained

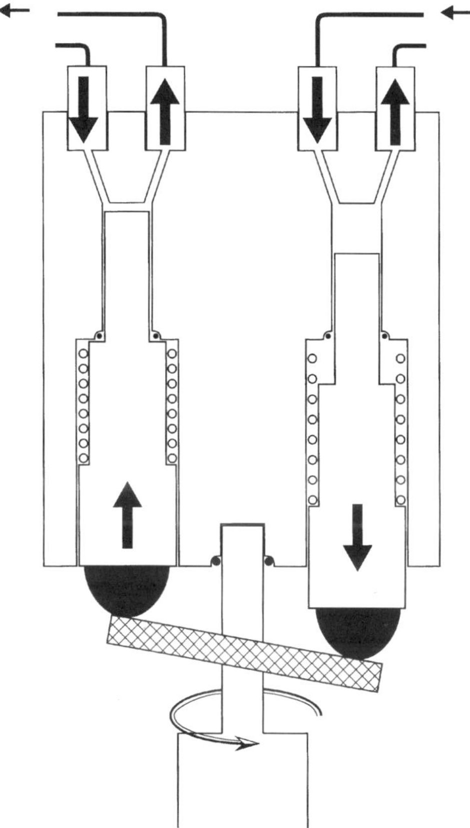

Figure 2.1b Pumping technology: Piston pump. (see text for details)

solutions. These effects can be reduced by thermally insulating the pumps or by increasing the flow rate.

2.2.1.4 Solenoid pumps

Solenoid pumps (Lee Company) are an other interesting and promising alternative for the following reasons: built with inert material (PEEK polyetheretherketone, FCR Fluorocarbon rubber), maintenance free, small size, light weight, self priming, dispense variable or fixed volume with increment as small as 0.1 µL, low power consumption. A solenoid driven-piston presses and releases a diaphragm which is integrated with a dual check valve, resulting in a pulsating flow (Figure 2.1d). The flow rate was proportional to the frequency of the pulses applied to the solenoid. Depending on the application, the pulsating flow can be drawback. Jannasch *et al.* (1994) used a solenoid pump to inject standard solution during the calibration operation of the osmotic *in situ* analyser where a single stroke (50 µL) was enough to realise one hydrodynamic injection. Daniel *et al.*

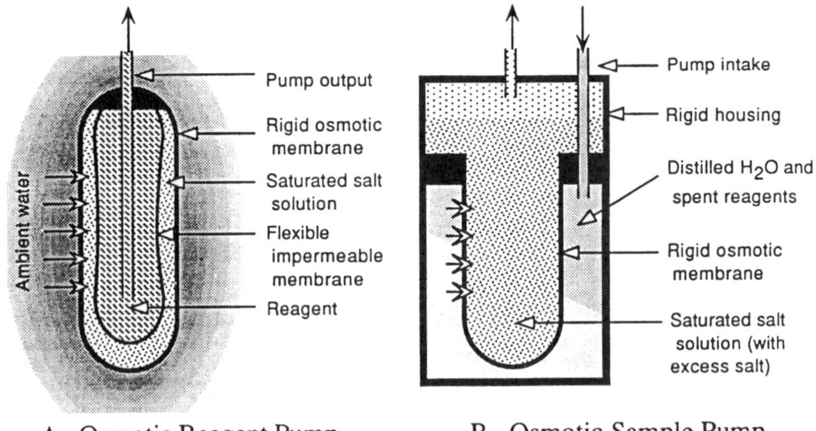

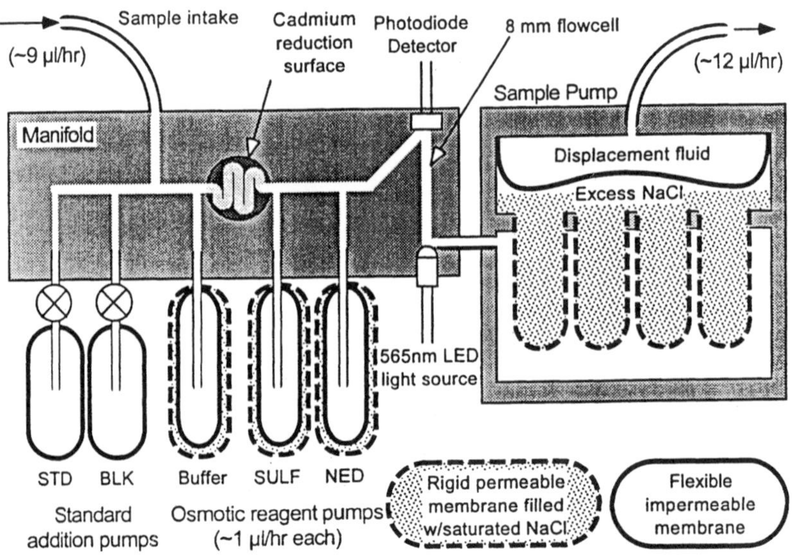

Figure 2.1c Pumping technology: Osmotic pump. Upper panel A and B are schematics of the reagent pump and the sample pump. The lower panel shows the whole osmotically pumped nitrate analyser. (with the permission of Analytical Chemistry)

(1997) describing a sequential flow analyser with Adsorptive Cathodic Stripping Voltametry (ACSV) for ship-board trace metal determination. The pulsating flow was not a drawback in this case because detection with ACSV requires a pre-concentration step which integrates many strokes. Weeks *et al.* (1996) evaluated the quantitative performances of the pump incorporated in a classical FIA analyser

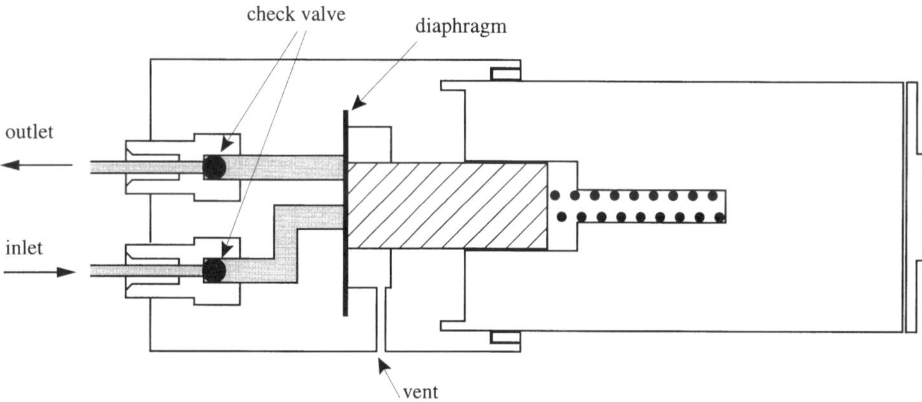

Figure 2.1d Pumping technology: Solenoid pump. (see text for details)

for nitrite determination. The authors concluded that solenoid pumps can produce acceptable performances in a typical FIA systems, but that the pulsation of the flow stream requires the chemist to pay careful attention to the optimisation of the manifold. Careful adjustment of the flow rates, length and diameter of the tubing, length and diameter of the flow cell, and digital signal processing are needed to obtain performances similar to the typical FIA analyser. An *in situ* analyser using solenoid pumps has been built and successfully tested (Figure 2.2).

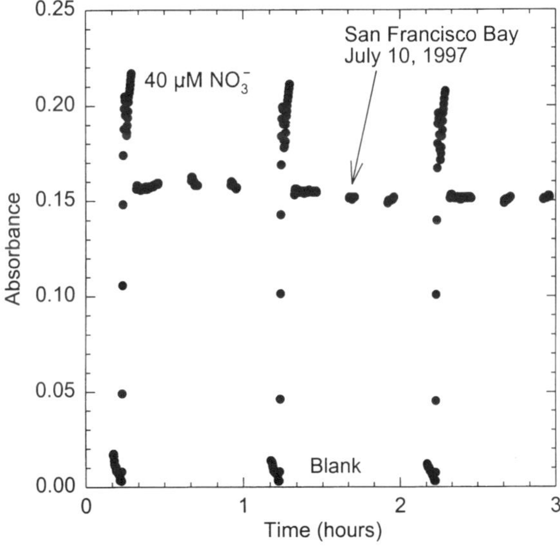

Figure 2.2 Nitrate measurement provided by an *in situ* analyser equipped with solenoid pumps. Blank, standard (40 µM) and San Fransisco Bay sea water were alternatively analysed demonstrating the reliability of the new device (K. Johnson, MLML).

2.2.1.5 Other systems

Piezoelectrically driven membrane pumps (Van Lintel *et al.*, 1988) or capillary based electro-osmotic pumps (Daguspta and Liu, 1994) have also been developed to propel sample, standards and reagents through micro flow analysers (Verpoorte *et al.*, 1994). With regard to *in situ* analysis, the main drawback of such pump, is the high voltage required for operation.

2.2.2 Manifold

As shown in the Figure 2.3 the manifold is the hydraulic circuit from the sample inlet to the waste outlet. Adapted from FIA technology (Ruzicka and Hansen, 1975; Ruzicka and Hansen, 1988), *in situ* manifolds are usually built with chemically inert PTFE tubing (with the exception of pumping tubing – sea above). The choice of the inner diameter was very important to ensure rapid radial mixing and avoid smearing of the samples along the manifold. The range of 0.5–0.8 mm has often been accepted as a good compromise. Daniel *et al.* (1995a) found that PTFE tubing tend to contract and harden when expose to low temperatures ($<10\,^{\circ}C$). PEEK exhibits the best performance under pressure, but this material is not very flexible and can only be used in parts of the manifold with low curvature which excludes (knitted) mixing coils. The manifold tubing are interconnected using fittings for high pressure liquid chromatography (HPLC). Glass bead columns have been also used to ensure complete mixing of the sample with the reagents (Clark *et al.*, 1989; Brooks and Dorsey, 1990).

Ruzica (1983) considering "the spaghetti-like heap of tubing" resulting from all the interconnections in the hydraulic circuit proposed to design the manifold as integrated microconduits. He made integrated manifolds by pressing (cold-forging) preformed wires into PVC blocks and sealing the path with a gasket and cover. Liquid connections were usually made on the back side of the block. This method was also used by Jannasch *et al.* (1994) to miniaturize the manifolds for the osmotically pumped analyzer, but has now been replaced by numerically controlled milling for repeatability.

2.2.3 Valves

The *in situ* calibration of the chemistry probably carries the greatest advantage of the chemical analyser over chemical sensor. However, this advantage is gained at the price of more complicated technology. The need for a pumping system has already been discussed, but there is also a need for valves to select between sample, blank and standard solutions. Solenoid valves present the most obvious technology. Mounted in an oil-filled pressure compensated vessel, six pinch solenoid valves with Viton tubing have been used in the *in situ* chemical analyser, SCANNER (Johnson *et al.* 1986a).

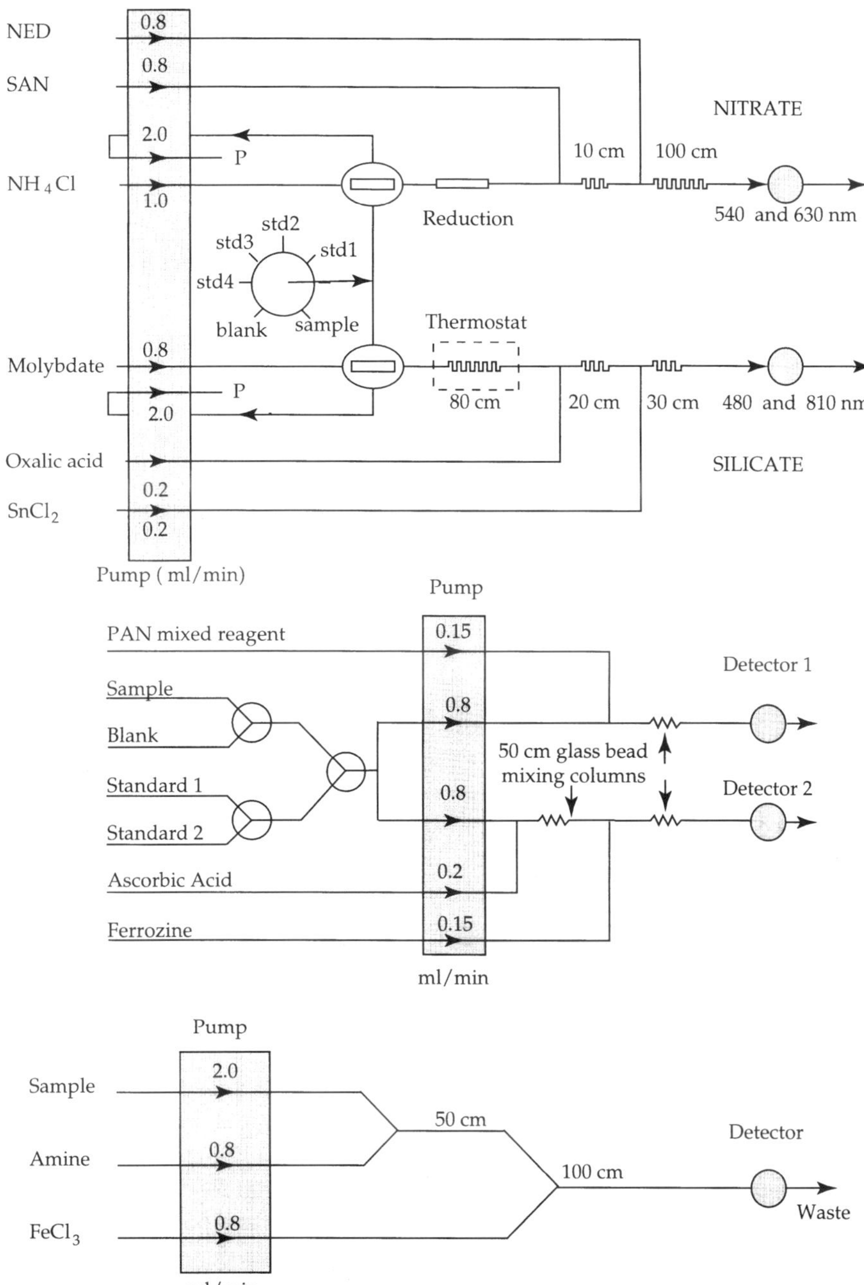

Figure 2.3 Examples of manifold used with *in situ* chemical analysers. The nitrate/silicate manifold was implemented on the FIA analyser (Floch *et al.*, 1998). The manganese-iron and sulfide manifold were implemented on the SCANNER.

Most of the *in situ* chemical analysers are based on continuous flow analysis (CFA). They are sometimes referred to as flow injection analyser (FIA), because they use unsegmented flow. In addition, they do not inject the sample into the reagent stream. Since FIA requires an injection valve, Johnson and co-workers wrote in 1986 "in submersible systems, the mechanical complexity of sample injection outweigh any advantages that are gained". However ten years later the technical challenge has been taken up and a rotary HPLC valve was adapted to *in situ* operation (Daniel *et al.*, 1995a). This new technology uses a step motor actuated rotor and optical encoding to control its position and status. A multi-way selection valve has been built on the same principle.

2.2.4 Colorimetric Detectors

Colorimetric detection is based on Beer's law which links the solution absorbance (A) of the light by the solution to the analyte concentration (C) by the relationship $A = \varepsilon CL$, where ε is coefficient of molar absorption related to the chemical structure of the product and function of the wavelength, and L is length of the flow cell. Although Beers' law is rigorously followed when the light is monochromatic, light measured in real systems is characterised by a finite bandwidth. In this case, the best sensitivity of the method and the larger range of linearity for the calibration curve are closely related to the following conditions: (i) coincidence of the measurement wavelength with the peak absorption wavelength of the product and (ii) a detector bandwidth lower than the width of the product absorption band.

In general, a colorimetric detector contains a light source, a wavelength selector (optional), a flow cell and a receptor transforming the light into an electrical signal.

Most of the *in situ* chemical analysers (Johnson *et al.*, 1986a; Jannasch *et al.*, 1994; Gamo *et al.*, 1994) have used light-emitting diodes (LED) as the light source. This low cost and pressure resistant optoelectronic component is small, has a low electrical power consumption, and works reliably *in-situ*. However, the wavelength of the maximum emission is not necessarily the same as the maximum absorption. For example the nitrate analyser uses a LED with maximum emission at 559 to 565 nm whereas the maximum absorption of the coloured complex is 540 nm. In addition, the wavelengths available with LED does not covered the entire visible spectrum despite the advent of "rainbow" LEDs (combining three LEDs, blue, green and red).

A more complicated system has been tested by Daniel *et al.* (1995b) to permit dual wavelength measurement. An incandescent light source produced a path of white light guided to the cell by an optic fibre. A simultaneous measurement of the light intensity at two wavelengths was performed using a multimode coupler and interference filters (bandwidth 10 nm). This design permits different detector chemistries simply by changing the excitation filters and permits future implementation of more powerful detectors such as photo-diode array (PDA) detectors.

Whatever the nature of the light source, two photoreceptors are generally used. One measures the light transmitted through the solution to estimate the chemical concentration while the other monitors the short-term fluctuations and long-term drift of the light source. These detectors are generally phototransistors.

2.3 CHEMISTRY

2.3.1 Colorimetric Chemical Reactions

Among the large variety of dissolved chemical species present in sea water, only a small number have been measured *in situ* by chemical analysers: nitrite, nitrate, phosphate, silicic acid, hydrogen sulfide, hydrogen peroxide, manganese and iron.

Nitrite determination, for example, is based on the formation of an azo dye. In the first step, nitrite reacts with an aromatic amine (sulfanilamide hydrochloride, SAN) to form a diazonium ion. Second, the diazonium compound couples with an other aromatic amine (N-(1-naphthyl)-ethylenediamine dihydrochloride, NED) to produce an azo dye. The peak of absorbance is at 540 nm. Nitrate was measured using the same method after reduction to the nitrite ion. There is an extensive literature describing various procedures for the reduction step. The most efficient and applied method is the heterogeneous reduction by copper-coated cadmium. This step is pH sensitive and a buffer (ammonium chloride) is needed. This method has been well described for manual, segmented flow analysis (Brewer and Riley, 1965; Hager *et al.*, 1972) or FIA (Johnson and Petty, 1983; Clinch *et al.*, 1987). It has been successfully implemented on *in situ* analysers, sometimes with minor modification, Imidazol replacing ammonium chloride (Johnson *et al.*, 1989; Jannasch *et al.*, 1994). Cadmium tubing or sinusoidal microconduits printed in a solid disk of cadmium have also been used in place of filings loaded in the column. The stability of the reagents and standards is an important aspect of long term monitoring. The SAN solution is stable for many months (Strickland and Parsons, 1972). The NED should be renewed every month and sooner when brown coloration due to oxidation develops. However no deterioration of the yield of the dye formation was observed for periods of several months when the reagent was vacuum-degassed and kept out of direct light (Jannasch *et al.*, 1994). Concerning the stability of the standards, Aminot and Kerouel (1991) demonstrated that autoclaving is an efficient method for the nitrate standard storage for up to four months.

Silicic acid (silicate) determination is based on the formation of the yellow β silicomolybdic acid reduced in the silicomolybdenum blue (Truesdale and Smith 1975; Strickland and Parson, 1972). There are two isomers of the silicomolibic complex (α and β). The β is the more suitable for spectrophotometric determination. Its formation is favoured by pH adjustment in the range 1.4–1.8. The molybdate is also known to form complexes with phosphate and arsenate. These interferences are eliminated by mixing oxalic acid with the sample. The same result can also be obtained if the addition occurs after the molybdo-complex formation. The most common reductant is a mixture of metol (p-methylamino-phenol sulphate) and sulfite. However, for *in situ* measurements, stannous chloride has been chosen as reductant in place of metol sulfite or ascorbic acid (Johnson *et al.*, 1986a) because it enhances the reaction rate and the sensitivity of the method. The activity of the reductant dramatically decrease with time, however degassing and storage without any air contact increase its life time.

Dissolved manganese was analysed using 1-(2-pyridylazo)-2-naphthol (PAN) (Chin *et al.*, 1992). The manifold is shown in Figure 2.3. Interference from dissolved iron can be masked by the iron-specific chelating agent desferriox-amineB. The detection limit ranges from 15–13 nM.

Iron determination used the formation of a coloured complex between fer-rozine and Fe(II) (Stookey 1970). Fe(III) is determined by the same method after reduction by ascorbic acid. The average detection limit is 25 nM

Colorimetric catalytic methods may be very useful for *in situ* determination because of its high sensitivity. For example, Hirayama and Unohara (1988) developed a method for the spectrophotometric catalytic determination of Fe(III) based on the catalytic oxidation of N,N-dimethyl-p-phenylenediamine by hydrogen peroxide. This method was modified by Measures *et al.* (1995) for flow injection analysis. Resing and Molt (1992) also reported such a catalytic method for manganese determination in sea water.

The determination of hydrogen sulfide in sea water is based on the reaction of dimethyl-p-phenylenediamine with sulfide in the presence of Fe(III) in an acidic media. The product of the reaction is methylene blue (Strickland and Parson, 1972). A similar method has been also reported for automated segmented flow analysis (Grasshoff and Chan, 1971). Sakamoto *et al.* (1986) adapted this method to FIA and continuous unsegmented flow analysis for *in situ* use. Due to the lower temperature expected in the deep sea the concentrations of the reagent differed in two manifolds. Depending on the manifold, two different sensitivities were obtained: 1.2 to 200 μM or 0.12 to 75 μM.

2.3.2 Influence of Environmental Parameters

When chemical reactions are carried out *in situ*, they can be significantly affected by various environmental parameters. For example, equilibrium reactions are controlled by temperature and pressure, whereas kinetic reactions are highly temperature dependant. Environmental changes may also alter the measurement by modifying the physical properties of the fluid (viscosity) and the mechanical properties of the analyser (diameter of the tubing and dimension of the flowcell), both contributing to change the flow rate, mixing conditions through the manifold and detector stability. Also, many parameters can vary widely, such as temperature (e.g. in a thermocline or arround hot vents) and salinity (e.g. in estuaries).

From an experimental point of view, the study of the influence of the pressure on the extent of a chemical reaction is often difficult because of restricted access to high pressure test facilities. Johnson *et al.* (1986a) tested the influence of hydrostatic pressure on silicic acid and sulfide colorimetric determination. The main conclusion was that pressure has a greater influence on reactions that do not go to completion. For both determinands absorbance increases with pressure. On the other hand, the formation of the coloured nitrate complex is pressure independent (Johnson *et al.*, 1989; Daniel *et al.*, 1995a), probably due to complete reaction.

The effect of temperature on the extent of a chemical reaction is easier to study than pressure effect and is therefore better understood. These studies are fundamental because of important temperature variations occurring in different parts of the water column (e.g. seasonal thermocline). The silicic acid colorimetric determination provides a very instructive example of the possible temperature effects and the various solutions proposed to overcome such difficulties. The reaction requires several minutes to go to completion, whereas the reaction mixture in a chemical analyser have a residence time of roughly one minute. The slow reaction rate is caused by the two slow steps involved in the reaction (Thompsen *et al.*, 1983), the formation of the molybdosilicate complex and its reduction. Johnson *et al.* (1986a) reported a 5 to 6 fold increase of the absorbance of a Si standard when the temperature increase from measured 5 °C to 19 °C. In FIA analyser, Floch *et al.* (1998) report a dramatic decrease of the sensitivity with temperature and a modification of the shape of the peak. (Figure 2.4). The influence of temperature can be reduced by increasing the reaction rate. This goal has been reached by using stannous chloride as reductant (Johnson *et al.*,

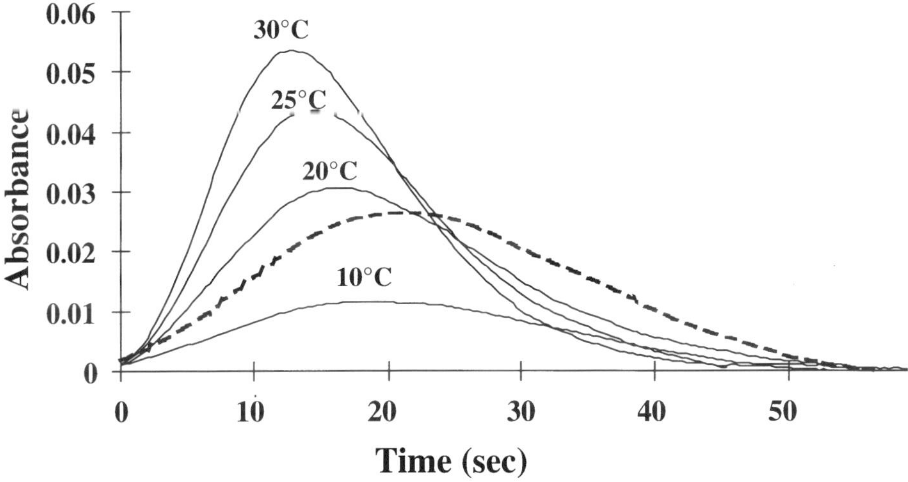

Figure 2.4 Influence of temperature on the FIA signal in the determination of silicic acid. The silicic acid concentration is 10 μM. The continuous lines are for the signals provided when the whole system (manifold, detector, standards and reagents) at the different temperatures. The dotted line shows the signal when the manifold is kept at 0 °C and the silicomolybdate complex formation area is thermostated at 30 °C.

1986a) or by thermostating the molybdosilicate complex formation reaction zone at 30 °C (Floch *et al.*, 1998).

2.3.3 *In situ* Calibration

The wide variation in environmental conditions that *in situ* analyzers encounter demonstrates a need for frequent *in situ* calibration. Fortunately, most analyzers can easily be equipped to automatically analyze a set of known standards at preset intervals. The resulting absorbance measured during their analysis is then used to calibrate the analyzer for those particular conditions (pressure and temperature). Most *in situ* analyzers use only two standard concentrations. The biggest drawback to frequent calibrations is that data cannot be collected simultaneously. Analyzers driven by peristaltic pumps usually use a selection valve to introduce sample, standard or blank, whereas osmotically pumped analyzers use micro solenoid pumps to backflush the intake tubes with standards.

2.4 EXAMPLES OF DEPLOYMENT AT SEA

2.4.1 High Resolution Analysers

In situ analysers are required when large variations in the concentration of chemical species are expected. This is especially important for environments around point sources such as hydrothermal vents and river inputs. These environments are often a source of concentrated dissolved chemical species which rapidly disperse. This process leads to large chemical gradients with a high potential for significant biogeochemical process. So it is not surprising that the first *in situ* chemical analyser have been used around hydrothermal vents. Johnson *et al.* (1986b) described the large changes in sulfide and silicate based on more than 10,000 measurements gathered by the SCANNER mounted on the submersible Alvin. The large number of data facilitated the study of the non conservative processes occurring during the mixing, demonstrating significant sulfide removal within clumps of mussels. Such conclusions must have been more speculative if only based on discrete samples analysis for two reasons: first, the small number of measurements available and secondly, the uncertainties due to the oxidation of sulphide before analysis. Thus, this first deployment clearly highlighted two of the main advantages of an *in situ* chemical analyser over usual discrete sampling and ship-based (or land-based) analysis. Manganese and iron (II + III) have also been measured in hydrothermal plumes with *in situ* analysers producing a two dimensional map of the chemical concentration (Coale *et al.*, 1991; Chin *et al.*, 1994) (Figure 2.5). Based on the SCANNER's principle, Massoth and collaborators developed a next-generation chemical analyser (Massoth *et al.*, 1995) to probe the submarine hydrothermal environment. The SUAVE (Submersible System used to Assess Vented Emissions) has been fitted to research

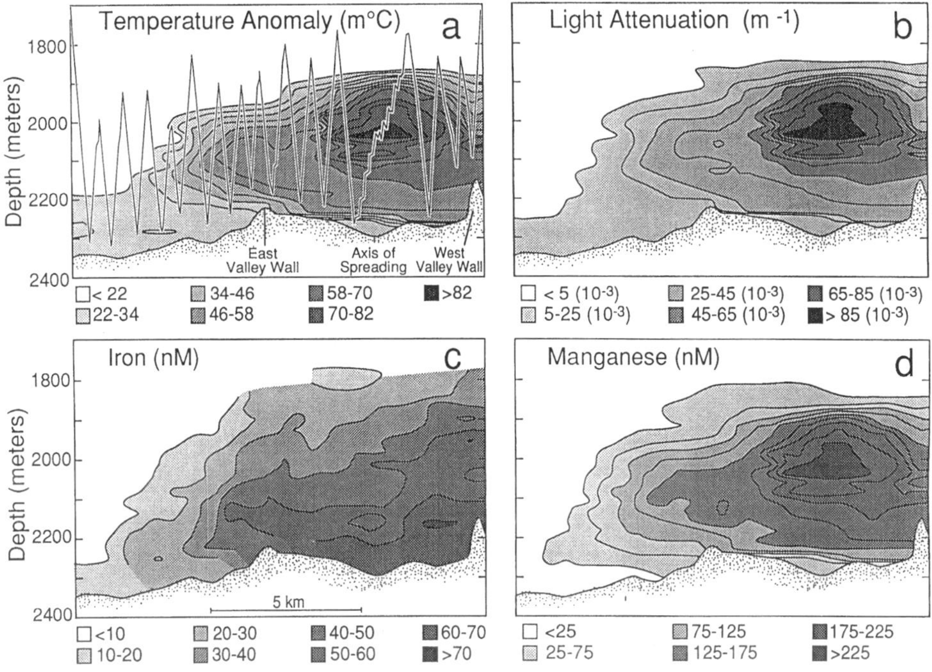

Figure 2.5 Two dimensional distributions deduced from *in situ* measurements of iron and manganese by the SCANNER secured to a CTCT/rosette package and towed across the Cleft segment at the southern end of the Juan de Fuca ridge (Pacific Ocean 130 °W, 45 °N) (Courtesy Kenneth Coale, MLML).

submersibles, on board Remote Operated Vehicles (ROV), and on CTD profiling packages (casts and Tow-yo's). Since 1991, 226 deployments with 1500 hours of *in situ* analysis time have been accumulated (Massoth and Milburn, 1997).

The thermocline which generally separates, nutrient-depleted surface water from nutrient rich deep water, is another environment with high chemical gradients. This region is of high biogeochemical interest because primary production is mainly driven by nutrient input from deep to surface water. These inputs result from a continuous diffusion process, as well as from episodic events. In both cases, *in situ* chemical analysis is especially useful. Using the SCANNER, Johnson *et al.* (1989), (Figure 2.6) described the fine scale structure of nitrate profiles. A vertical span of 5 m could be detected at lowering rates up to 40 m min^{-1}.

2.4.2 Time Series Analysers

Long-term *in situ* analyzers are needed to continuously monitor the concentrations of dissolved chemicals, and give us the best access to chemical data from locations or times when ships cannot be present (e.g. during storms, periodic

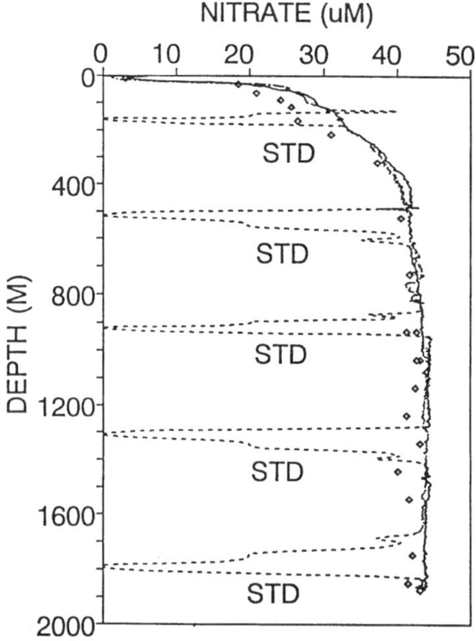

Figure 2.6 Vertical nitrate profiles: comparison between *in situ* measurements (continuous line) and ship board analysis (diamond). The dotted line shows the signal of the blank and 40 µM standard. (provided by Ken Johnson MLML)

upwelling events, or planktonic blooms). Long-term time-series measurements require autonomous *in situ* analyzers with very low electrical power and reagent volume requirements. The *in situ* osmotically pumped chemical analyzer developed by Jannasch *et al.* (1994) attempts to fulfill these requirement. The improved resolution of data collected by a long term *in situ* analyzer over standard shipboard time series analyses is clearly shown with data from a mooring in Monterey Bay (Figure 2.7). These analyzers have also been deployed on a regularly serviced deep ocean mooring located 80 kilometers south west of Bermuda (Dickey *et al.*, 1997). Data from these analyzers show that sporadic nutrient injections can supply the mixed layer with significant amounts of nutrients during short periods usually missed by ship cruises (Figure 2.8).

2.5 LIMITS OF *IN SITU* ANALYSERS

Comparing the number of analysis carried out by an *in situ* chemical analyser with the data provided by the usual hydrocasts it is obvious that the resolution of the chemical gradients in sea water has been greatly improved. However it is also important to keep in mind some of their limitations. The high resolution

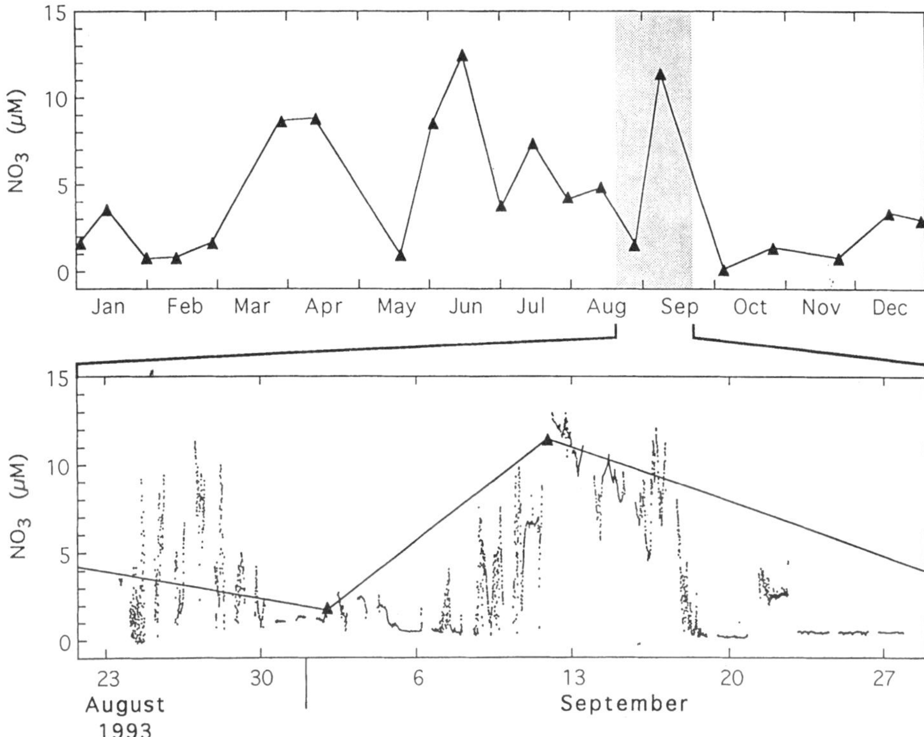

Figure 2.7 Time variation of the nitrate concentration in the Monterey bay: comparison between results of continuous monitoring by an osmotic analyser and monthly determinations. (provided by Ken Johnson, MLML)

chemical analyser has a response time of the order of (at least) a few minutes. During this time, the CFA continuously pumps the sea water sample. Some mixing of successive samples can occur leading to smearing of the chemical gradient. A modeling approach has been developed (Floch, 1998) to define a parameter characterising each analyser. This parameter is deduced from the characteristic of the manifold and the shape of the signal. Using this parameter it is possible to compare a real chemical profile to the profile measured by the *in situ* analyser. Figure 2.9 highlights the artefacts which may result when moving an *in situ* chemical analyser at different lowering rates through a strong chemical gradient (e.g. seasonal thermocline). In contrast, the FIA *in situ* analyser does not produce any smearing of the gradient but it only provides discontinuous measurements leading to "gaps" along any profile. The depth interval between each measurement grows with an increase in lowering rate of the analyser. Whatever the sort of analyser, the only way to reduce these artefacts or to fill the gaps in the profile, is to decrease the response time of the analysers.

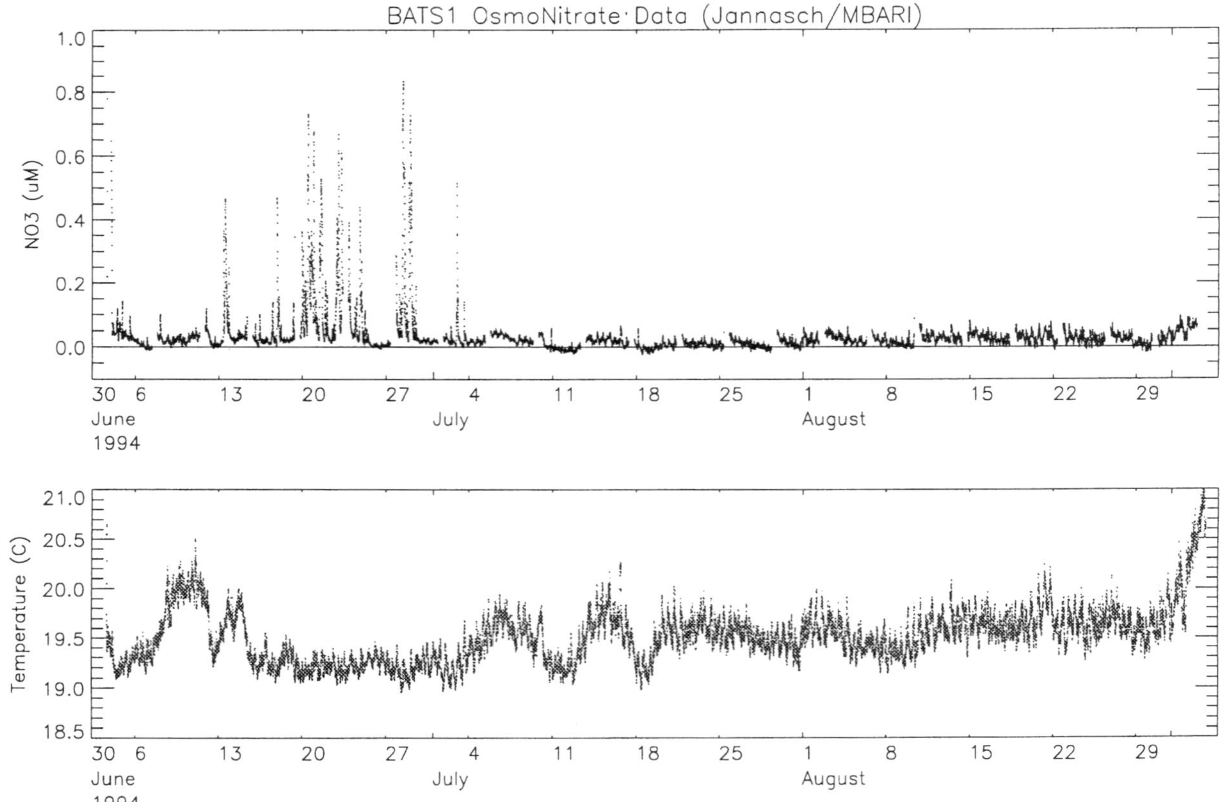

Figure 2.8 Results from the deployment (80 m depth) of the long term osmotic analyser at the Bermuda Times Series Station. Very short nitrate pulses (several hours) can be observed. Such pulses was expected but not observable using conventional sampling and analytical procedures. (provided by Hans Jannasch, MBARI)

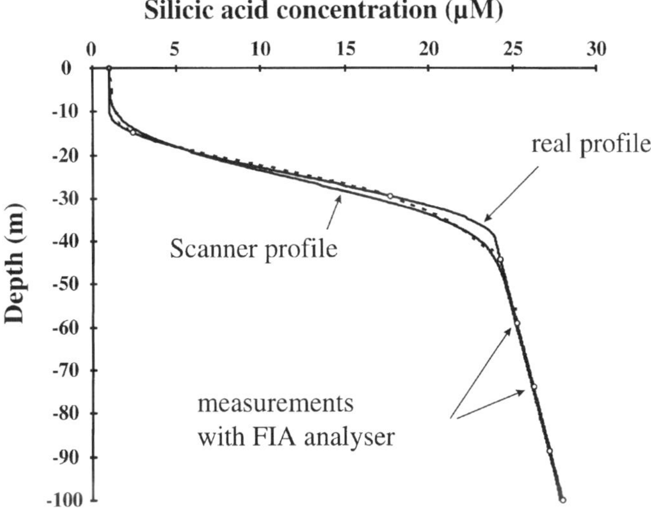

Figure 2.9 Comparison between a real chemical profile (dotted line) and modelled data. The model calculates at a fixed deepening speed the profile measured by an *in situ* chemical analysers. Open circles represent the measurements provided by FIA *in situ* analyser lowered at the same speed 0.25 m s^{-1}.

The other limitations of *in situ* chemical analysers are linked to long term use. In addition to the main problems of fouling and power consumption, the stability of the reagent and the standards and the consumption of reagents are major issues. Both have already been addressed in this chapter. Although technological solutions are available more work is still needed before most of the chemistries met these requirements.

Finally, in regard with the number of chemical species present in the ocean, work with *in situ* chemistries is simply beginning. The wider use of *in situ* analysers by chemical oceanographers requires a multitude of instruments. This implies that more laboratory-made prototypes be prototyped and manufactured. The future commercial availability of *in situ* analyzers will be severely limited by the relatively small oceanographic market, and will more likely be controlled by the overall market and demand for environmental monitoring.

References

Aminot, A. and Kérouel, R. (1991). Autoclaved sea water as a reference material for the determination of nitrate and phosphate in sea water, *Analytica Chimica Acta*, 248, 277–283.

Atkins, W.R.G. (1923a). The phosphate content of fresh and salt waters in its relationship to the growth of the algual plankton, *Journal of Marine Biologist Association U.K.*, 14, 119–150.

Atkins, W.R.G. (1923b). The silica content of some natural waters and culture media, *Journal of Marine Biologist Association U.K.*, 13, 151–159.

Brewer, J.P.G. and Riley, J.P. (1965). The automatic determination of nitrate in sea water, *Deep Sea Research*, 12, 765–772.

Brooks, S.H. and Dorsey, J.G. (1990). Moment analysis for the evaluation of flow injection manifolds, *Analytica Chimica Acta*, 229, 35–46.

Chin, C.S., Johnson, K.S. and Coale, K.H. (1992). Spectrophotometric determination of dissolved manganese in natural waters with 1-(2-pyridylazo)-2naphtol: application to analysis *in situ* in hydrothermal plumes, *Marine Chemistry*, 37, 65–82.

Chin, C.S., Coale, K.H., Elrod, V.A., Johnson, K.S., Massoth, G.J. and Baker, E.T. (1994). *In situ* observations of dissolved iron and manganese in hydrothermal vent plumes, Juan de Fuca Ridge, *Journal of Geophysical Research,* 99(B3), 4969–4984.

Clinch, J.R., Worsfold, P.J. and Casey, H. (1987). An automated spectrophotometric field monitor for water quality parameters, *Analytica Chimica Acta*, 200, 523–531.

Cline, J.D. (1969). *Limnology and Oceanography*, 14, 454

Clark, G.D., Hungerford, J.M. and Christian, G.D. (1989). Optimization of confluent mixing in flow injection analysis, *Analytical Chemistry*, 61, 973–979.

Coale, K.H., Chin, C.S., Massoth, G.J., Johnson, K.S. and Baker, E.T. (1991). *In situ* chemical mapping of dissolved iron and manganese in hydrothermal plumes, *Nature*, 352, 325–328.

Dasgupta, P.K. and Liu, S. (1994). Electroosmosis: A reliable fluid propulsion for flow injection analysis, *Analytical Chemistry*, 66, 1792–1798.

Daniel, A., Birot, D., Blain, S., Tréguer, P., Menut, E. and Leilde, B. (1995a). A submersible flow-injection analyser for the *in-situ* determination of nitrite and nitrate in coastal waters, *Marine Chemistry*, 51, 67–77.

Daniel, A., Birot, D., Lehaitre, M. and Poncin, J. (1995b). Characterisation and reduction of interferences in flow-injection analysis for the *in situ* determination of nitrate and nitrite in sea water, *Analytica Chimica Acta*, 308, 413–424.

Daniel, A., Baker, A.R. and Van Den Berg, C.M.G. (in preparation) Sequential flow analysis coupled with ACSV for on-line monitoring of trace elements in the marine environment.

Deacon, G.E.R. (1933). *Discovery report*, 7, 173.

Dickey, T.D., Frye, D., Jannasch, H.W., Boyle, E. and Knap, A.H. (1997). Bermuda sensor system tesbed, *Sea technology*, 81–86.

Floch, J., Blain, S., Tréguer, P. and Birot, D. (1998). *In situ* determination of silicic acid in sea water based on FIA and colorimetric dual wavelength measurements, *Analytica Chimica Acta*, 377, 157–166.

Floch, J. (1998) Instrumentation *in situ* et modelisation pour l'étude des variations haute fréquence des sels nutritifs dans la colonne d'eau. PhD thesis, 127.

Gamo, T., Sakai, H., Nakayama, E., Ishida, K. and Kimoto, H. (1994) A submersible flow-trough analyse for *in-situ* colorimetric measurement down to 2000 m depth in the ocean, *Analytical sciences*, 10, 843–848.

Grasshoff, K.M. and Chan, K.M. (1971). An automatic method for the determination of hydrogen sulfide, pp. 73–80 in Grasshoff, K. *et al.* (eds), *methods of seawater analysis*, Verlag Chemie.

Hager, S.W., Atlas, E.L., Gordon, L.I., Mantyla, A.W. and Park, P.K. (1972). A comparison at sea of manual and Autoanalyser analysis of phosphate, nitrate, and silicate, *Limnology and Oceanography*, 17, 931–937.

Harvey, H.W. (1926). *Journal of Marine Biologists Association U.K.*, 14, 72.

Hyrayama, K. and N. Unohohara (1988). Spectrophotometric catalytic determination of an ultratrace amount of iron(III) in water based on the oxidation of N,N-dimethyl-p-pheylenediamine by hydrogen peroxide, *Analytical Chemistry*, 60, 2573–2577.

Jannasch, H.W., Johnson, K.S. and Sakamoto, C.M. (1994). Submersible, osmotically pumped analysers for continuous determination of nitrate *in situ*, *Analytical Chemistry*, 66, 3352–3361.

Johnson, K.S. and Petty, R.L. (1983). Determination of nitrate and nitrite in seawater by flow injection analysis, *Limnology and Oceanography*, 28(6), 1260–1266.

Johnson, K.S., Beehler, C.L. and Sakamoto-Arnold, C.M. (1986). A submersible flow analysis system, *Analytica Chimica Acta*, 179, 245–257.

Johnson, K.S., Beehler, C.L. and Sakamoto-Arnold, C.M. (1986). *In situ* measurements of chemical distributions in a deep-sea hydrothermal vent field, *Science*, 231, 1139–1141.

Johnson, K.S., Sakamoto Arnold, C.M. and Beehler, C.L. (1989). Continuous determination of nitrate concentration *in situ*, *Deep sea Research*, 36, 1407–1413.

Massoth, G.J., Baker, E.T., Feely, R.A., Butterfield, D.A., Embley, R.E., Lupton, J.E., Thompson, R.E. and Cannon, G.A. (1995). Observations of manganese and iron at the Coaxial seafloor eruption site, Juan de Fuca ridge, *Geophysical Research Letters*, Vol. 22, 151–154.

Massoth, G.J. and Milburn, H.B. (1997). SUAVE (Submersible System Used to Assess Vented Emissions): A diverse tool to probe the submarine hydrothermal environment in: *proceedings of Marine Analytical Chemistry for Monitoring and Oceanographic Research workshop*, Ed. Blain, S., 80–87.

Measure, C.I., Yuan, J. and Resing, J.A. (1995). Determination of iron in seawater by flow injection analysis using in-line preconcentration and spectrophotometric detection, *Marine Chemistry*, 50, 1–10.

Resing, J.A. and Mottl, M.J. (1992). Determination of manganese in seawater using flow injection analysis using on-line preconcentration and spectrophotometric detection, *Analytical Chemistry*, 64, 2682–2687.

Ruzicka, J. and Hansen, E.H. (1975). Flow injection analyses part I; a new concept of fast continous flow analysis, *Analytica Chimica Acta*, 78, 145–157.

Ruzicka, J. (1983). Flow injection analysis from test tube to integrated microconduits, *Analytical Chemistry*, 55(11), 1040A-1053A.

Ruzicka, J. and Hansen, E.H. (1988). Flow injection analysis, Wiley, New York, 2nd.

Sakamoto-Arnold, C.M., Johnson, K.S. and Beehler, C.L. (1986). Determination of hydrogen sulfide in seawater using flow injection analysis and flow analysis, *Limnology and Oceanography*, 31(4) 894–900.

Stookey, L.L. (1970). Ferrozine-a new spectrophotometric reagent for iron, *Analytical Chemistry*, 42, 779–781.

Strickland, J.D. and Parson, T.R. (1972). A practical handbook of sea water analysis Bulletin 167 Fisheries Research Board of Canada, 310.

Taylor, C.D., Howes, B.L. and Doherty, K.W. (1993). Automated Instrumentation for time-series measurement of primary production and nutrient status in production platform-accessible environments, *Marine Technology Science Journal*, 27(2), 32–44.

Theeuwes, F. and Yum, S.I. (1976). Principles of the design and operation of generic osmotic pumps for the delivery of semisolid or liquid drug formulations, *Annual Biomedical Engineering*, 4, 343–353.

Thompsen, J., Johnson, K.S. and Law, R. (1983). Determination of reactive silicate in sea water by flow injection analysis, *Analytical Chemistry*, 55, 2378–2382.

Truesdale, V.W. and Smith, C.J. (1976). The automatic determination of silicate dissolved in fresh water by means of procedures involving the use of either α- and β- molybdosilicic acid, *Analyst*, 101, 19.

Van Lintel, T.G., Van der Pool, F.C.M. and Bouwstra, S. (1988). *Sensors ans Actuators*, 15, 153.

Verpoorte, E.M.J., Van der Schoot, B.H., Jeanneret, S.A. Manz and de Rooij, N.F., Silicon-based chemical microsensor and microsystems in *interfacial design and chemical sensing*, 245–254.

Weeks, D.A. and Johnson, K.S. (1996). Solenoid pumps for flow injection analysis, *Analytical Chemistry*, 68, 2717–2719.

3. FLOW INJECTION WITH CHEMILUMINESCENCE DETECTION FOR THE SHIPBOARD MONITORING OF TRACE METALS

PAUL J. WORSFOLD[a], ERIC P. ACHTERBERG[a], ANDREW R. BOWIE[a,b], RICHARD C. SANDFORD[a,b] and R. FAUZI C. MANTOURA[b]

[a]*Department of Environmental Sciences, Plymouth Environmental Research Centre, University of Plymouth, Drake Circus, Plymouth, PL4 8AA, UK and* [b]*Plymouth Marine Laboratory, Centre for Coastal and Marine Sciences, Citadel Hill, Plymouth, PL1 2PB, UK*

3.1 MARINE BIOGEOCHEMISTRY OF TRACE METALS

Trace metals in seawater are of concern because of their potential impact on marine life in estuaries and coastal waters, which receive increased inputs from domestic, industrial and agricultural waste disposal. Interest in trace metals in open ocean waters has focused in recent years on the involvement of elements like Fe, Cu, Co and Zn in primary productivity in remote oceanic regions.

Progress in chemical oceanography is determined to a large extent by developments in analytical methods. Many of the trace metals of interest occur at low concentrations in seawater and the complex saline matrix poses many challenges for the marine chemist. Advances in sampling and analytical techniques over the last 25 years have greatly improved our knowledge of the distribution of dissolved metals in the oceans and the oceanic behaviour of many trace metals is now thought to be reasonably well understood. The setting-up of GEOSECS (Geochemical Ocean Sections Study) in the late 1960s was a key step towards our current understanding of oceanic trace metal behaviour. The trace metal concentrations in the vertical oceanic profiles determined for GEOSECS were considerably lower than any previous measurements and required a re-think of oceanic metal behaviour. The main differentiation between oceanic trace metals is based upon their tendency to undergo transfer from solution to solid phases, which is described as particle reactivity, or geochemical reactivity. A common approach to compare the particle reactivity of elements is by use of the mean oceanic residence time (MORT). The MORT of water molecules in the ocean with respect to evaporation, precipitation over continents, and return to the oceans via rivers is about 40,000 years. Elements with MORT longer than that of water become concentrated in sea water, whereas an element with a MORT shorter than water are depleted in sea water relative to the concentration in which they are supplied in river water. MORT of a constituent in seawater is commonly defined by the equation

$$\text{MORT} = A/(\mathrm{d}A/\mathrm{d}t), \tag{3.1}$$

where A is the amount of constituent dissolved in the ocean and $\mathrm{d}A/\mathrm{d}t$ is the amount introduced or removed each year. Estimates of MORT are commonly based on river inputs to the ocean. The definition of MORT assumes the existence of steady state conditions and a well mixed reservoir. Highly particle reactive trace metals are removed rapidly from the water column and as a result have a short MORT. Such elements are called scavenged elements and examples include Mn (890 y), Pb (1100 y) and Al (370 y) (MORT values between brackets (Martin and Whitfield, 1983)). Their distributions are strongly determined by their external inputs (see Figure 3.1a), and their concentrations decrease markedly away from sediment-water and air-sea interfaces. The assumption of a well mixed oceanic reservoir is not fulfilled for scavenged elements and their calculated MORT's should therefore be treated as nominal values. Conservative elements have long MORT's ($> 10^6$ y), because they lack reactivity. As a consequence, concentrations of conservative elements are constant relative to salinity and their distributions are governed by water mixing processes (see Figure 3.1b). Examples of such elements include U (4.9×10^5 y), Cs (4.2×10^5 y) and Li (5.5×10^5 y) (MORT values between brackets (Martin and Whitfield, 1983)). Trace metals with an intermediate MORT are influenced by both their reactivity and the water mixing processes in the ocean. Their vertical distributions tend to mimic nutrients, indicating an association with biogenic particles. Such elements are termed nutrient-type or recycled trace metals, and include Ni (1.4×10^4 y), Zn (1.2×10^2 y), Cd (1.8×10^4 y) and Cu (2.4×10^3 y) (MORT values between brackets (Martin and Whitfield, 1983)). Lowest concentrations of nutrient-type trace metals are observed in the surface ocean, coinciding with enhanced primary productivity. These elements are taken up by organisms, either because of adsorption onto the cell (Cd) or because they are essential micro-nutrients (e.g. Cu and Zn). Upon decomposition and dissolution of organisms deeper in the water column or at the sediment-water interface, the nutrients and nutrient-type trace metals are released back into the water column. As a result, nutrient-type trace metals show vertical oceanic profiles with surface depletion and enrichment at depth (see Figure 3.1c) (Chester, 1990).

Although the oceanic profiles for the conservative, nutrient-type and scavenged trace metals can be explained within a broad biogeochemical context, many questions still remain about the exact workings of the trace metal-particle and trace metal-organism interactions. In this context, the importance of trace metal speciation becomes apparent. Speciation can be defined as a description of the individual physico-chemical forms of an element which together make up its total concentration. The different trace metal species include free aqueous ions, ion pairs, complexes (both inorganic and organic), colloids and particles (see Figure 3.2). In addition, trace metals can have different oxidation states in the marine

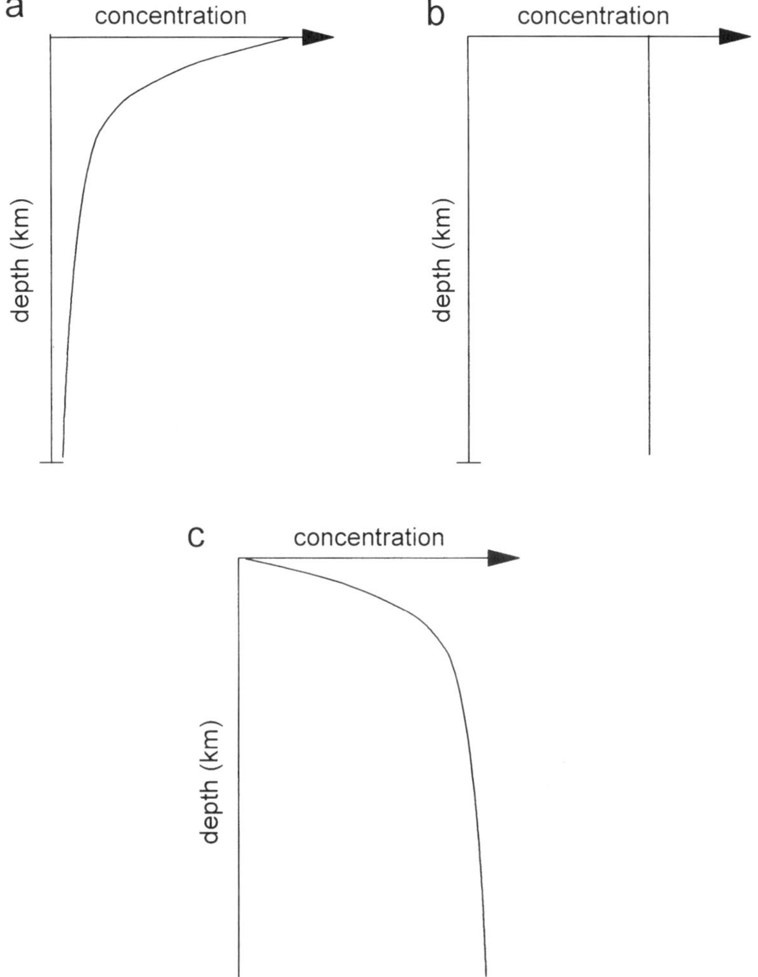

Figure 3.1 Schematic depth profiles representing the 3 general types of oceanic trace metal behaviour: (a) scavenged; (b) conservative; and (c) nutrient-type or recycled. The conditions which cause the shape of the profiles are discussed in the text.

environment. Studies have shown that the different chemical trace metal species exhibit differences in availability for uptake by organisms, toxicity and particle reactivity. For example, the behaviour of dissolved Cr in seawater is strongly influenced by its redox chemistry, with the reduced form (Cr(III)) being highly particle reactive, and the oxidised form (Cr(VI)) behaving more conservatively (Achterberg and van den Berg, 1997). In the case of Cu, free aqueous Cu ions appear to be the available form for the dinoflagellate *Gonyaulax tamarensis*, with enhanced Cu^{2+} concentrations resulting in toxic effects ($[Cu^{2+}] > 10^{-10.4}\ mol\,l^{-1}$)

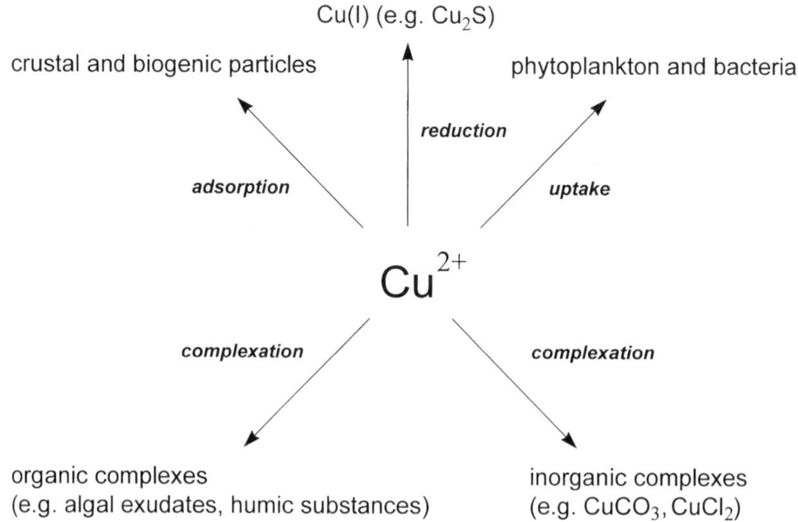

Figure 3.2 Schematic diagram of the processes affecting the speciation of Cu in sea water. This is a generic diagram, the physico-chemical speciation of many other trace metals in seawater is determined by similar processes. Reduction of Cu(II) to Cu(I) in seawater can for example be found in anoxic deep waters of fjords, the Black Sea and the Baltic Sea.

and low levels resulting in Cu deficiency with reduced growth ($[Cu^{2+}] < 10^{-12.5}$ $mol\,l^{-1}$). (Schenck 1984; Anderson and Morel, 1978).

Although the distributions of most trace metals have been known for 15–20 years, there are still some which offer extra analytical challenges. Iron is one of these trace metals which, because of its ubiquitous presence in the terrestrial environment and its low concentration in the oceanic environment, has only relatively recently been investigated in detail. This recent attention originates from the role of Fe as a limiting micro-nutrient in High Nutrient Low Chlorophyll (HNLC) regions. Dissolved Fe exhibits a nutrient-type behaviour with low levels (< 0.1 nM) in oceanic surface waters and enhanced constant values at greater depth (ca. 0.6 nM) for stations in the North Atlantic and North Pacific (Johnson, Gordon and Coale, 1997). A better knowledge of the oceanic distribution of Fe will improve our understanding about the various Fe inputs to the ocean, and the global importance of HNLC regions.

Improved analytical techniques need to be developed for the accurate determination of different Fe species in seawater, as the available Fe species for algae are thought to be free ferric or ferrous ions. In addition to these ions, it is thought that an important fraction of dissolved iron in the ocean occurs in the form of colloids. Also, it has been implied that organic ligands (perhaps including bacterial exudates like siderophores) are involved in complexing Fe, a mechanism which has been suggested to keep Fe in solution by preventing insoluble Fe-oxyhydroxide formation (Johnson, Gordon and Coale, 1997). As a result of the transient nature of the chemical equilibrium between the various dissolved Fe forms, it is important that speciation measurements are performed immediately upon sampling. The application of automated ship-board flow-injection based analytical systems could therefore make a significant contribution to high-resolution contamination-free oceanic Fe speciation measurements.

Much effort has concentrated in the past on open ocean trace metal work, but a lot of uncertainty still exists about trace metal behaviour in enclosed basins (e.g. Western Mediterranean) and coastal seas. Further investigations into trace metal biogeochemistry in such systems are important because of the pollution threat to these waters, which are often close to population centres. Dissolved trace metals in these waters do not appear to mimic nutrient behaviour. In some cases (e.g. Western Mediterranean) this may be caused by a low productivity and therefore low rate of nutrient cycling, leading to a decoupling of the trace metals and nutrient cycles. The result of this is that trace metal retention occurs in the surface waters of the Western Mediterranean and concentrations of nutrient-type trace metals (e.g. Ni, Cr and Cu) are significantly higher than in open ocean surface waters (Morley *et al.*, 1997). A better knowledge of the interaction between metals and primary producers in systems like the Western Mediterranean is important to discriminate between these natural biogeochemical processes and potential anthropogenic perturbations (Achterberg and van den Berg, 1997).

In other waters, such as the coastal North Sea, concentrations of nutrient-type metals are not significantly influenced by enhanced primary productivity during the summer periods. The concentrations of elements like Ni and Cu do not show a decrease during the summer period as is the case for nitrate and phosphate. Radiotracer studies have indicated that biological uptake of nutrient-type trace metals does occur in these coastal waters (Turner *et al.*, 1992). The decoupling between nutrients and nutrient-type trace metals in the North Sea has been explained by the high metal inputs into the system (Tappin *et al.*, 1995). As a consequence, any biological uptake of nutrient-type trace metals does not result in a significant decrease in metal concentrations because of the enhanced trace metal levels.

Redox-sensitive trace metals like Mn, Fe and Co, show a distinct seasonal variation in their dissolved concentrations in the North Sea with enhanced levels during the summer period. The increase in concentration of these trace metals has been explained by porewater inputs and reduced mixing of the coastal waters upon thermal stratification (Tappin *et al.*, 1995). Enhanced primary productivity during the summer period results in an increased amount of sinking detritus in the North Sea. Bacterial mineralisation of the detritus at the sediment-water interface causes a reduction of the redox-potential and hence an increased diffusion of redox-sensitive trace metals from porewaters to overlaying waters. Phytoplankton blooms invoked by enhanced nutrient inputs into coastal seas therefore may result in mobilisation of redox-sensitive trace metals, linking eutrophication to enhanced metal concentrations in marine systems. In the autumn period, storms erode the water column stratification and oxygenate the water resulting in removal of redox sensitive metals.

Many questions still remain about the decoupling of nutrients and nutrient-type trace metals in coastal waters, and about the importance of sediment-water exchange of redox-sensitive trace metals. Because of the strong temporal and spatial variability in coastal waters, pollution and biogeochemical studies require high-resolution data. The best approach for future studies therefore will be the application of automated shipboard and *in-situ* analytical instrumentation. A better understanding of coastal trace metal behaviour will aid prediction of effects of industrial discharges, and will improve the forecasting of the influence of more stringent legislation with respect to metal and nutrient emissions.

3.2 FLOW INJECTION (FI)

FI was described by Ruzicka and Hansen (1981) as an unsegmented flow technique in which a volume of liquid sample is inserted into a moving liquid carrier stream, whereupon it undergoes physical dispersion as it is transported to a flow-through detector for measurement. The transient response is in the form of a peak, the height of which is usually directly related to analyte concentration. The degree of sample dispersion is controlled by factors such as flow rate, tubing length and diameter and manifold geometry, and is highly reproducible. The technique is now widely used in analytical laboratories for the automation of wet chemical methods and has considerable potential for use on board ship and in submersible analysers (Andrew *et al.*, 1994).

A simple FI manifold is shown schematically in Figure 3.3a, and typically consists of a propulsion unit (e.g. a peristaltic pump), a six-port rotary sample

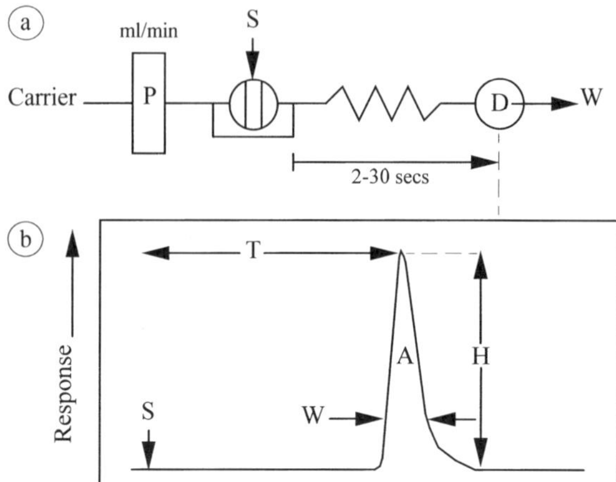

Figure 3.3a Schematic diagram of a single channel FI manifold; P is the pump, S is the injection valve, D is the flow-through detector and W is waste. The chart recorder output shows the recording starting at S, H is the peak height, W is the peak width, A is the peak area and T is the residence time corresponding to the peak height.

injection valve and a flow-through detector (e.g. a spectrophotometer). PTFE tubing is used for sample and reagent transport, with tightly-wound coils often included to aid mixing. The manifold shown in Figure 3.3a is a single-channel system, in which the carrier stream, which can also contain a reagent, transports the sample to the detector. If the method requires more than one reagent, additional streams can be merged with the carrier stream at suitable points in the manifold, as shown in Figure 3.3b (McCormack, 1996).

Reagent consumption is generally low in FI systems (an important factor for shipboard applications), and can be reduced still further by use of a reagent injection manifold, whereby a discrete volume of reagent is injected into a continuously flowing sample stream. This option is suitable for applications in which the sample is in abundant supply (as in many marine situations), and is particularly useful when expensive reagents are necessary. Simultaneous FI determinations can be performed by designing split-line manifolds in which the sample is injected into more than one flow channel, undergoing different reaction chemistries in each. Other components which are often incorporated into FI systems include gas dialysis units (for the diffusion of a gaseous analyte from a carrier (donor) stream through a microporous membrane into a reagent (acceptor) stream) and solid phase reaction columns (in which the injected sample reacts with or is retained by a column packed with solid material). Examples of such manifolds are shown in Figure 3.3b (McCormack, 1996).

3.3 CHEMILUMINESCENCE

Chemiluminescence (CL) can be defined as the production of electromagnetic radiation (ultraviolet, visible or infrared) by a chemical reaction and bioluminescence as visible wavelength CL produced by living organisms or chemical systems derived therefrom. Such processes are uncommon as most chemical reactions release energy as heat but CL can be observed in solid-, gas- and liquid-phase reactions and all three have been used as the basis of analytical methods. The potential of CL-based analytical techniques in biomedical, toxicological, environmental and related areas is attested to by the number of monographs and review articles covering various aspects of its application (see e.g. Robards and Worsfold, 1992; Bowie, Sanders and Worsfold, 1996). In solution, CL finds many applications in analytical chemistry for the determination of metal ions, inorganic anions, biomolecules, carcinogens and drugs in a variety of environmental and clinical matrices. The advantages of CL for analysis includes: high sensitivity and a wide linear dynamic range, both of which can generally be achieved with relatively simple instrumentation. The high sensitivities characteristic of such procedures depend on a number of factors, but especially on the absence of a need for a radiation (or emission) source, which reduces or eliminates Raman and Rayleigh scattering and noise associated with the source. This allows the detector to be operated at extreme photomultiplier voltages with a significant improvement in the signal-to-noise ratio relative to conventional fluorescence

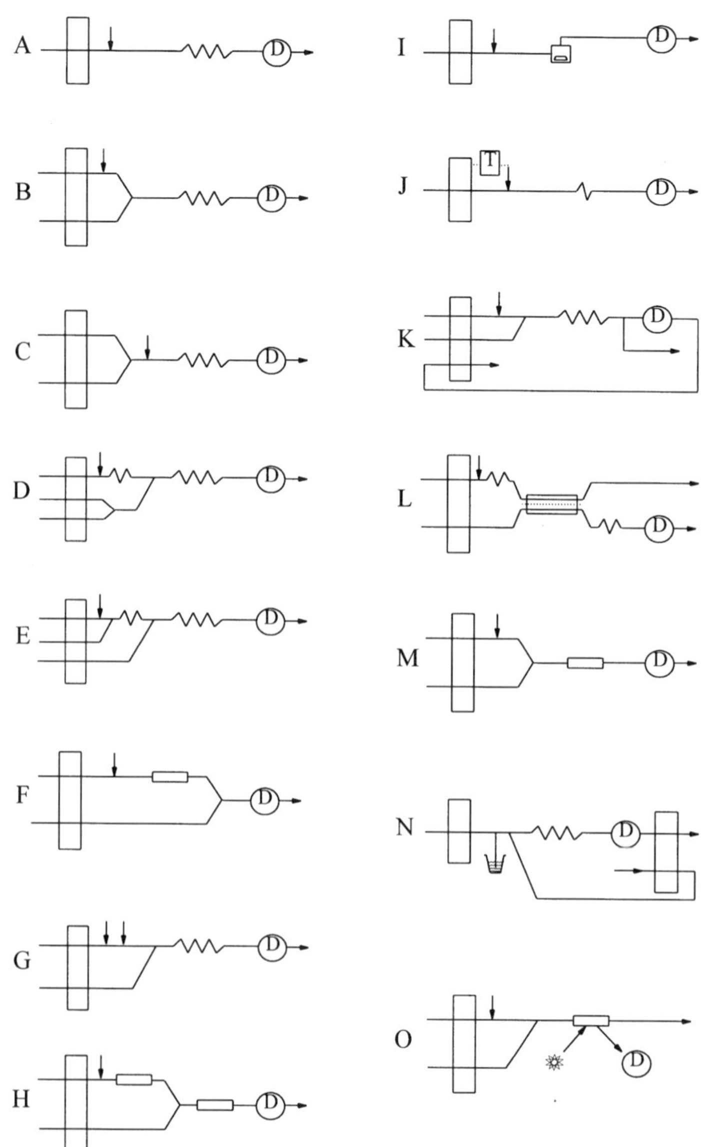

Figure 3.3b Types of FI manifold. A, single line; B, two line with a single confluence point; C, reagent premix into a single line; D, two line with a single confluence point and reagent premix; E, three line with two confluence points; F, packed reactor in line; G, double injection and single confluence (zone penetration); H, sequential reactors (with immobilised enzymes); I, single line with mixing chamber; J, single line, stopped flow; K, solvent extraction; L, dialysis, ultrafiltration, or gas diffusion; M, two lines, single confluence, incorporating a packed reactor; N, hydrodynamic injection; O, two line, one confluence, optosensing on solid surface.

detection. Femtomole detection limits are not uncommon with attomole levels detected for some compounds; as little as 120 molecules have been detected in the case of certain enzymes. Furthermore, calibration responses in CL analysis are often linear over several orders of magnitude. For analytical applications, the rapid and transient nature of solution phase CL emission requires a means of rapidly and reproducibly mixing the sample and reagent streams. Therefore, FI is an ideal technique for sample delivery when using CL detection.

There are many inorganic and organic chemical reactions that are known to produce light. Chemiluminescent reactions have been defined as direct or indirect depending on the origin of the CL. The latter is also described as sensitised or energy transfer chemiluminescence. With direct CL, the reaction generates the primary excited-state molecule which is then responsible for light emission. In contrast, in "sensitised" reactions the excited product is not the light emitter, but rather transfers its energy to a fluorescent acceptor which then emits light. The discrete energy transfer step in sensitised CL has distinct advantages in the design of analytical systems, as optimum structural features can be designed and incorporated into separate chemical reactant and fluorophore molecules. For instance, unlike in direct CL, the spectrum of the chemiluminescent emission is not determined by the consumed chemiluminescent reagent but instead by the fluorophore. Hence, the fluorophore in sensitised CL is selected to provide the optimum compromise between excitation and emission wavelengths, chemical stability and quantum yield. This is in contrast to direct CL, wherein reactivity and fluorescence properties reside in a single molecule.

Chemiluminescent emission can be characterised by the four parameters of colour, intensity and rates of production and decay of intensity. Reaction time and duration of chemiluminescent reactions vary from very rapid and/or short-lived (< 1 s) to very slow and/or long-lasting (> 1 day). Apart from the obvious role of the substrate, reaction conditions have a significant effect on the progress of the chemiluminescent reaction. The wide range of reaction rates observed places considerable demands on the development of instrumentation for monitoring CL reactions. However, from the analytical viewpoint, the CL emission intensity (I) has most impact on the application of a chemiluminescent reaction, as it is I_{CL} that is measured (either as an integral over the lifetime of the emission or as a transient response) and this is dependent on the rate of reaction, its efficiency at generating molecules in an excited state (expressed as the quantum yield or efficiency) and, in sensitised CL, on the fluorophore. The quantum yield of a reaction is the product of the efficiencies of the excitation and emission steps and values ranging from 10^{-15} to nearly 1 have been observed (Campbell, 1988). Higher values are usually associated with bioluminescent reactions whereas the chemiluminescent reactions commonly exploited in analysis have typical quantum yields of 0.001–0.1. However, the almost complete absence of background emission means that even very inefficient systems with much lower quantum yields (< 0.001) can be monitored and exploited for analysis. Further, the colour and quantum yield can be affected markedly by the polarity of the solvent, the solution temperature and, under the influence of secondary physical processes, by quenching of the excited state. For example, the quantum yield of luminol oxidised in dimethyl sulphoxide is 0.05 compared with 0.01 in water and the respective colours are blue-violet (λ_{max} 425 nm) and blue-green (λ_{max} 480–502 nm).

Because I_{CL} is dependent on the rate of the chemiluminescent reaction, measurement of emission intensity can be used as the basis of quantification for any species whose concentration determines this reaction rate. The analytical versatility of CL depends on this and the fact that reaction conditions can be adjusted so than any one of the reaction components, i.e. chemiluminescent substrate, oxidant, "catalyst", cofactor or sensitiser, can be made rate determining by adjusting concentrations so that the analyte is the limiting reactant, that is, all other reactants are present in excess.

For a reaction to be chemiluminescent, an excited state molecule must be produced during the course of the reaction. The observed emission stems from the ejection of a photon from this excited state. Three essential features are associated with these reactions:

(i) The reaction must be exothermic in order to generate sufficient energy for formation of the electronically excited state (for emission in the visible region the minimum energy requirement is 180 kJ mol^{-1}).
(ii) A pathway must exist for the formation of the electronically excited state.
(iii) The excited state must be capable of deactivation by the emission of radiation or energy transfer to a fluorophore.

There are general empirical guidelines for predicting chemiluminescent behaviour. For instance, if an analyte itself or an oxidation product is fluorescent then there is the possibility that oxidation of the molecule will produce CL. However, there are many exceptions to this principle and, in many cases, chemiluminescent reactions cannot be predicted. Nonetheless, many solution phase CL reactions involve the oxidation of aromatic substances. One of the most commonly used, particularly for the determination of trace metals (see below) and hydrogen peroxide (Price *et al.*, 1994) in seawater is the luminol reaction. This was first reported in 1928 and involves the oxidation of luminol (5-amino-2,3-dihydrophthalazine-1,4-dione or 5-aminophthalhydrazide), an acyl hydrazide, in basic solution (pH 10–11) (see Figure 3.4). The reaction is catalysed by haeme-containing enzymes (e.g. peroxidase) and a number of metal ions of which cobalt(II), copper(II) and iron(II) are particularly effective. In addition, hexacyanoferrate(III) can act as catalyst/co-oxidant in the reaction. Due to the rapid and transient CL emission this reaction is ideally suited to incorporation within an FI manifold.

3.4 ON-LINE ANALYTE PRECONCENTRATION

As stated above, most solution phase CL methods suffer from poor selectivity. For example, many trace metal ions (e.g. Fe(II), Co(II), Cu(II), Mn(II)) are known to

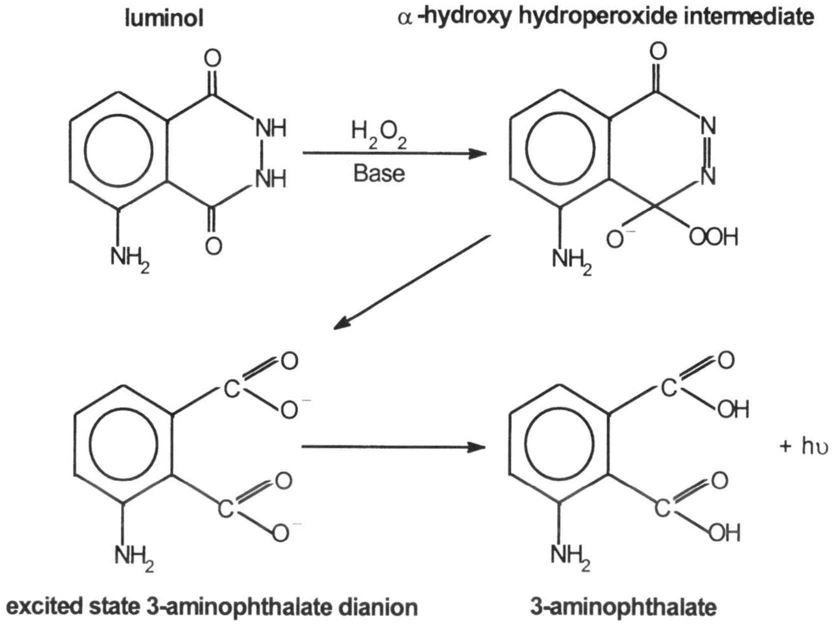

Figure 3.4 Luminol chemiluminescence reaction scheme.

catalyse the oxidation of luminol in the absence of hydrogen peroxide (Klopf and Nieman, 1983). These interfering transition metals act to alter the apparent quantitative relationship between observed luminescent power and analyte concentration. Furthermore, the signal generated in many CL reactions is substantially suppressed for trace metal standards prepared in various solutions containing major seawater cations (e.g. Ca^{2+}, Mg^{2+}) compared with pure water standards, and solutions containing halide ions showed significant CL signal increases.

Elimination of the sea-salt matrix prior to introduction of the buffered CL reagents is essential to prevent any precipitation of such major seawater ions in the FI manifold. Many solution phase CL reactions are optimum at high alkaline pH levels. Additionally, preconcentration of the analyte is generally required to perform open-ocean seawater analyses, where trace elements are often only present at sub-nanomolar concentrations.

Chelating resins have entertained extensive interest in recent years for the preconcentration of trace metals from seawater (Nickson *et al.*, 1995). The method of Landing, Haraldsson and Paxeus (1986) has been most commonly applied for the synthesis of an 8-hydroxyquinoline (8-HQ) microcolumn used in FI-CL determinations. 8-HQ is selective towards transition and heavy metal cations relative to alkali and alkaline-earth cations. Previous methods have involved the immobilisation of 8-HQ onto silica substrates, which offer the advantages of good mechanical strength, resistance to swelling and rapid overall

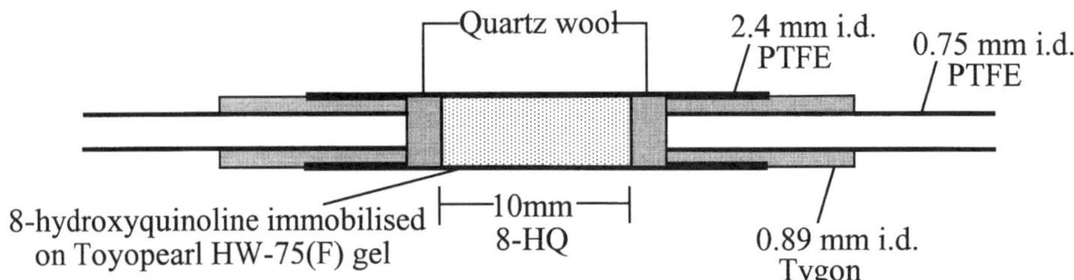

Figure 3.5 Schematic diagram of the 8-HQ chelating resin preconcentration column.

exchange kinetics in column applications. Such immobilisations, however, are unstable at high pH. The method of Landing used the highly porous, mechanically and chemically stable, hydrophilic organic resin gel Toyopearl-TSK as the solid support. Toyopearl-TSK consists of intertwined vinyl polymer agglomerates, which offer stability, high porosity, and high hydrophilicity due to the presence of ether linkages and hydroxyl groups. The polymer itself exhibits no cation exchange capacity and does not concentrate dissolved organic species. A TSK-8HQ microcolumn has been used in a FI-CL manifold for the analysis of Fe (Elrod, Johnson and Coale, 1991; Powell, King and Landing, 1995; Bowie *et al.*, 1998), Mn (Chapin, Johnson and Coale, 1991), Co (Sakamoto-Arnold and Johnson, 1987) and Cu (Coale *et al.*, 1992).

A typical 8-HQ microcolumn is shown in Figure 3.5. The columns were constructed from 2.4 mm i.d. PTFE tubing and the ends fabricated using 0.89 mm i.d. Tygon tubing connected to the 0.75 mm i.d. PTFE manifold tubing. A plug of quartz wool was placed at either end in order to retain the resin inside the column. Such columns can easily be incorporated in-line into a FI manifold, and allow analyte preconcentration/matrix elimination to take place upstream of the CL reaction mechanism. Careful buffering of the seawater as it passes across the column can also allow for extra selectivity to be obtained between both different transition metals and redox species. Furthermore, partitioning between different chemical forms of one metal species (e.g. between colloidal, leachable, organically bound fractions) is possible because 8-HQ may show selectivity towards the different metal species and complexes.

3.5 IRON

Iron is the fourth most abundant element in the Earth's crust, present at a concentration of about 6%. However, like other reactive trace metals, dissolved iron is only present in open-oceanic waters at sub-nanomolar levels. Iron is an essential micro-nutrient, limiting primary production in remote oceanic regions (Martin and Fitzwater, 1988; Martin *et al.*, 1990; Coale *et al.*, 1996) such as the Equatorial and North East Pacific and the Southern Ocean, where macronutrients are plentiful. Such limitation may have important ramifications for the atmosphere – ocean CO_2 flux and global carbon cycles (Martin, 1990; Cooper, Watson and Nightingale, 1996). Seeding experiments during Iron-Ex I and II (Martin *et al.*, 1994; Coale *et al.*, 1996) expeditions in the Equatorial Pacific, using $FeSO_4$, resulted in enhanced primary productivity, and in the case of Iron-Ex II, also in a noticeable, enhanced draw-down of CO_2. Iron exists in two forms in marine systems. Fe(III) is the thermodynamically stable form in oxygenated seawater, existing primarily as insoluble oxy-hydroxides or colloidal matter (Stumm and Morgan, 1981; Hudson and Morel, 1990; Turner *et al.*, 1981). Fe(II) is a transient species in surface oxic waters, existing via chemical or photochemical Fe(III) reduction (O'Sullivan *et al.*, 1991; Johnson *et al.*, 1994; Gledhill and van den Berg, 1995) or via atmospheric deposition (Millero *et al.*, 1995; Zhuang *et al.*, 1990; Martin and Gordon, 1988), and is oxidised rapidly by O_2 and H_2O_2 species at seawater pH (King *et al.*, 1995). Recently, organic complexation has been thought to occur to a significant extent in marine systems (Gledhill and van den Berg, 1994; Rue and Bruland, 1995; van den Berg, 1995). Major sources of iron to the world's oceans include: aeolian deposition, fluvial transport, hydrothermal venting, continental shelf regeneration and upwelling of Fe-enriched subsurface waters. In order to assess Fe input and removal mechanisms, and to enhance the understanding of redox and chemical speciation of Fe associated with various operationally defined fractions, there is a need for new shipboard analytical methodologies to determine ultratrace levels of iron.

FI-CL techniques have attracted interest in recent years for the shipboard mapping of iron. Two CL reactions systems have been exploited for this purpose. Elrod, Johnson and Coale (1991) used the brilliant sulfoflavin reagent to measure Fe(II) species. The relatively poor sensitivity of the CL reaction precludes its usage except in hydrothermal plume mapping or in waters with enhanced iron concentrations. Other workers (Obata, Karatani and Nakayama, 1993; Powell, King and Landing, 1995) have utilised the well documented luminol CL reaction to determine Fe at sub-nanomolar levels.

Bowie *et al.* (1998) have recently deployed a luminol CL system along the Atlantic Meridional Transect, mapping Fe levels in the upper water column from 50 °N to 50 °S. The FI-CL manifold used for this work is shown in Figure 3.6. After Fe(III) reduction by sulphite, total dissolved (II + III) iron levels are determined via in-line matrix elimination/analyte preconcentration on a 8-HQ chelating microcolumn. The preconcentrated iron is then stripped from the column using a weak HCl eluent. The analyte is then brought to the detection unit where it mixes with the buffered (pH 12.2) luminol reagent stream, catalysing the reaction and generating a signal in the form of a narrow peak. The analytical performance for the FI-CL system are given in Figure 3.7. As shown, such techniques are highly sensitive and rapid (a seawater sample can be quantified within 20 min.) which is ideal for the mapping of Fe in fertilisation experiments. The manifold also has the potential for determining the partitioning between different chemical forms of iron in seawater. By careful pretreatment, FI-CL techniques are capable of differentiating between the colloidal, leachable and particulate fractions, as well as assessing the effect naturally

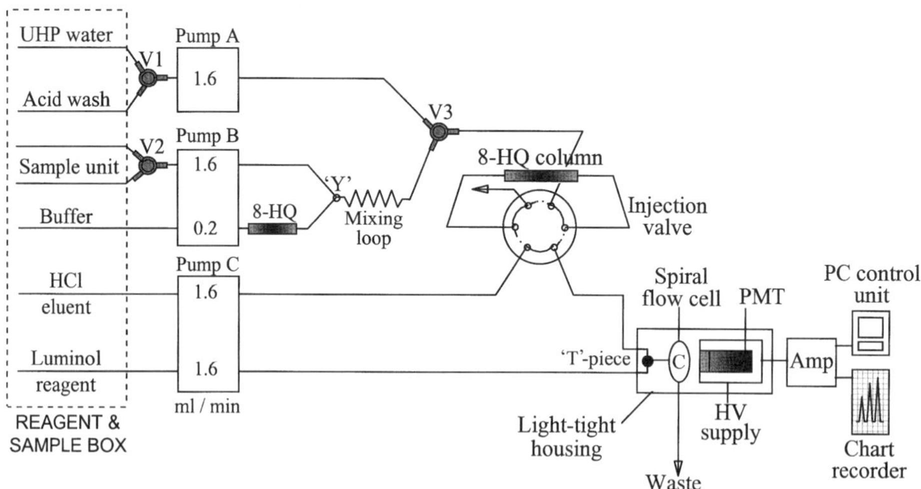

Figure 3.6 Flow injection manifold used for the chemiluminescence based determination of Fe in seawater: PMT = photomultiplier unit; C = spiral flow cell; V1, V2 & V3 = switching valves; A = wash pump; B = sample pump; C = reagents pump.

Limit of detection (3s)	40 pM
Linear dynamic range	0.04 – 10 nM
R squared	0.9986
Reproducibility	2 – 5 %
Sample analysis rate	3 per hour

Figure 3.7 Figures of merit for the FI-CL system used to determine Fe in seawater.

occurring organic ligands have on the recovery of iron by the chelating resin (Obata *et al.*, 1997). Furthermore, the manifold can be easily automated which will lead to future generations of standalone and *in-situ* monitors.

A vertical distribution of total (II + III) dissolvable, unfiltered Fe in unfiltered samples collected in the upper water column (7–200 m), as determined by FI-CL, is shown in Figure 3.8(a). The corresponding chlorophyll fluorescence, temperature and salinity profiles for this station are given in Figure 3.9(b). The location for the sampling station was in the sub-tropical eastern Atlantic Ocean, at 24 °N, 21 °W. The analysis of unfiltered samples using the FI-CL method is likely to have resulted in the detection of an important fraction of colloidal and labile particulate Fe which dissolved during acidification. Enhanced surface water concentrations at the 7 m depth (1.2 nM) were possibly due to Fe laden atmospheric deposits. Levels decreased with depth through the euphotic zone and reached a minimum (0.4 nM) coinciding with the fluorescence maximum. Fe concentrations showed an increase below 100 m depth which may have been caused by

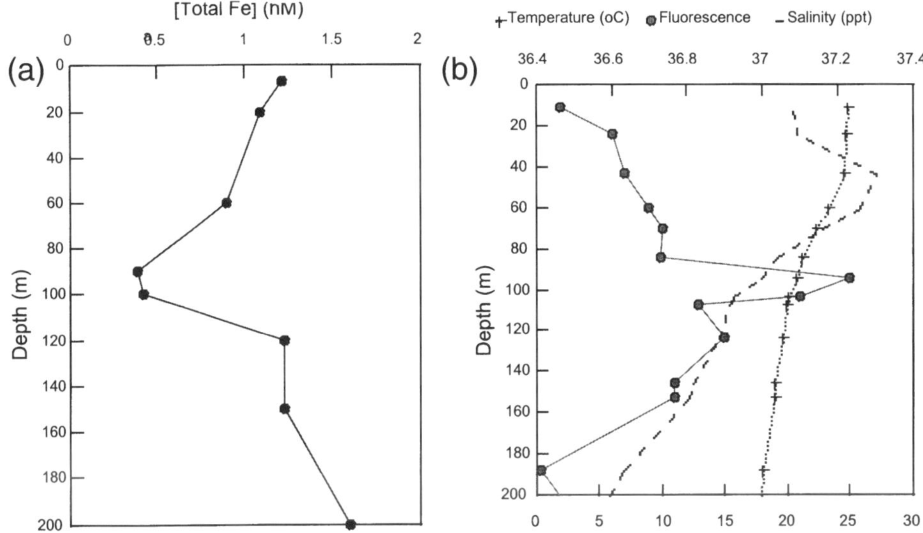

Figure 3.8 (a) Vertical profile of dissolved Fe(II + III) distribution in NE Atlantic seawater sample; (b) Corresponding temperature, salinity and chlorophyll fluorescence profiles (measured by Neil Brown CTD sensors); AMT-3 CTD station 8, SDY 273, 24 °N, 21 °W.

release upon breakdown of Fe containing detritus. The fluorescence measurements are indicative of chlorophyll levels and therefore the Fe minimum at ca. 100 m depth is directly related to photosynthetic uptake.

Future analytical developments of FI-CL methodologies will involve the quantification of the separate redox species via selective microcolumn preconcentration. By incorporating two flow lines and microcolumns, one side can measure Fe(II) concentration (using no Fe(III) reduction), whilst the other line can be used to determine the reducible or "total" dissolved fraction. Furthermore, using cross-flow ultrafiltration instrumentation, present FI-CL technologies can be employed to investigate different size fractions of Fe (e.g. truly dissolved, colloidal, particulate) present in marine waters.

3.6 MANGANESE

Manganese is a geochemically reactive element, and is found in the oceans at concentrations ranging from < 0.1 to 3 nM (Donat and Bruland, 1995). Manganese(II) is removed from the dissolved phase by oxidation to insoluble manganese(IV) oxide, which acts as a scavenger for other trace elements. Particulate Mn(IV) in sediments is diffused into the water column under mild reducing conditions, such as the oxygen minimum layer and near-shore anoxic sediments. Sources of manganese to the oceans include aeolian dust, coastal, sediment release, riverine and hydrothermal inputs, and the use of Mn as a tracer has improved our understanding of such input mechanisms. The ability to map Mn accurately in plumes has enhanced the understanding of hydrothermal activity at mid-ocean ridges and its role in the geochemical cycling of trace elements.

Several FI-CL methods have been reported for the determination of manganese in seawater. Chapin *et al.* (1991) presented a method based on the Mn(II)-catalysis of the oxidation of 7,7,8,8-tetracyanoquinodimethane (TCNQ) in alkaline solution, resulting in the emission of light. By incorporating a mini-column containing the 8-hydroxquinoline chelating resin, a limit of detection of 0.1 nM was achieved. This method has been widely used at sea where profiles were in good agreement with data measured by GFAAS in a shore-based laboratory. The reaction was also exploited by Bowie *et al.* (1995) in order to determine manganese in potable waters. The manifold for this method is illustrated in Figure 3.9. No chelating resin mini-column was incorporated into the FI system since the nature of the matrix and the enhanced Mn levels meant the reaction was sensitive enough without in-line preconcentration. Bowie *et al.* (1995) investigated the reaction mechanism in detail and the CL emission was sensitised using eosin Y and further enhanced by means of didodecyldimethylammonium bromide (DDAB) bilayer vesicles.

Nakayama *et al.* (1989) developed an automated chemiluminescence method for the determination of manganese in seawater. The technique was based on the combination of selective electrolytic preconcentration using a glassy carbon electrode, followed by CL detection using the alkaline luminol reaction. Interfering metal ions were removed by selective extraction chromatography using a chelex-100 column. The limit of detection (0.9 nM) allowed the method to be

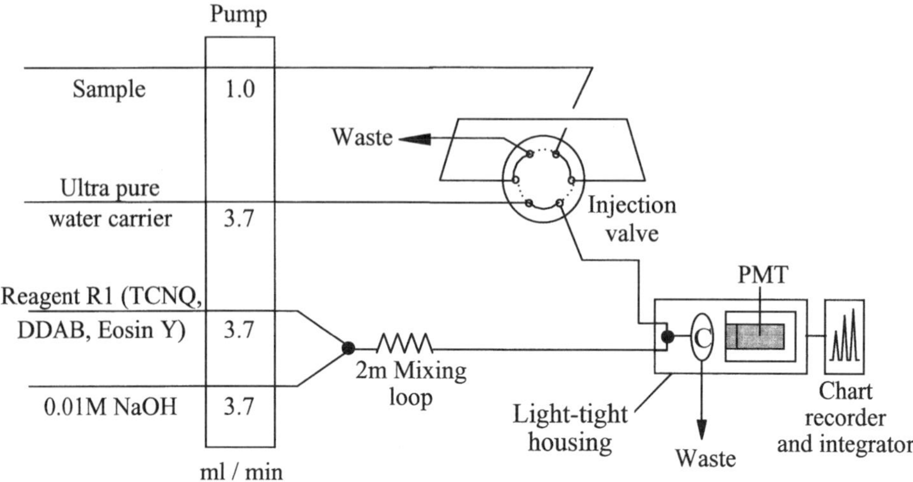

Figure 3.9 Flow injection manifold used for the chemiluminescence based determination of Mn in natural waters: PMT = photomultiplier unit; C = spiral flow cell.

applied to samples collected through the 0–1750 m water column in seas adjacent to Japan, where hydrothermal activity was noted. The Mn(II) catalysis on the alkaline oxidation of luminol has also been reported in synthetic media (Burguera and Townshend, 1981; Lin *et al.*, 1993). Both methods have thoroughly investigated the effect of potential interfering metal ions on the reaction, and one can conclude that extraction of the analyte from its seawater matrix would be required before using the reaction schemes for shipboard FI-CL determinations.

3.7 COPPER

Copper (Cu) is an essential micronutrient for many marine organisms, but may be toxic at enhanced concentrations. The principal redox state in oxygenated seawater is Cu(II), but the occurrence of Cu(I) has been reported in anoxic bottom waters (e.g. in fjords) (Apte and Batley, 1995). The principal inputs of Cu to marine systems are: atmospheric deposition, fluvial inputs, and industrial and urban waste water discharges. The application of Cu in anti-fouling paints has particularly affected Cu concentrations in estuaries and coastal waters.

The physico-chemical form of Cu is an important factor in determining its role in marine ecosystem processes, with free cupric ions (Cu^{2+}) reported to be the most available/toxic Cu form for marine species (Gledhill *et al.*, 1997). The free cupric ion constitutes only a small fraction ($< 1\%$) of the total dissolved Cu concentration, with organically complexed Cu forming the major dissolved fraction (90–99%) (Buckley and van den Berg, 1986). The remainder exists as inorganically complexed Cu. Concentrations of Cu^{2+} as low as 10^{-13} M have been reported for oceanic waters by Coale and Bruland (1988) and Moffet and Zika (1987). Ambient concentrations of total dissolved Cu are typically in the range 0.5–6 nM in the open ocean (Coale and Bruland, 1988), 10–20 nM in coastal seas (e.g. North Sea, Tappin *et al.*, 1995) and in the μM range in perturbed estuaries such as the Fal in Cornwall, UK (Rijstenbil *et al.*, 1991).

CL and voltammetric techniques are able to determine Cu speciation directly in seawater, whereby an operational differentiation is made between "labile" and total dissolved Cu. Labile Cu includes Cu^{2+} ions and inorganic and weak organic complexes. This labile Cu fraction is thought to be representative of the bioavailable concentration of Cu in seawater. A shipboard method for dissolved Cu would ideally be capable of determining both sub-nM levels in open ocean waters and enhanced (high nM) levels in coastal waters with a high spatial and temporal resolution. A good approach is to use the FI-CL technique described above in conjunction with the Cu catalysed 1,10-phenanthroline CL reaction (Federova *et al.*, 1982), a simplified scheme of which is shown in Figure 3.10. 1,10-Phenanthroline combines with Cu(II) to form a chelate $[Cu\,phen_n]^{2+}$ ($n = 1$ or 2). The bound copper catalyses the decomposition of hydrogen peroxide to produce the superoxide radical which cleaves the 1,10-phenanthroline ring at the $C_5=C_6$ position to form a peroxide radical ion. This breaks down to form a dioxetane,

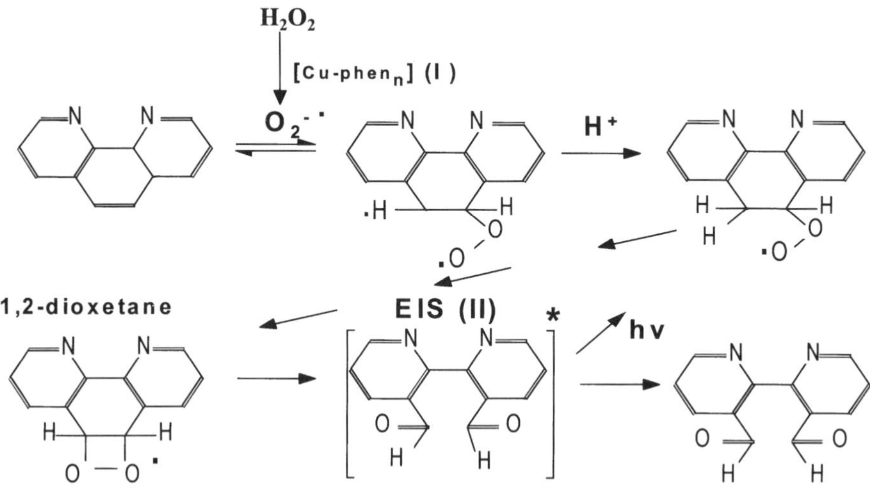

Figure 3.10 Copper-1,10-phenanthroline chemiluminescence reaction scheme.

which in turn decomposes to an electronically excited 3,3′-diformyl-2,2′-dipyridyl species, which decays back to the ground state resulting in CL emission.

Typical analytical figures of merit for the FI-CL manifold shown in Figure 3.11, without preconcentration, are a sample throughput of $30\,h^{-1}$, a linear range of 0.1–60 nM ($R^2 = 0.999$), reproducibility of 1–3 % (n = 5) and a limit of

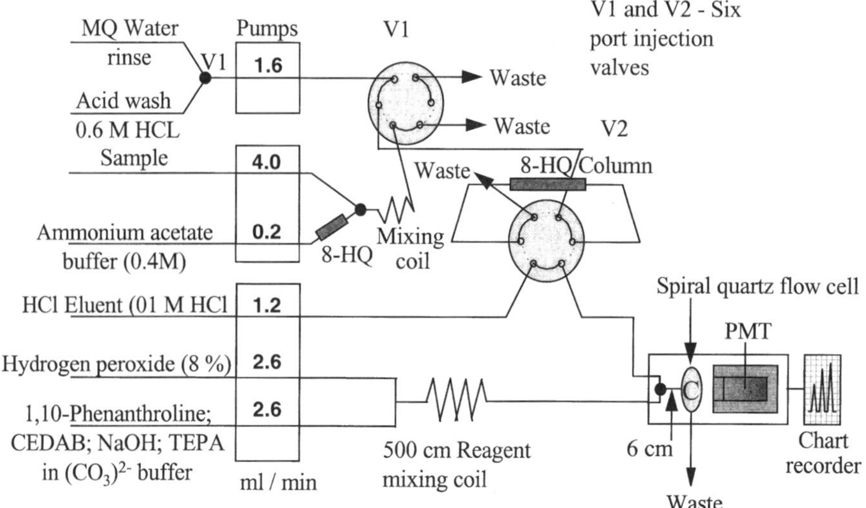

Figure 3.11 Flow injection manifold used for the chemiluminescence based determination of Cu in seawater.

detection (3s) of 0.5 nM in a water matrix. The incorporation of a preconcentration microcolumn of 8-HQ immobilised on an inert Fractogel backbone reduces the limit of detection and increases selectivity via analyte separation from the seawater matrix (Coale *et al.*, 1992). An in line photo-oxidation unit for breakdown of Cu complexing organic ligands would also facilitate total dissolved Cu measurements (Achterberg and van den Berg, 1994).

3.8 COBALT

Cobalt (Co) is a micronutrient for marine organisms and exists principally as Co(II) in oxygenated waters. Co is a geochemically reactive metal, occurring at enhanced levels in surface oceanic waters (100–200 pM) and showing a decrease in concentration with depth (to < 40 pM) (Sakomoto-Arnold and Johnson, 1987; Vega and van den Berg, 1997). Co levels in coastal waters of the central North Sea are reported to range from 160 pM in winter to > 800 pM in summer (Tappin *et al.*, 1995).

The CL reaction for Co is based on the oxidation of gallic acid in the presence of hydrogen peroxide, with Co acting as the catalyst. Sakamoto-Arnold and Johnson (1987) reported a limit of detection of 8 pM using a manifold similar to that shown in Figure 3.13. This was achieved in a seawater matrix using

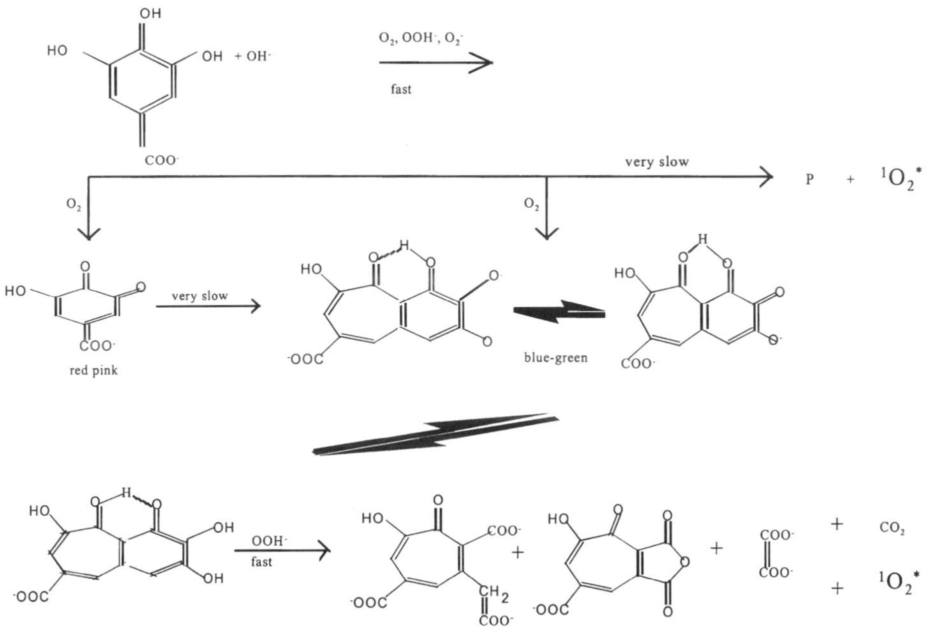

Figure 3.12 Cobalt-gallic acid chemiluminescence reaction scheme.

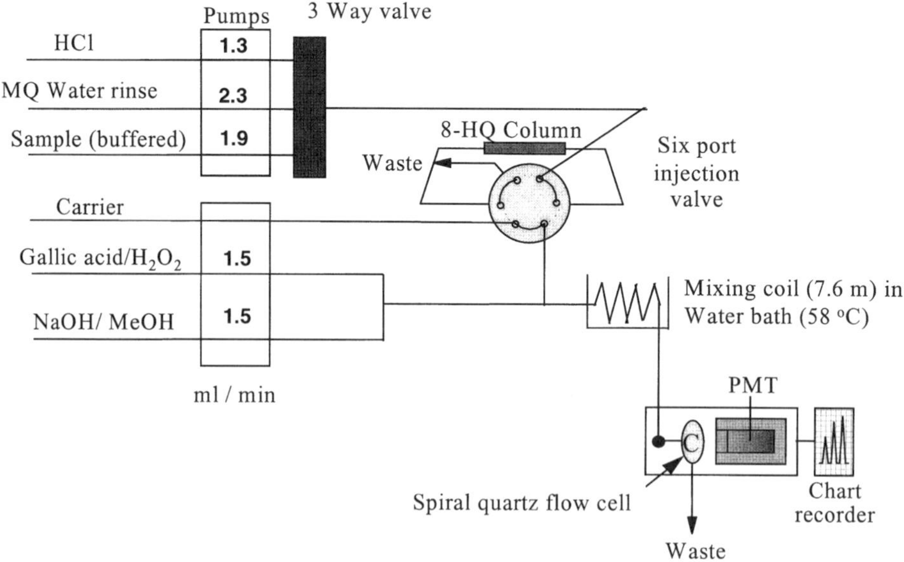

Figure 3.13 Flow injection manifold used for the chemiluminescence based determination of Co in seawater.

immobilised 8-HQ to preconcentrate Co(II) with a 4 minute sample load time. The complete analytical cycle took 8 minutes and the reproducibility was $< 5\%$.

3.9 OTHER SPECIES

Other trace metals can be determined using a variety of reactions (Robards and Worsfold, 1992; Bowie, Sanders and Worsfold, 1996) incorporated within FI-CL manifolds. These include the determination of chromium species, Ag(I), Au(III), Rh(III) and Ti(IV) using modifications of the luminol reaction. This reaction has also been used to determine hydrogen peroxide, which is often used as the oxidant, in the presence of an excess of Co(II) as the catalyst. The method was successfully deployed on board ship in the Mediterranean Sea, with a limit of detection of 5 nM and a linear range of 5–500 nM (Price, Worsfold and Mantoura, 1994).

3.10 CONCLUSIONS

The combination of flow injection with chemiluminescence detection is well suited to shipboard (and submersible) deployment for the sensitive determination of trace metals and other species in seawater. Solid phase microcolumns containing chelating reagents can be incorporated within the flow injection manifold to provide on-line preconcentration of the analyte and effective matrix removal. Precise control of the chemical environment within the flow injection manifold provides good reproducibility and allows for additional on-line physical and chemical treatment of the sample. This approach to shipboard monitoring therefore offers considerable scope to the chemical oceanography community as a means of providing high quality analytical data with good temporal and spatial resolution.

Acknowledgements

P.J.W., E.P.A. and A.R.B. would like to thank the EU MAST programme for funding their MEMOSEA project (contract MAS3-CT97-0143) in partial support of this work. A.R.B. and R.C.S. would like to thank the University of Plymouth and the CCMS Plymouth Marine Laboratory for funding their PhD studentships.

References

Achterberg, E.P. and van den Berg, C.M.G. (1994). In-line ultraviolet-digestion of natural water samples for trace metal determination using an automated voltammetric system, *Anal. Chim. Acta*, 291, 213–232.

Achterberg, E.P. and van den Berg, C.M.G. (1997). Chemical speciation of chromium and nickel in the Western Mediterranean, *Deep-Sea Research II*, 44, 693–720.

Anderson, D.M. and Morel, F.M. (1978). Copper sensitivity of Gonyaulax tamarensis, *Limnol. Ocean.*, 23, 283–295.

Andrew, K.N., Blundell, N.J., Price, D. and Worsfold, P.J. (1994). Flow injection techniques for water monitoring, *Anal. Chem.*, 66, 916A–922A.

Apte, S.C. and Batley, G.E. (1995). Trace metal speciation: non electrochemical approaches. In: Tessier, A. and Turner, D. R. (1995) *Metal Speciation and Bioavailability in Aquatic Systems*, Chichester: Wiley.

Bowie, A.R., Achterberg, E.P., Mantoura, R.F.C. and Worsfold, P.J. (1998). Determination of sub-nanomolar levels of iron in seawater using flow injection with chemiluminescence detection, *Anal. Chim. Acta*, 361, 189–200.

Bowie, A.R., Fielden, P.R., Lowe, R.D. and Snook, R.D. (1995). Sensitive determination of manganese using flow-injection and chemiluminescent detection, *Analyst*, 120, 2119–2127.

Bowie, A.R., Sanders, M.G. and Worsfold, P.J. (1996). Analytical applications of liquid phase chemiluminescence reactions – A review, *J. Biolumin. Chemilum.*, 11, 61–90.

Buckley, P.J.M. and van den Berg, C.M.G. (1986). Copper complexation profiles in the Atlantic Ocean, *Mar. Chem.*, 19, 281–296.

Burguera, J.L. and Townshend, A. (1981). Determination of manganese(II) by a chemiluminescence reaction, *Talanta*, 28, 731–735.

Campbell, A.K. (1988). *Chemiluminescence: Principles and applications in biology and medicine*, pp. 1–608, Chichester: Ellis Horwood.

Chapin, T.P., Johnson, K.S. and Coale, K.H. (1991). Rapid-determination of manganese in sea-water by flow-injection analysis with chemiluminescence detection, *Anal. Chim. Acta*, 249, 469–478.

Chester, R. (1990). *Marine geochemistry*, p. 698, London: Unwin Hyman.
Coale, K.H. and Bruland, K.W., (1988). Copper Complexation in the North Pacific, *Limnol Oceanogr.*, 33, 1084–1101.
Coale, K.H., Johnson, K.S., Fitzwater, S.E., Gordon, R.M., Tanner, S., Chavez, F.P., *et al.* (1996). A massive phytoplankton bloom induced by an ecosystem-scale iron fertilization experiment in the equatorial Pacific Ocean, *Nature*, 383, 495–501.
Coale, K.H., Johnson, K.S., Stout, P.M. and Sakamoto, C.M. (1992). Determination of copper in sea-water using a flow-injection method with chemiluminescence detection, *Anal. Chim. Acta*, 266, 345–351.
Cooper, D.J., Watson, A.J. and Nightingale, P.D. (1996). Large decrease in ocean-surface CO_2 fugacity in response to *in-situ* iron fertilization, *Nature*, 383, 511–513.
Donat, J.R. and Bruland, K.W. (1995). Trace Elements in the Oceans. In Salbu, B. and Steinners, E. (Eds) Trace Metals in Natural Waters, 11, 247–281.
Elrod, V.A., Johnson, K.S. and Coale, K.H. (1991). Determination of subnanomolar levels of iron(II) and total dissolved iron in seawater by flow-injection analysis with chemiluminescence detection, *Anal. Chem.*, 63, 893–898.
Federova, O.S., Olkin, S.E. and Berdnikov, V.M. (1982). The chemiluminescence mechanism in 1,10-phenanthroline oxidation during the catalytic decomposition of hydrogen peroxide, *Z. Phys. Chemie, Leipzig*, 3, 529–549.
Gledhill, M., Nimmo, M., Hill, S.J. and Brown, M. (1997). The toxicity of copper species to marine algae with particular reference to macroalgae, *J. Phycol.*, 33, 2–11.
Gledhill, M. and van den Berg, C.M.G. (1994). Determination of complexation of iron(III) with natural organic complexing ligands in seawater using cathodic stripping voltammetry, *Mar. Chem.*, 47, 41–54.
Gledhill, M. and van den Berg, C.M.G. (1995). Measurement of the redox speciation of iron in seawater by catalytic cathodic stripping voltammetry, *Mar. Chem.*, 50, 51–61.
Hudson, R.J.M. and Morel, F.M.M. (1990). Iron transport in marine phytoplankton: Kinetics of cellular and medium coordination reactions, *Limnol. Oceanogr.*, 35, 1002–1020.
Johnson, K.S., Coale, K.H., Elrod, V.A. and Tindale, N.W. (1994). Iron photochemistry in seawater from the equatorial Pacific, *Mar. Chem.*, 46, 319–334.
Johnson, K.S., Gordon, R.M. and Coale, K.H. (1997). What controls dissolved iron concentrations in the world ocean? *Mar. Chem.*, 57, 137–161.
King, D.W., Lounsbury, H.A. and Millero, F.J. (1995). Rates and mechanism of Fe(II) oxidation at nanomolar total iron concentrations, *Environ. Sci. Technol.*, 29, 818–824.
Klopf, L.L. and Nieman, T.A. (1983). Effect of iron(II), cobalt(II), copper(II), and manganese(II) on the chemiluminescence of luminol in the absence of hydrogen peroxide, *Anal. Chem.*, 55, 1080–1083.
Landing, W.M., Haraldsson, C. and Paxeus, N. (1986). Vinyl polymer agglomerate based transition metal cation chelating ion-exchange resin containing the 8-hydroxyquinoline functional group, *Anal. Chem.*, 58, 3031–3035.
Lin, Q.X., Guiraum, A., Escobar, R. and Delarosa, F.F. (1993). Flow-injection chemiluminescence determination of cobalt(II) and manganese(II), *Anal. Chim. Acta*, 283, 379–385.
Martin, J.H. (1990). Glacial-interglacial CO_2 change: The iron hypothesis, *Paleoceanography*, 5, 1–13.
Martin, J.H., Coale, K.H., Johnson, K.S., Fitzwater, S.E., Gordon, R.M., Tanner, S.J. *et al.* (1994). Testing the iron hypothesis in ecosystems of the equatorial Pacific Ocean, *Nature*, 371, 123–129.
Martin, J.H. and Fitzwater, S.E. (1988). Iron-deficiency limits phytoplankton growth in the northeast Pacific sub-arctic, *Nature*, 331, 341–343.

Martin, J.H. and Gordon, R.M. (1988). Northeast Pacific iron distributions in relation to phytoplankton productivity, *Deep-Sea Research Part A*, 35, 177–196.
Martin, J.H., Gordon, R.M. and Fitzwater, S.E. (1990). Iron in Antarctic waters, *Nature*, 345, 156–158.
Martin, J.M. and Whitfield, M. (1983). The significance of the river input of chemical elements to the ocean. In Wong, C.S. and Bruland, E. (Eds) *Trace metals in sea water*, pp. 265–296, New York: Plenum Press.
McCormack, T., Flow injection chemistries for the *in situ* monitoring of nutrients in seawater, PhD Thesis, University of Plymouth.
Millero, F.J., Gonzalez-Davila, M. and Santana-Casiano, J.M. (1995). Reduction of Fe(III) with sulfite in natural waters, *Journal of Geophysical Research*, 100, D4. 7235–7244.
Morly, N.H., Burton, J.D., Tankere, S.P.C. and Martin, J.-M. (1997). Distribution and behaviour of some dissolved trace metals in the western Mediterranean Sea, *Deep-Sea Research II*, 44, 675–691.
Moffet, J.W. and Zika, R.G. (1987). Solvent extraction of copper acetylacetonate in studies of copper(II) speciation in seawater, *Mar. Chem.*, 21, 301–313.
Nakayama, E., Isshiki, K., Sohrin, Y. and Karatani, H. (1989). Automated determination of manganese in seawater by electrolytic concentration and chemiluminescence detection, *Anal. Chem.*, 61, 1392–1396.
Nickson, R.A., Hill, S.J. and Worsfold, P.J. (1995). Solid phase techniques for the preconcentration of trace metals from natural waters, *Anal. Proc.*, 32, 387–395.
O'Sullivan, D.W., Hanson, A.K., Miller, W.L. and Kester, D.R. (1991). Measurement of Fe(II) in surface water of the equatorial Pacific, *Limnol. Oceanogr.*, 36, 1727–1741.
Obata, H., Karatani, H. and Nakayama, E. (1993). Automated determination of iron in seawater by chelating resin concentration and chemiluminescence detection, *Anal. Chem.*, 65, 1524–1528.
Obata, H., Karatani, H., Matsui, M. and Nakayama, E. (1997). Fundamental studies for chemical speciation of iron in seawater with an improved analytical method, *Mar. Chem.*, 56, 97–106.
Powell, R.T., King, D.W. and Landing, W.M. (1995). Iron distributions in surface waters of the South Atlantic, *Mar. Chem.*, 50, 13–20.
Price, D., Worsfold, P.J. and Mantoura, R.F.C. (1994). Determination of hydrogen peroxide in seawater by flow injection analysis with chemiluminescence detection, *Anal. Chim. Acta*, 298, 121–128.
Rijstenbil, J.W., Merks, A.G.A., Peene, J., Poortvliet, T.C.W. and Wijnholds, J.A. (1991). Phytoplankton composition and spatial distribution of copper and zinc in the Fal estuary (Cornwall, UK), *Hydrobiological Bulletin*, 25, 37–44.
Robards, K. and Worsfold, P.J. (1992). Analytical applications of liquid-phase chemiluminescence, *Anal Chim. Acta*, 266, 147–173.
Rue, E.L. and Bruland, K.W. (1995). Complexation of iron(III) by natural organic ligands in the Central North Pacific as determined by a new competitive ligand equilibration/adsorptive cathodic stripping voltammetric method, *Mar. Chem.*, 50, 117–138.
Ruzicka, J. and Hansen, E.H. (1981). *Flow Injection Analysis*. pp. 1–207, Chichester: Wiley.
Sakamoto-Arnold, C.M. and Johnson, K.S. (1987). Determination of picomolar levels of cobalt in seawater by flow-injection analysis with chemiluminescence detection, *Anal. Chem.*, 59, 1789–1794.
Schenck, R.C. (1984). Copper deficiency and toxicity in Gonyaulax tamarensis (Lebour), *Mar. Biol. Lett.*, 5, 13–19.

Stumm, W. and Morgan, J.J. (1981). *Aquatic Chemistry*. p. 5, New York: Wiley.

Tappin, A.D., Millward, G.E., Statham, P.J., Burton, J.D. and Morris, A.W. (1995). Trace metals in the Central and Southern North Sea, *Estuar. Coastal Shelf Sci.*, 41, 275–323.

Turner, A., Millward, G.E., Bale, A.J. and Morris, A.W. (1992). The solid-solution partitioning of trace metals in the southern North Sea – *in situ* radiochemical experiments, *Continental Shelf Research*, 12, 1311–1329.

Turner, D.R., Whitfield, M. and Dickson, A.G. (1981). The equilibrium specistion of dissolved components in freshwater ans seawater at 25 °C and 1 atm. pressure, *Geochim. Cosmochim. Acta*, 45, 855–882.

van den Berg, C.M.G. (1995). Evidence for organic complexation of iron in seawater, *Mar. Chem.* 50, 139–157.

Vega, M. and van den Berg, C.M.G. (1997). Determination of cobalt in seawater by catalytic adsorptive stripping voltammetry, *Anal. Chem.*, 69, 874–881.

Zhuang, G., Duce, R.A. and Kester, D.R. (1990). The dissolution of atmospheric iron in surface seawater of the open ocean, *J. Geophys. Res. C. Oceans.*, 95, 16207–16216.

4. USE OF A WET CHEMICAL ANALYSER FOR THE *IN SITU* MONITORING OF NITRATE

DAVID J. HYDES[a], PAUL N. WRIGHT[a,b]
and MARK B. RAWLINSON[c]

[a]*Southampton Oceanography Centre, Southampton, SO14 3ZH, UK,*
[b]*Present address: Maritime Faculty, Southampton Institute, East Park Terrace, Southampton, SO14 0YN, UK and*
[c]*WS Ocean System Ltd, Unit 4, Omni Business Centre, Omega Park, Alton, Hampshire, GU34 2QD, UK*

4.1 INTRODUCTION

Eutrophication is a major environmental issue within coastal waters world-wide. It is considered to be driven by the supply of nutrients from rivers draining urbanised and intensively farmed hinterlands (Nixon, 1995). Consequently, monitoring of nutrients in coastal environments is becoming increasingly important due to the need to understand the relationship between anthropogenic discharge of nutrients and the occurrence of harmful algal blooms.

Traditional techniques of nutrient monitoring may involve collection of samples on estuarine surveys, at tidal stations or regular sampling from prearranged locations. Monitoring schemes in which only a single sample per river is collected, often no more frequently than weekly, can fail to identify transients, such as storms events, which can pass in the matter of hours. It has been postulated that the inputs of nutrient rich storm waters to coastal zones can have a dramatic effect upon primary production (Galegos *et al.*, 1992). Similarly, if such transient signals are not detected estimates of total discharge load are severely under estimated. Models of nutrient uptake by algae need to be validated against data sets collected at temporal resolutions that are difficult to match using conventional sampling techniques. Additionally, there is still debate as to how samples should be preserved between collection and analysis in the laboratory (Dore *et al.*, 1996). Whilst the consensus of opinion suggests that nitrate levels are rarely effected by sample freezing, it has been shown that concentrations of other nutrients, particularly silicate, can be altered by freezing. Mercuric chloride is frequently added to samples as a preservative (Kremling and Wenk, 1986). Use of mercuric chloride is not only undesirable from a safety aspect, but also can give rise to problems with the subsequent analysis. It may effect the sensitivity of the standard nitrate reduction technique by changing the properties of the cadmium reduction column (Nydal, 1976), and during the determination of ammonia it can prevent the formation of the coloured complex.

In order to eliminate these problems in traditional monitoring of nutrients, the need exists for autonomous, *in situ*, techniques that measure nutrients at a high temporal resolution (hourly) yet can be left within the environment for a period of time giving continuous, longer term (weeks to months) data. This chapter covers the use of a chemical analyser for the determination of concentrations of nitrate in water which has been designed for such *in-situ* use. The instrument described here is the NAS-2, *in-situ* nutrient analyser manufactured by WS Ocean System Ltd. The original design concept was developed at the Scottish Office Marine Laboratory, Aberdeen.

4.2 THE NAS-2 ANALYSER

4.2.1 Method

The chemical method used to determine concentrations of nitrate follows well documented techniques (Wood *et al.*, 1967; Grasshoff *et al.*, 1983). The method requires the reduction of nitrate to nitrite. Chemical reduction is achieved by passing the sample over a reduction column of copper coated cadmium wire (Stainton, 1974; Nydal, 1976). The column may loose reduction efficiency over time. During use, the column is charged with the ammonium chloride buffer solution containing copper to preserve the efficiency of the column between each sampling. The resulting concentration of nitrite is then determined by formation of an azo dye, the intensity of whose purple colour is determined colorimetrically.

4.2.2 Reagents

Sulphanilamide: Dissolve 10 g in 200 ml of 50% HCl, dilute to 1 l with distilled water, then add 2 ml BRIJ-35 wetting agent (30% solution).

Napthylethelynedihydrochloride: Dissolve 1 g in 1 l distilled water.

Ammonium chloride: Dissolve 20 g and 0.2 g copper sulphate in 1 l of distilled water.

For marine applications the On Board Standard (OBS) is prepared in Low Nutrient Seawater (LNS). LNS is an aged surface sea water containing less than 0.1 μM of nitrate plus nitrite. It is supplied by Ocean Scientific International Ltd. An amount of a primary standard containing 10.00 mM of nitrate as potassium nitrate in distilled water is added to the LNS to give a nitrate concentration at the upper end of the range expected for the observations to be carried out.

4.2.3 Main Components and their function

The NAS-2 nitrate sensor has five major components. These are: an eight way rotary valve, a motor driven syringe, a colorimeter, reagent housing, and an electronic controlling unit. The layout of a NAS-2 is shown in Figure 4.1. The

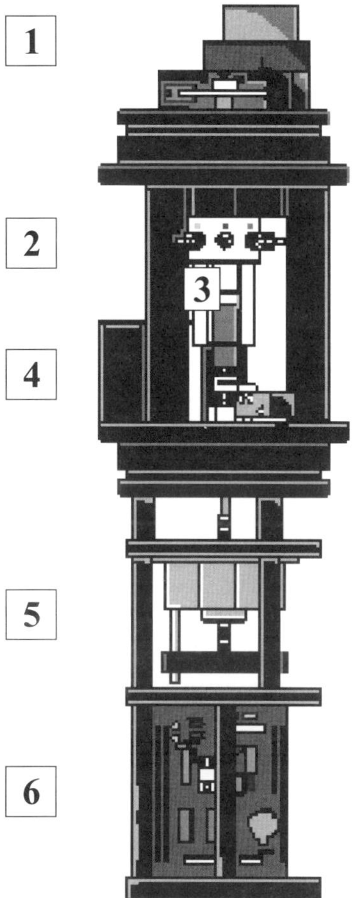

Figure 4.1 Diagram of mechanical structure of a NAS-2: 1. Stepper motor; 2. Muti-port valve; 3. Syringe; 4. Colorimeter; 5. Syringe drive motor and 6. Control electronics.

rotary valve and syringe are both driven by stepper motors. The chemicals used in the analysis are stored in plastic "transfusion" bags in an upper housing attached to the top of the unit as it is shown in Figure 4.1. The unit is about 80 cm long, 21 cm in diameter, and weighs approximately 10 kgs. The chemical system is essentially pressure balanced as all the chemical circuitry at ambient pressure, external to the electronic instrumentation. The depth limitation is 200 m, which is the pressure rating of the housing for the electronics. The likelihood of bio-fouling of moving components is minimised as all moving parts are contained within housings.

Where the turbidity of the water is high, problems could arise from high blanks which are difficult to measure reproducibly, or particles foulling the valves and contaminating the cadmium reduction column. To combat the introduction of foreign matter into the unit, a filter is fitted at the end of the sampling tube.

A water sample is taken by aligning the rotary valve to the inlet port then drawing down the syringe plunger. This blank aliquot is then injected into the colorimeter and the value recorded. After expulsion of the blank, another aliquot is taken. The sample is then pushed out of the syringe onto the reduction column. After reduction from nitrate to nitrite has occurred, the sample is drawn back into the syringe. The valve moves around to the next ports and draws in the two reagents in turn. By moving the piston up and down 4 times the sample and regents are effectively mixed by the vortexes set up in the liquid each time it re-enters the syringe cylinder. Two minutes are allowed for colour formation then the solution is pushed into the colorimeter where the extinction of the solution is measured. The colorimeter consists of a narrow linear capillary tube with a LED source at one end and a photodiode detector at the other. An additional detector is placed adjacent to the source to monitor LED intensity directly.

The concentration of nitrate is calculated by comparing the absorbance of the sample to the value obtained for an "on board standard" (OBS) after correcting the measured absorbance by the value obtained for the blank. As mentioned above blanks are determined prior to every sample taken. OBS values are determined at a rate of one per six samples. The repeated analysis of the OBS allows the user to monitor how the unit has drifted over the time. Drift is likely due to changes in the functioning of the colorimeter circuitry and because the amount of colour developed in sample will vary as the temperature of the instrument changes in response to changes in the surrounding water. In the case of the determination of nitrate this temperature drift is likely to be relatively small as the colour forming reaction is rapid and the reaction will be near to its end point before it is introduced into the colorimeter. For the determination of nitrate the major cause of drift in the values obtained for the OBS and samples is likely to be shifts in the efficiency of the cadmium reduction column.

Control and retrieval of data can be done by direct connection of PC-computer through a RS232 communications port on the instrument, or can be done remotely by means of telemetry. Instructions can be given either by a menu driven program, or by means of a command language. The software allows for control of these protocols by the user. For example, longer periods between the collection of samples can be set up or the details of the procedure can be altered – such as the relative volumes of samples and reagents, or the duration of the reduction step in the copper-cadmium column.

4.3 RESULTS

Results from three deployments are summarised in Table 4.1. These deployments were off the French coast in March and April 1995, and in the more turbid waters of Southampton Water, an estuary on the south coast of the UK in August–October 1996, and March–April 1997. The deployments in South-

Table 4.1 Summary of 3 deployments of the NAS-2 nitrate analyser.

Survey	*Brest*	*SONUS 1*	*SONUS 2*
Start	21/3/95	22/8/96	5/3/97
End	18/4/95	17/10/96	9/4/97
No. of samples	561	1343	697
No. of OBS	114	224	138
OBS concentration (μM)	47.5	30	50
Conc. ranges samples (μM)	5–30	33–5	10–60
% Std Deviation Blank	0.7	4.0	N/A
% Std Deviation OBS	5.3	8.9	13.4

ampton Water were carried out as part of the Southern Nutrient Study (SONUS). Testing of the NAS-2 during this study was linked to the development of a data buoy system (Wright *et al.*, 1997). At both sites the analyser was moored in waters in which the salinity was likely to be in the range 30 ± 3, anthropogenic discharges of nutrients in both areas are relatively high.

In Figure 4.2, the data from the first week of the deployment off the French coast are plotted. During this deployment a boat was used each working day when weather and other conditions permitted to collect a sample using a standard

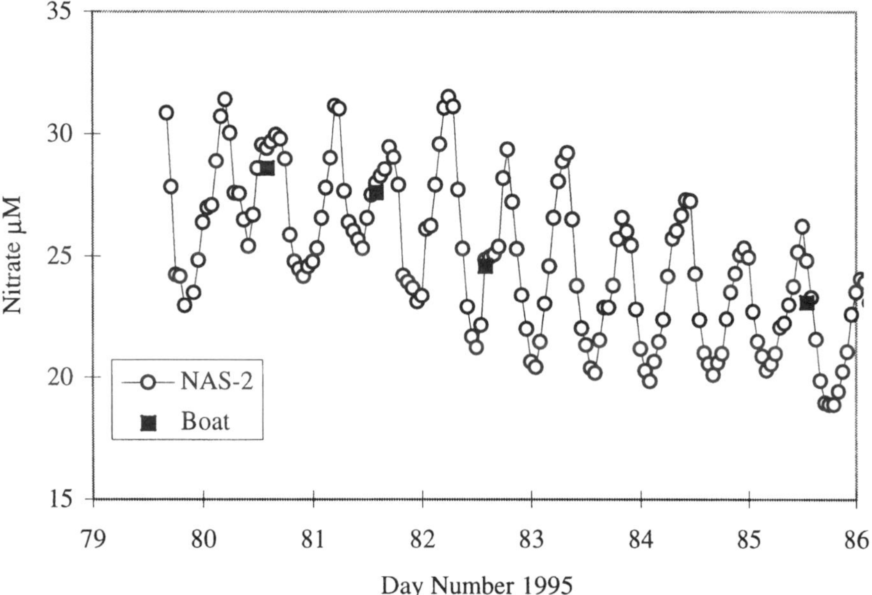

Figure 4.2 Detailed plot of first weeks observations made on the deployment off the French coast.

water sampling bottle. This sample was collected to coincide with the sample collected by the NAS-2 close to 1300 hours each day. The sample was then returned to the laboratory and its nitrate concentration was determined using a standard auto-analyser method. Figure 4.2 shows that the NAS-2 data is providing a record of the variation in the concentration as the water conditions around the analyser change with the state of the tide. The accuracy of the measurements compared to the bottle samples appears to be good.

In total 18 samples were collected by boat during the first deployment. In Table 4.2 the concentrations of nitrate determined in these samples are compared with the value recorded by the NAS-2 closest to the time of collection of the boat sample. The mean difference over all 18 samples is equivalent to 1.2 µM of nitrate a percentage difference of 6.7 %. This result compares favourably with the degree of accuracy that is usually reported, when different laboratories report determination of nitrate in inter-comparison samples (Aminot and Kirkwood, 1994). A potential problem with the determinations carried out on the NAS-2 is that the calibration of the NAS-2 samples is only done relative to a single standard

Table 4.2 Comparison of results from NAS-2 with those from conventionally collected samples.

Day 1995	*Nitrate* µM	*Nitrate* µM	*Difference*	*% Difference*
	NAS-2	*Boat*		
81	29.4	28.6	0.8	2.7
82	28.0	27.6	0.4	1.3
83	24.8	24.6	0.2	1.0
86	24.8	23.1	1.7	6.9
87	23.8	22.9	0.9	3.6
88	26.2	25.3	0.9	3.3
89	31.0	30.1	0.9	3.0
90	24.4	23.8	0.6	2.6
93	21.9	19.8	2.1	9.5
94	19.4	18.1	1.3	6.6
95	19.1	17.1	2.0	10.6
96	18.3	16.3	2.0	10.8
100	12.9	10.8	2.1	16.2
101	14.7	14.0	0.7	4.6
102	15.2	13.6	1.6	10.6
103	12.7	11.0	1.7	13.3
104	10.8	9.3	1.5	13.5
108	5.6	5.5	0.1	1.2
			mean	mean
			1.2	6.7

solution and a blank value. For this to give accurate results the instrument must be set up so that all the samples measured will be with in the linear working range of the colorimeter at the wavelength of the analysis. A relatively wide dynamic range of sample concentrations were encountered during the first deployment as it extended into the period of the phytoplankton bloom which reduced observed concentrations of nitrate from 29 μM at the start of the observations to 5 μM at the end. The plot of the NAS-2 results against the boat results (Figure 4.3) shows that there appears to be no significant difference between the linearity of the NAS-2 and auto-analyser determinations over this concentration range.

A critical problem that has to be over come with all determinations of nitrate in sea water is maintaining a reproducible response from the cadmium reduction column for periods of time of a few months. The design of the reduction column on the NAS-2 follows the suggestion by Stainton (1974) that cadmium wire is a suitable form to use the cadmium in. The reactivity of the wire is enhanced by coating it with precipitated copper (Nydal, 1976), following the procedure given by Stainton (1974). In order to retain the reactivity of the column between sample determination the column is flushed at the end of each processing cycle with ammonium chloride buffer solution containing copper sulphate.

In Figure 4.4 all the absorbances measurements recorded for the On Board Standard are plotted for the deployment off the French coast and the first SONUS deployment. The variation seen in this plot is due to changes in the condition of the cadmium column, but will also be influenced by other factors such as the ambient temperature and the condition of the chemical reagents.

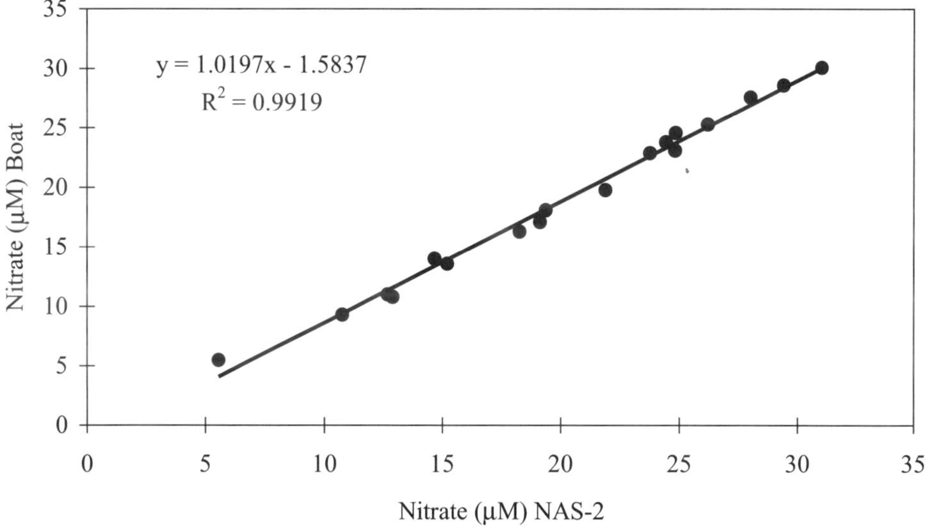

Figure 4.3 Plot of the NAS-2 data against data from laboratory based auto-analyser measurements on samples collected by boat. Results of linear regression analysis of the data are shown.

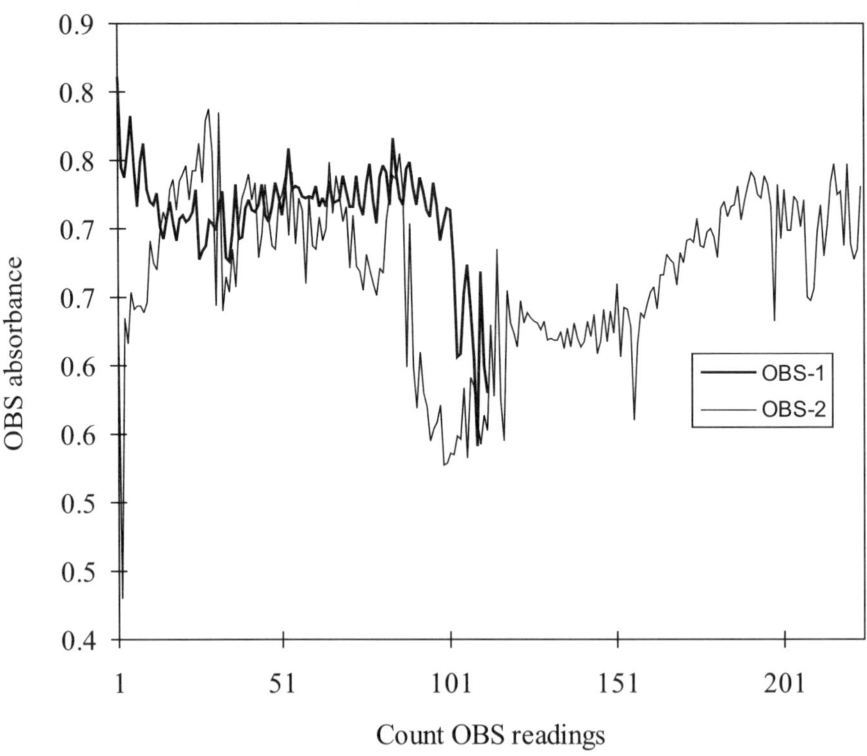

Figure 4.4 Plot of the variation with time of the absorbance measured for the On Board Standard during the deployment off the French coast – OBS-1 and the first SONUS deployment – OBS-2.

In Figure 4.5, for comparison, we present data which records the equivalent change in the calibration coefficient for the determination of nitrate after reduction to nitrite using a Stainton type cadmium wire reduction column on an AA-II type auto-analyser during an oceanographic cruise (RRS Discovery 216). This data was obtained by the analysis of four standard solutions in duplicate at the start of each analytical run before the determination of samples. Omitting the first 3 runs at the start of the cruise the percent standard deviation was 4.1%.

4.4 EXAMPLE RESULTS

The first deployment in Southampton Water started on 22 August 1996 and lasted for 56 days. Figure 4.6 show variations in nitrate levels that occur with respect to the daily tidal signal and on longer time scales. On a daily time scale, higher nitrate concentrations correspond to the presence of a larger component of fresh water rich in nitrate at low water. There is also cyclicity in the data which

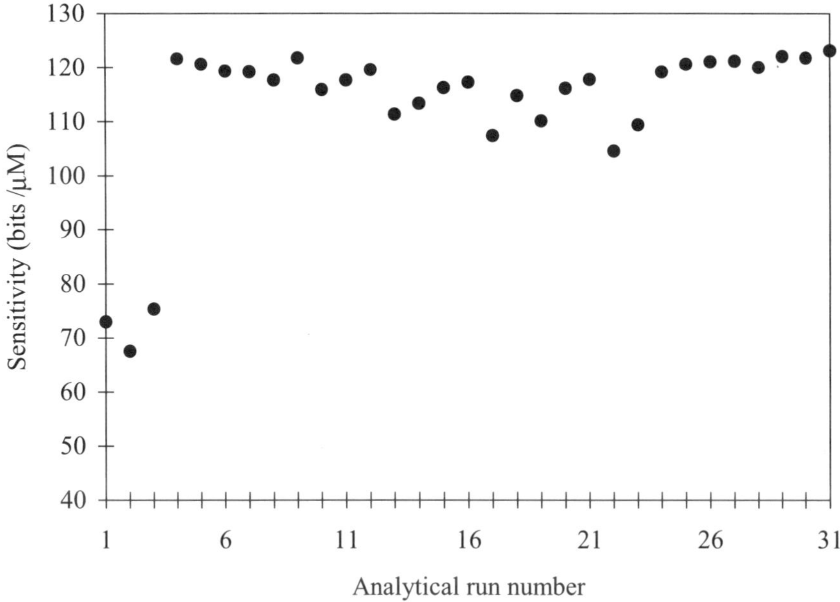

Figure 4.5 Plot of the variation in the calibration coefficient obtained on a AA-II type auto-analyser for nitrate determinations over 20 days on RRS Discovery cruise 216.

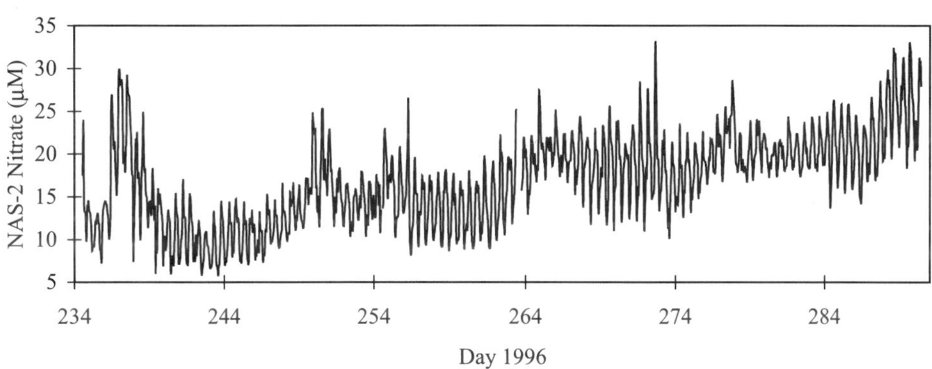

Figure 4.6 Results of NAS-2 determinations of nitrate in Southampton Water between 22 August and 17 October 1996.

correspond to the spring neap tidal cycle. Again this is due to the variation in the fraction of fresh water at the mooring site. The general increase over the whole period is due to the increase in concentrations of nitrate in open sea water in autumn at the end of the phytoplankton production season.

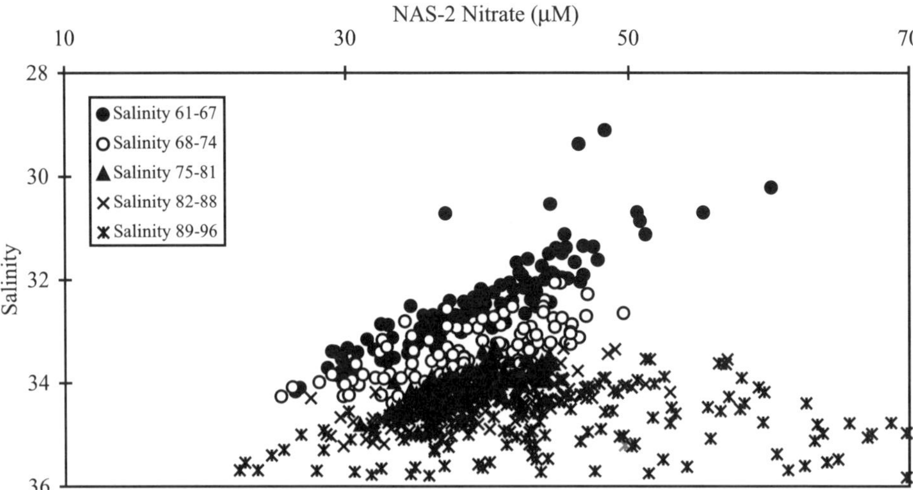

Figure 4.7 Results from the second SONUS deployment starting on the 5 March 1997 (Day 61 1997). Data are grouped into weekly sets starting with days 61 to 67.

During the second SONUS deployment, the NAS-2 was deployed in conjunction with a wider array of sensors. From this deployment we have available corresponding data from a conductivity sensor. This data can used to calculate the salinity of the water. All the nitrate data and corresponding salinity data are plotted against each other in Figure 4.7. On inspecting this data it could be seen that while the nitrate results stay with in similar range through the course of this deployment, the salinity values drift to values which are unreasonably high for Southampton Water. The gradual progression in this drift has been indicted in Figure 4.7 by plotting the data in weekly groups. The cause of the drift in salinity measurements is probably fouling of the sensor by the growth of plant matter around the conductivity sensor head. Logistical difficulties prevented servicing of the mooring during this deployment.

4.5 CONCLUSIONS

The data presented here show that the NAS-2 wet chemical analyser, can produce results for the determination of nitrate *in-situ* which are comparable in quality to those obtainable in a laboratory using more conventional automated systems. This quality of data can be achieved for periods of over one month. The data plotted in Figure 4.7, suggests that where problems arise from fouling analysers like the NAS-2 which are self calibrating may be more reliable than other instruments even well established ones such as conductivity sensors. The instrument works well within coastal areas, identifying both short term and long term variations which are difficult to observe during more traditional surveys.

Acknowledgements

This work was supported in part by the UK Department of the Environment, Transport and the Regions funding for the SONUS project (DoETR Contract PECD 1/9/36). We are very grateful to Christian Le Gall of the IFREMER laboratory for making available their results from tests of the NAS-2.

References

Aminot, A. and Kirkwood, D.S. (1994). The 1993 QUASIMEME laboratory performance study: nutrients in seawater and standard solutions, *Marine Pollution Bulletin*, 29, 159–165.

Dore, J.E., Houlihan, D.V., Tien, G., Tupas, L. and Karl, D.M. (1996). Freezing as a method of sample preservation for the analysis of dissolved inorganic nutrients in seawater, *Marine Chemistry*, 53, 173–185.

Gallegos, C.L., Jordan, T.E. and Correll, D.L. (1992). Event scale response of phytoplankton to watershed inputs in a subestuary: timing, magnitude, and the location of blooms, *Limnology and Oceanography*, 37, 813–828.

Grasshoff, K.M., Erhardt, K.M. and Kremling, K. (1983). Methods of seawater analysis, *Verlag-Chemie*, 419.

Kremling, K. and Wenk, A. (1986). On the storage of dissolved inorganic phosphate, nitrate, and reactive silicate in Atlantic Ocean water samples, *Meersforeschung*, 31, 69–74.

Nixon, S.W. (1995). Coastal marine eutrophication: A definition, social causes, and future concerns, *Ophelia*, 41, 199–219.

Nydal, F. (1976). On the optimum conditions for the reduction of nitrate to nitrite by cadmium, *Talanta*, 23, 349–357.

Stainton, M.P. (1974). A simple effective reduction column for use in the automated determination of nitrate in seawater, *Analytical Chemistry*, 46, 1616.

Wright, P.N., Hydes, D.J., Lauria, M.-L., Sharples, J. and Purdie, D. (1997). Results from data buoy measuremnts of processes related to phytoplankton production in a temperate latitude estuary with high nutrient inputs, Southampton Water U.K., *Deutsches Hydrographisches Zeitschrift*, 49, 210–210.

Wood, E.D., Armstrong, F.A.J. and Richards, F.A. (1967). Determination of nitrate in seawater by cadmium-copper reduction to nitrite, *Journal of the Marine Biological Association of the UK*, 47, 23–31.

5. SENSING OF NITRATE CONCENTRATION BY UV ABSORPTION SPECTROPHOTOMETRY

CHARLES H. CLAYSON

Southampton Oceanography Centre, European Way,
Southampton S014 3ZH, UK

5.1 INTRODUCTION

Measurements of nitrate distribution over the world's oceans are urgently required by marine biologists. Nitrate is one of the main nutrients involved in the growth of phytoplankton at the start of the marine food chain. We need to know its distribution, allied to the distributions of other measured physical and chemical variables and solar energy input, in order to understand the processes of primary production and to assist in the development of models of these processes. Levels of nitrate in the open oceans are generally small, of the order of a few tenths of a milligram of nitrate (NO_3^-) per litre, so that instruments must be very sensitive and stable to measure the distribution with sufficient accuracy.

Environmental protection agencies also have a considerable interest in monitoring the dispersion of nitrate in river outflows around coasts; the nitrate originates largely from agricultural fertiliser run-off into the rivers. In this case, the nitrate levels are much higher, typically up to a few tens of milligrams of nitrate (NO_3^-) per litre.

Spot measurements of nitrate concentration have been possible for many years, using chemical analysis of water bottle samples in the laboratory and, more recently, by using *in-situ* wet-chemical colorimetric nitrate analysers. However, the lack of an instrument with a response speed suited for use in towed undulating vehicles, or on vertical wire casts, has severely limited the mapping of nitrate distributions.

The adoption of UV absorption spectrophotometry, a technique used by the water industry for many years in monitoring nitrate concentration in natural waters, allows an increase in sampling rate of about 2 orders of magnitude to around 1 sample/second; this is an adequate rate for rapid surveying in conjunction with other instruments such as Conductivity, Temperature, Depth and Oxygen probes, Chlorophyll Sensors (fluorometers), Transmissometers, etc.

As a result of experimental work with a prototype, self-contained, *in-situ*, UV spectrophotometer, a production instrument has been developed and the principles, calibration and limitations of this instrument are described below.

5.2 THE BASICS OF ABSORPTION SPECTROPHOTOMETRY

Sea water displays selective absorption of radiation in the ultraviolet and visible spectrum (the UV/VIS spectrum); that is to say, the absorption of radiation is dependent upon the wavelength of the radiation. Absorption is the conversion of radiation energy into molecular energy in a medium, resulting in a reduction of the transmitted radiation flux. This is distinct from scattering, where radiation is scattered by particles over a range of directions, also resulting in a reduction of the directly transmitted flux.

The quantised internal energy of a molecule is the sum of its electronic, vibrational and rotational energies. An incident photon having the appropriate frequency, ν, and hence energy, E ($= h\nu$, where h is Planck's constant), can change the molecule between its ground state and an excited state, its energy being transferred to the molecule. The rotational energy levels are quite close together, resulting in an absorption band consisting of a number of closely spaced lines. As discussed by Jaffe and Orchin (1964), these lines overlap in the case of liquids, so that a continuous band results in the absorption spectrum. Different molecules or ions have differing absorption spectra and this is the basis for absorption spectrophotometry.

In the 18th century, Bouguer proposed a law of absorption which states that for radiation passing through a given thickness of a given isotropic medium, a fraction of the incident flux is absorbed which is independent of the incident flux, Φ. The fraction absorbed is proportional to the thickness, so that for an infinitesimal thickness, dx:

$$\frac{\mathrm{d}\Phi}{\Phi} = -\alpha \mathrm{d}x,$$

where α is the absorption coefficient.

Integrating over a thickness, b, of the medium, we have:

$$\int_{\Phi_0}^{\Phi_b} \frac{\mathrm{d}\Phi}{\Phi} = -\alpha \int_0^b \mathrm{d}x,$$

so that the flux, Φ_b, after transmission through the distance b, is related to the incident flux, Φ_0, by:

$$\log_e(\Phi_b) - \log_e(\Phi_0) = -\alpha \cdot b$$

We can rewrite this as:

$$\Phi_b = \Phi_0 e^{-A}$$

or

$$A = \log_e\left(\frac{\Phi_0}{\Phi_b}\right),$$

where A ($= \alpha \cdot b$) is the *absorbance*. This is normally referred to as Lambert's Law, although it stems from Bouguer's original work. It is important to note that, whereas the absorption coefficient is solely a property of the medium, the absorbance refers to flux reduction over a specified path length, b.

If there are uniformly distributed scattering particles present, an additional exponential reduction in flux results and the above expression can be modified to include this by the inclusion of a scattering coefficient, β:

$$\Phi_b = \Phi_0 e^{-(\alpha+\beta)b}$$

Beer studied the dependence of the absorption coefficient upon the concentration of the absorbing substance, c, and found this to be linear. The combination of Lambert's and Beer's findings resulted in the Beer-Lambert law, sometimes referred to as Beer's law:

$$\Phi_b = \Phi_0 e^{-\varepsilon c b}.$$

The absorbance is now given by:

$$A = \varepsilon c b,$$

where ε is the (wavelength-dependent) molar *absorptivity coefficient*, which has units of $M^{-1}cm^{-1}$, M being the concentration in units of molarity (moles/litre).

In practical measurements, the absorbance (as determined from Φ_b and Φ_0) is found to be linearly related to concentration only over a limited range. This is due to a number of factors which may include:

stray light
scattering by particulates
fluorescence
saturation due to depletion of ground state molecules

However, for absorbances of less than about 3, the linearity is normally adequate for analytical work.

As noted above, the absorptivity coefficient for a given molecule, or ion, is wavelength dependent $= \varepsilon(\lambda)$. Also, the wavelength dependent absorbance, $A(\lambda)$, due to the combination of a number of molecules or ions in solution, is the sum of the component absorbances, i.e.

$$A(\lambda) = (\varepsilon_1(\lambda) \cdot c_1 + \varepsilon_2(\lambda) \cdot c_2 + \varepsilon_3(\lambda) \cdot c_3 \ldots)b$$

The absorptivity spectrum $\varepsilon(\lambda)$ due to a dissolved species may be temperature sensitive; this phenomenon is termed thermochromism and results from changes in the equilibrium between reacting species.

5.3 UV ABSORPTION BY SEA WATER

UV absorption techniques have been used for many years by water authorities as a standard technique for monitoring nitrate (NO_3^- ion) concentrations in "fresh" water supplies (see, for example, American Public Health Association *et al.*, 1985).

It is important to differentiate the absorption due to NO_3^- ions from that due to the other ions normally present in sea water; of these the halide ions (Cl^-, Br^-, I^-) are dominant, although SO_4^{2-}, HCO_3^-, HPO_4^{2-}, and NO_2^- ions are always present in lower concentrations. Other "interfering" organic compounds may be present which further complicate the data analysis, as will be discussed below.

In the case of so-called low-nutrient sea water† (LNS), the absorbance in the far UV range of the spectrum, A_{LNS}, results mainly from the Cl^- ions and is of the form shown by the solid line in Figure 5.1, below.

These absorbances are relative to triple-distilled water‡ and were measured using a laboratory spectrometer with a path length, *b*, of 10 mm. The absorbance for LNS of salinity (s) 35 peaks at 2.438 at a wavelength of 205 nm and the

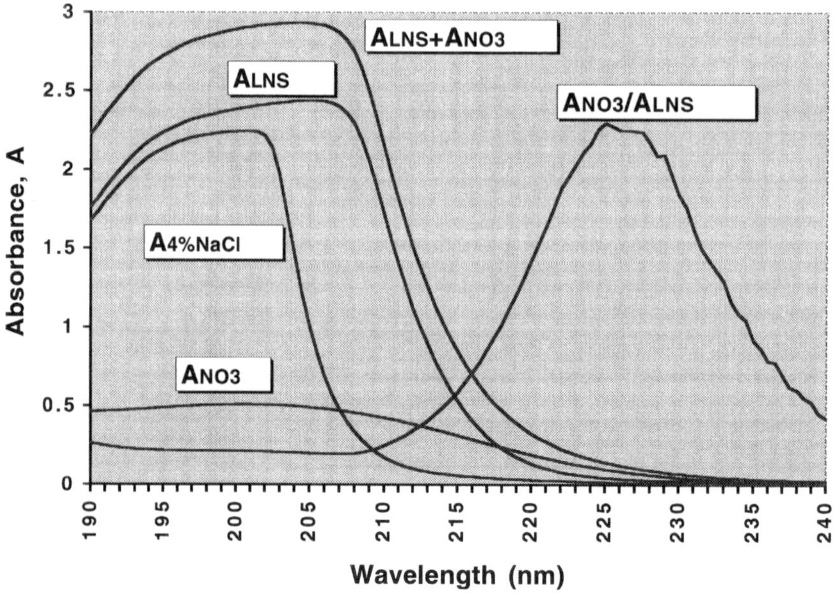

Figure 5.1 Absorbance Spectra of Sea Water and Nitrate.

---

†Low-nutrient sea water is collected from regions far from land and, after processing to remove micro-organisms, is sold commercially.

‡In practice, the absorbance is calculated from the natural logarithm of the ratio of the transmitted flux with the cell (length b) filled with a blank of distilled water, to the transmitted flux with the cell filled with the measurand.

absorbance then falls off sharply with increasing wavelength. The absorbance for s = 40 NaCl peaks at only 2.245 at a wavelength of 201 nm, demonstrating the significant effect of the other ions in the LNS.

In contrast, the absorbance for 50 μM nitrate in distilled water, A_{NO3}, (a high concentration which would normally be found only in coastal waters) peaks at only 0.510 at a wavelength of 200.5 nm. However, the absorbance for the nitrate solution falls off more gradually that that for LNS, so that it exceeds the latter above a wavelength of 218 nm. The ratio of A_{NO3} to A_{LNS} is maximum at a wavelength of about 227 nm, although both absorbances are then rather small.

As noted above, sea water unfortunately contains varying concentrations of other substances which absorb significantly in this spectral region; in particular, varying "cocktails" of organic molecules are present in the vicinity of river outflows and, locally, further out at sea. These give rise to an "interfering" absorption spectrum which is, if a wide range of molecules is present, relatively smooth in comparison with the absorption spectra of sea water and nitrate. By taking absorption measurements over a number of wavelengths in the vicinity of 220 nm, one can at least infer the presence of such interference, if not eliminate it precisely. This problem is discussed further in section 5.8, below.

A number of sea water samples from a number of locations were analysed by Hydes (1995), using a laboratory spectrophotometer (more correctly, if clumsily, termed a UV spectroabsorptiometer) and an auto analyser. The good correlation of the 220 nm absorbance with the nitrate concentration as measured by the auto analyser implied a low extent of apparent interference. This encouraged Hydes to develop a sea-going (*in-situ*) UV spectroabsorptiometer – brief details of which are given below to demonstrate some of the technological problems which had to be overcome.

5.4 TECHNOLOGICAL CONSIDERATIONS FOR AN *IN-SITU* INSTRUMENT

From the data shown in Figure 5.1, the absorbance of a 10 mm path length, *b*, through LNS at 220 nm is approximately 0.134 (relative to distilled water) and this is increased to 0.326 by 50 μM of nitrate.

With a path length of 200 mm, an absorbance range of about 2.69 to 3.453 would cover the range 0 to 10 μM nitrate in sea water of salinity 35. With a path length of 66.7 mm, an absorbance range of about 0.896 to 3.447 would cover the range of about 0 to 100 μM nitrate in sea water of salinity 35. As explained above, an instrument covering a larger range of absorbance, whilst possibly giving better resolution, might suffer from non-linearity.

By employing a retro-reflecting prism, set on spacing pillars above a window in the instrument pressure housing, we can readily alter the path length in the measurand by changing the pillar lengths. This method of construction also has a considerable advantage over the two-housing form of construction in that it only requires one pressure housing (no linking electrical connections) and one window.

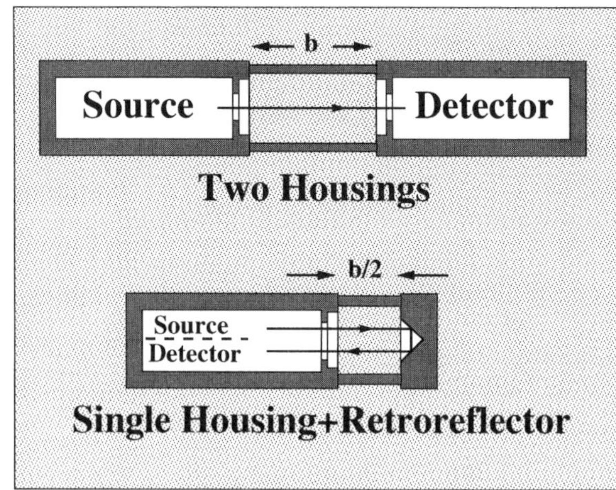

Figure 5.2 Alternative Configurations for Instrument.

The choice of the UV light source is critical in this application. Continuous Deuterium gas discharge lamps have a stable output which is rich in the UV range of interest (e.g. 200 to 280 nm), but they typically have a power consumption of the order of 30 W. This poses not only supply problems for self-contained battery-powered marine instruments, but also problems of heat dissipation. They also suffer from a fairly short operating life (typically 500–1000 hours to half intensity, although recently introduced devices can achieve 4000 hours). Quartz-tungsten-halogen (QTH) filament lamps produce only small amounts of light below 300 nm wavelength unless over-run, when their life is severely reduced. Mercury and Xenon continuous arc sources, whilst having a high UV output, require between 50 and 100 watts of electrical power, which can not be supplied in this application.

Pulsed Deuterium sources can be operated on a very low duty cycle but still require an appreciable average power level to maintain a minimum operating temperature. This is achieved by keeping the tube "simmering" between pulses; typical average power input levels are 20 W.

Pulsed Xenon sources do not give such high UV levels as Deuterium sources and they do not offer such good pulse-to-pulse stability. However, at the (practically useful) flash rate of 8 Hz, they can be operated at average power input levels of order 1 W, with lifetimes of order 2500 hours; this was the main factor resulting in the selection of this type of source. The arc electrodes may be conical or prismatic and one or more trigger electrodes are normally incorporated to facilitate the initiation of the discharge and to improve its positional stability. In our experience, the conical type displays higher stability, although a burn-in period is necessary before the output becomes stable.

In the case of the Xenon flashtube used in the instrument described, the arc is approximately 1.6 mm × 0.1 mm. Whilst this is almost a point source in the plane

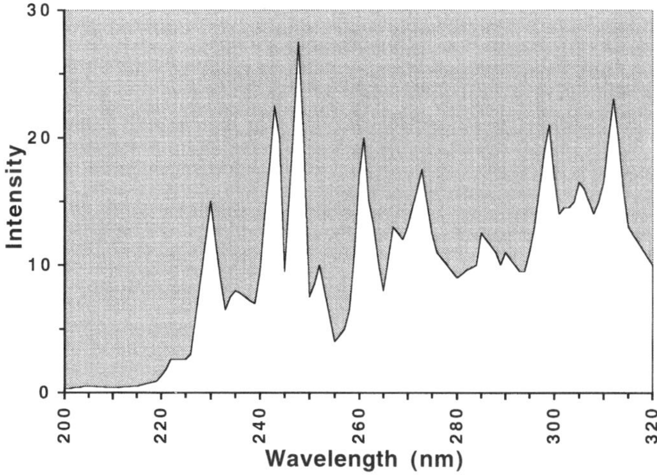

Figure 5.3 The output spectrum of a typical xenon flash tube.

normal to the axis of the arc, the length of the arc makes collimation in this plane difficult without the use of cylindrical lenses. Consequently, the arc axis is oriented in a plane parallel with the lines of the diffraction grating used in the spectrometer assembly. This prevents any undue increase in spectrometer bandwidth.

The output spectrum shows considerable variation with wavelength, as shown in Figure 5.3, above; furthermore, it drops off considerably below about 230 nm. Whilst this is not ideal, the spectrometer detector gains are software-controlled to compensate. This spectrum contrasts with the smooth output spectrum of a Deuterium source which, in fact, rises with decreasing wavelength over the range of interest. Unfortunately, the power consumption of a deuterium source could only be tolerated in a shipboard instrument.

UV-grade synthetic fused silica lenses, window and prism are used, so as to minimise absorption and scattering losses in these components; the use of UV-grade material also minimises fluorescence and it is less susceptible to solarisation – a process where photochemical reactions take place with impurities in the material, such as titanium, causing browning and greatly increased absorption loss. Synthetic sapphire was initially used for the pressure housing window, as it offers very high strength and good scratch resistance. However, only the highest purity grade of material was found to be free of solarisation problems and, even then, the absorption losses were higher than for a (thicker) UV-grade fused silica window of comparable pressure rating.

The pressure housing window is designed to withstand pressures of 100 bar (shallow version) and 600 bar (deep version); it is seated on a tapered stress-distributing ring within the pressure housing end cap as shown in Figure 5.4, below.

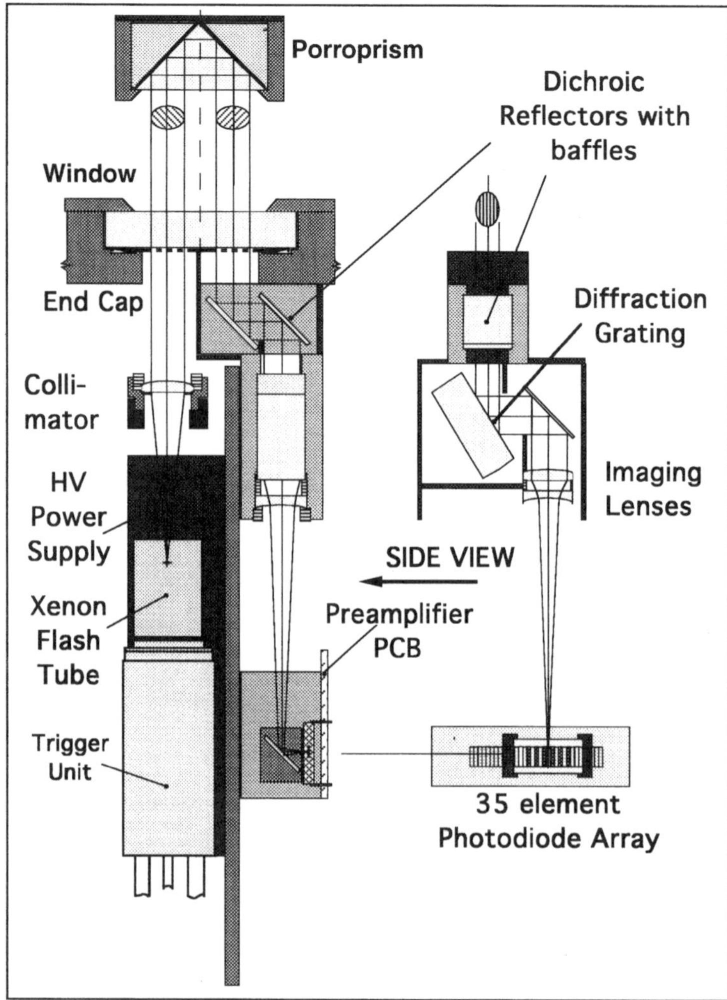

Figure 5.4 Schematic Diagram of Instrument.

The retro-reflecting prism is a standard 90° porroprism, coated with high-reflectivity aluminium on its two reflecting surfaces; these are protected from the sea water by encapsulation in a polyurethane potting material.

Dichroic reflectors are used to filter out visible light at the input to the spectrometer, whether originating directly from the source or as stray light from the environment. An interferometrically-produced (holographic) grating, optimised for 250 nm, is used to disperse the UV spectrum, *via* an imaging lens, onto a multi-element photo-detector array. A further dichroic filter is used to block any visible light from reaching the detector array.

Figure 5.5 The SUV6 (MkII) version of the instrument, courtesy Valeport Ltd.

The photo currents from each of six array elements are integrated, using a two-phase system to remove dark current (and any steady current due to extraneous light sources), and the resulting signals are digitised before processing by a micro-controller and subsequent output to external data acquisition systems. It is possible to normalise the signals with respect to the longest wavelength (transmissometer) signal, so as to reduce the effect of flash output level variations.

The spectrometer optics are designed to achieve spectral estimates at 15 nm spacing, with a bandwidth of 8 nm, in the range 205 to 280 nm.

The optics are set up by, first, adjusting the source collimator lenses for a near parallel beam (in the plane normal to the arc axis) at 220 nm, using a band-pass filter and using a UV fluorescent target for beam visualisation. The imaging lens adjustment is not critical, having a relatively small effect on the optical bandwidth of each detector channel. The grating angle is fixed at the nominal angle to the input beam axis for a 1st order diffracted beam at 90°, at a wavelength of 242.5 nm. The wavelength tuning is achieved by fine lateral adjustment of the detector array in the image plane, using narrow band filters in the water path.

The overall size of the instrument in its (shallow) pressure housing is about 400 mm long $\times$ 113 mm diameter; this was kept to a minimum to allow use of the instrument in towed undulators and on profiling moorings. An end chamber can be fitted over the external optics to allow use in a pumped system and for calibration purposes, although great care must be taken to avoid contamination of the measurand. O-ring lubricants, screw thread greases and cleaning solvents can cause significant contamination problems in a closed chamber.

5.5 CALIBRATION AND DATA PROCESSING

Calibration of the instrument can be achieved by filling the calibration chamber first with distilled water and then with low nutrient sea water, to which nitrate is progressively added in known quantities. *In situ* calibrations on vertical casts can be achieved by use in conjunction with a CTD/Rosette water bottle system or, in the case of a pumped system, in conjunction with a thermo-salinograph and manually drawn bottle samples. In both cases the water samples can be analysed using an auto analyser. It has been found that second order bi-variate fits can be derived from such calibration data which result in adequate accuracy of the derived nitrate level, c_{NO_3},

i.e.

$$c_{NO_3} = c_0 + k_1 \cdot A_{220} + k_2 \cdot A_{220}^2 + k_3 \cdot A_{235} + k_4 \cdot A_{235}^2,$$

and the salinity, S, is given by:

$$S = S_0 + k_5 \cdot A_{220} + k_6 \cdot A_{220}^2 + k_7 \cdot A_{235} + k_8 \cdot A_{235}^2,$$

where A_{220} and A_{235} are the measured absorbances at wavelengths of 220 nm and 235 nm and c_0, S_0, k_1 to k_8 are the fitted coefficients.

5.6 SOME PRACTICAL RESULTS

Figures 5.6 and 5.7, below, show how the use of this technique can yield accurate data; the data were obtained in a through flow pumped system on board ship and represent a period of 17 days continuous operation. In Figure 5.6, the 220 nm signal shows little relation to the auto analyser NO_3 level, due to salinity changes, but use of the 235 nm signal gives fitted NO_3 levels accurate to better than 0.2 μM standard error as shown in Figure 5.7.

The instrument has also been used on CTD casts, mounted in the CTD frame as shown in Figure 5.8, below.

The nitrate concentration profile, Figure 5.9, shows fine detail which correlates well with the fluorometer data; the details could not have been observed with conventional water sampling or wet-chemistry instrument techniques.

5.7 COMPARISON WITH OTHER TECHNIQUES

The advantages of the UV absorption technique over *in-situ*, wet-chemistry, instruments lie primarily in sampling rate and in the number of samples which

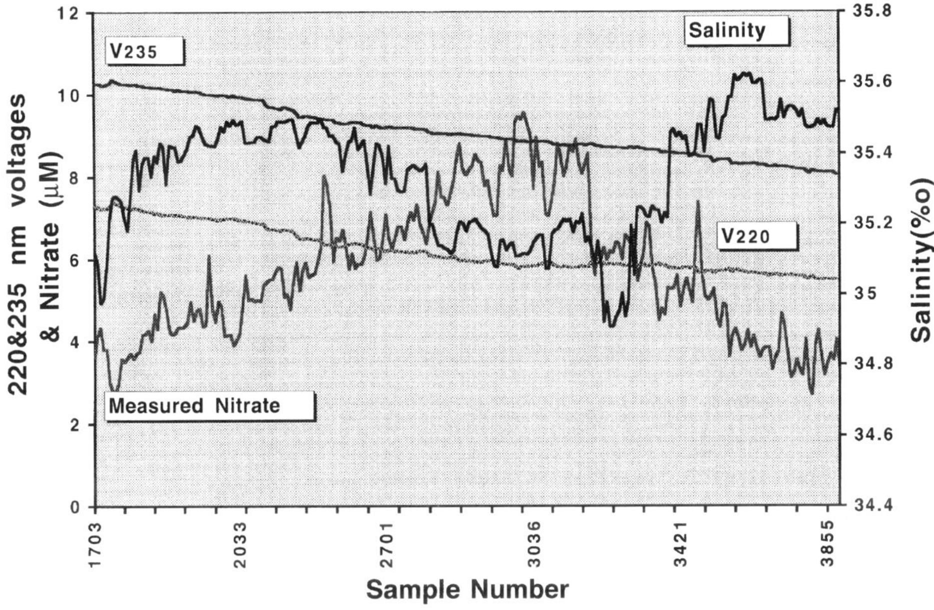

Figure 5.6 Sensor voltages, Auto analyser NO_3 and Salinity.

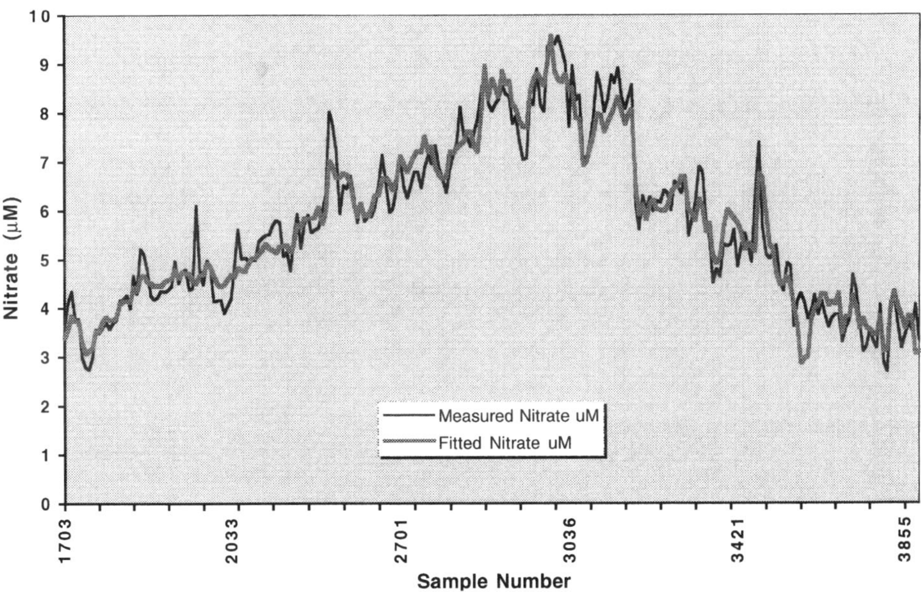

Figure 5.7 Measured (Auto analyser) NO_3 and fitted NO_3 calculated from a 2nd Order $A220$, $A235$ fit.

Figure 5.8 MKI sensor in CTD frame.

can be taken. The sampling rate is determined by the acceptable "noise" level, since the xenon flash tube source exhibits both short and long term fluctuations in output and a degree of smoothing is needed to reduce the effect of pulse-to-pulse output energy variation. A rate of 1 Hz can be achieved with a noise level of $\sim 1\ \mu M$, although this is dependent upon the individual flash tube used and upon its "burn-in" history. Wet-chemistry instruments, in contrast, have a sampling rate of order 0.03 Hz or less. Sensors incorporating the UV absorption technique can, therefore, be incorporated successfully in towed profilers (undulators) and in CTD packages for vertical profiling casts, whilst the wet-chemical sensors are only really suitable for use on moorings or for spot measurements. Ion-specific electrodes can also be used for sensing nitrate, but suffer from poor sensitivity and interference problems (American Public Health Association *et al.*, 1985).

In the absence of power supply limitations, the maximum number of samples is determined by flash tube life, which is of the order of 10^8 flashes to half the initial output energy or, typically, 3–4 months continuous operation. However, the calibration of the instrument must be checked at intervals over this life time,

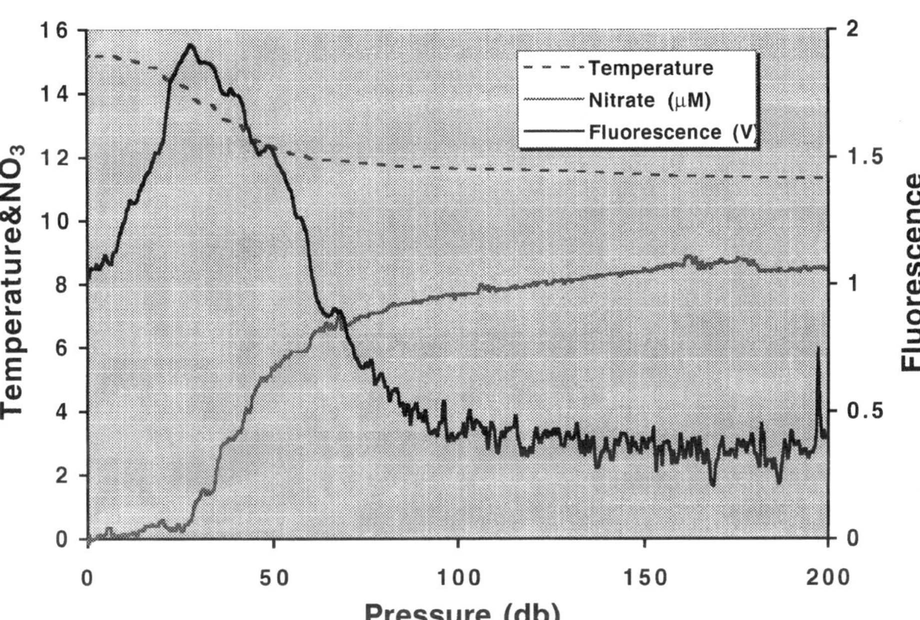

Figure 5.9 Correlation between NO_3^- from UV spectrometer data and Fluorescence (chlorophyll) data.

due to changes in the shape of the flash output spectrum. There is also the possibility of wavelength-selective fouling of the optical surfaces which are in contact with the sea water.

In the case of wet-chemistry instruments, the maximum number of samples is determined by the quantity of reagents incorporated in the instrument and upon its total pumping capacity. A general description of the technique and details of experimental equipment are given by McCormack (1996).

5.8 ELIMINATION OF INTERFERENCE

A problem of the UV absorption technique, especially noticeable in inshore waters where there is a high concentration of organic compounds, is the possibility of interference of the organic absorption spectra with the halide and nitrate absorption spectra. In the laboratory, it is possible to eliminate these compounds by chemical treatment of the sample prior to the absorption measurement; this is a standard technique used by the water industry. Such manipulation is clearly impractical for a fast response, *in-situ*, instrument.

Thomas and Gallot (1990) proposed multi-wavelength absorptiometry as a method for isolating the nitrate signal from that of other, interfering, species. As has been demonstrated above, it is possible to separate the salinity and the nitrate concentration – to a reasonable accuracy – by using absorption measurements at two wavelengths. This is relatively simple.

However, the absorption spectrum due to the potentially very large number of organic components (the interfering spectrum) is an unknown quantity. Thomas and Gallot overcome this difficulty by assuming that the interfering spectrum is relatively smooth, such that its $(p+1)$th derivative is negligible. The interfering spectrum, $Am(\lambda)$, can then be approximated by a polynomial of the form:

$$Am(\lambda) = a_0 + a_1 \cdot \lambda + a_2 \cdot \lambda^2 + \ldots + a_p \cdot \lambda^p$$

and the total absorption spectrum due to both the interfering components and the nitrate is then given by:

$$A(\lambda) = Am(\lambda) + \varepsilon_{NO_3}(\lambda) \cdot C_{NO_3} \cdot b,$$

where $\varepsilon_{NO_3}(\lambda)$ and C_{NO_3} are, respectively, the absorptivity coefficient and concentration of nitrate, and b is the path length, as described above. The absorptivity coefficient of nitrate as a function of wavelength, $\varepsilon_{NO_3}(\lambda)$, has to be known from previous measurements; this is assumed to be constant and unaffected by the presence of other species or by physical parameters such as temperature and pressure.

Measurement of the absorbance, $A(\lambda)$, at a number of wavelengths is followed by least-squares regression analysis, giving the coefficients $a_0 \ldots a_p$ and the desired nitrate concentration, C_{NO_3}. However, the method not only requires measurement at several wavelengths, but also accuracy. As might be expected, it is most effective when the absorption spectrum of the required species, nitrate in this case, is highly peaked in contrast to a relatively smoothly varying interfering background.

If this technique is not employed and if salinity is available from an independent sensor, this can be compared with the absorbance-derived salinity and, if different, one can infer that there is appreciable interference due to organics. In the open ocean, the level of organics is normally so small that this does not present a problem.

To put the degree of sensitivity into context, it is worth stressing that the nitrate levels in the deep ocean are generally less than 10 μM. This is an extremely low level in comparison with the safe level recommended by the US Environmental Protection Agency (1991) for drinking water of 700 μM. Colorimetric nitrate test kits for consumer use in home and farm water system testing typically have a detection limit of about 35 μM.

References

American Public Health Association *et al.* (1985). Standard Methods for examination of water and waste water, 16th edition, American Public Health Association, American Water Works Association, Water Pollution Control Federation.

Hydes, D.J. (1995). personal communication.

Jaffe, H. and Orchin, H. (1964). Theory and applications of Ultraviolet Spectroscopy, New York: John Wiley.

Jannasch, H.W. and Johnson, K.S. (1992). *In situ* instrumentation for dissolved nutrient and trace metal analyses in sea water, Proceedings of the MarChem '91 Workshop on Marine Chemistry. Martin, S.J., Codispoti, L.A. and Johnson, D.H. editors, ONR Report No. OCNR 12491–18.

Jannasch, H.W., Johnson, K.S. and Sakamoto, C.M. (1994). Submersible, Osmotically Pumped Analyzers for Continuous Determination of Nitrate *in situ*, *Analytical Chemistry*, Vol. 66, No. 20.

McCormack, T. (1996). Flow Injection Chemistries for the *In Situ* Monitoring of Nutrients in Sea Water, PhD. Thesis, University of Plymouth, UK, May 1996.

Thomas, O. and Gallot, S. (1990). Ultraviolet multiwavelength absorptiometry (UVMA) for the examination of natural waters and wastewaters, Part I: General considerations, *Fresenius' Journal of Analytical Chemistry*, 338(3), 234–237.

Thomas, O., Gallot, S. and Mazas, N. (1990). Ultraviolet multiwavelength absorptiometry (UVMA) for the examination of natural waters and wastewaters, Part II: Determination of nitrate, *Fresenius' Journal of Analytical Chemistry*, 338(3), 238–240.

United States Environmental Protection Agency, (1991). Is Your Drinking Water Safe?, WH-550, 570 9–91–005.

6. DEVELOPMENT OF AN OPTICAL CHEMICAL SENSOR FOR OCEANOGRAPHIC APPLICATIONS: THE SUBMERSIBLE AUTONOMOUS MOORED INSTRUMENT FOR SEAWATER CO_2

MICHAEL D. DEGRANDPRE, MATTHEW M. BAEHR[a] and TERENCE R. HAMMAR[b]

[a]*University of Montana, Department of Chemistry, Missoula, MT 59812, USA and*
[b]*Woods Hole Oceanographic Institution, Woods Hole, MA 02543, USA*

6.1 INTRODUCTION

Advances in marine chemistry are often driven by new technology that gives us the capability to measure an analyte more frequently and with greater sensitivity and selectivity. Improvements in mass spectrometry, for example, have made it possible for oceanographers to map ocean-wide helium, tritium and radiocarbon distributions, revealing patterns of ocean circulation (Jenkins, 1997; Jones *et al.*, 1994). Many more interesting questions could be addressed if more extensive spatial and temporal chemical data were available. Some progress has been made on the development of new automated chemical instrumentation on ships that can map surface distributions of many chemical properties (e.g. Dandonneau, 1995; Vodacek *et al.*, 1995) and provide real-time tracking of water masses (e.g. Watson *et al.*, 1994). Ideally though, marine chemists would like to obtain distributed chemical information over long time periods without spending large amounts of time and money at sea. Deploying chemical instrumentation on moorings, drogues, and autonomous vehicles that can remotely transmit data back to the lab is one possible solution to this problem. However, until recently, almost no automated chemical sensors or analyzers existed that could meet the criteria necessary for autonomous oceanographic measurements. These instruments should be simple, small, low power, and very stable or capable of automated calibration. Marine researchers have now pioneered a number of innovative sensors and analyzers that have been deployed on unmanned oceanographic platforms for extended periods (DeGrandpre and Bellerby, 1995).

One such instrument is the Submersible Autonomous Moored Instrument for CO_2, or SAMI-CO_2, developed at the Woods Hole Oceanographic Institution (DeGrandpre, 1993; DeGrandpre *et al.*, 1995a). Our objective was to design an autonomous instrument that could measure seawater pCO_2 (partial pressure of CO_2) for periods up to 6 months. Development of the first SAMI prototype began in early 1991 and it achieved its present form by early 1994. The evolution of the SAMI may typify the sort of lengthy, tortuous path required for the development of other autonomous chemical sensors. For this reason this chapter is devoted to a description of the various phases of the SAMI development. We first provide a brief background on fiber optic chemical sensors and CO_2 sensors. For those specifically interested in colorimetric based sensors, a detailed description of the operating principle and performance characteristics of the SAMI is given. We also describe problems that we have encountered in deploying these instruments at sea and past field results.

6.2 FIBER OPTIC CHEMICAL SENSORS

Chemists were quick to recognize that chemical sensors could be developed using the high quality inexpensive fiber optics that became available in the late 1970's. Like electrochemical sensors that use an electrical conductor to transmit information between the meter and the sensing mechanism, an optical chemical sensor could be made with fiber optics acting as a light conduit to and from an optically-based sensor. Fiber optics opened up the possibility of developing sensors based on established spectrochemical methods, including direct spectrophotometry, colorimetry, and photoluminescence. Considerable interest in chemical sensor applications for industrial, medical and environmental monitoring stimulated a large amount of research into these sensors, now called fiber optic chemical sensors (FOCS) or sometimes "optodes" (Seitz, 1988).

FOCS fall into two general categories, those that directly measure a spectroscopic property of the sample and those that utilize an analyte-selective reagent to transduce analyte concentration into an optical signal. While the former category has been successfully implemented in many real world applications (e.g. Cowles *et al.*, 1989; Melling *et al.*, 1991), the reagent-based FOCS have not met with wide success based on a survey of commercially available sensors. The multi-wavelength nature of FOCS was expected to enhance the sensor performance relative to their electrochemical counterparts but other limitations were encountered. Research over the past two decades on CO_2 FOCS provides an excellent example of the general evolution of FOCS; a brief review of this technology is therefore given.

There has been a longstanding interest in CO_2 sensors, primarily motivated by the need for blood CO_2 monitors. The first electrochemical CO_2 sensor was reported in 1958 (Severinghaus and Bradley, 1958). Fiber optic pCO_2 sensors first appeared in the early 1980's (Vurek *et al.*, 1983) and a large number of variations on the theme have since been reported (e.g. Zhujun and Seitz, 1984; Munkholm *et al.*, 1988; Wolfbeis *et al.*, 1988; Kawabata *et al.*, 1989; Parker *et al.*, 1993). These sensors are comprised of a pH indicator immobilized or entrapped within a gas-permeable membrane fixed on the fiber end(s). An aqueous bicarbonate solution is typically combined with the indicator at the fiber tips so that CO_2 can react with water to form carbonic acid; the subsequent formation of bicarbonate changes the pH-dependent fluorescence or absorbance. Early evaluations (e.g. Vurek *et al.*, 1983) revealed poor short-term stability and slow response times, both problems that originate from the fixed reagent chemistry. For example, it has been noted that changes in humidity or ionic strength causes water to diffuse into or out of the internal solution. The resulting change in solution composition leads to drift and loss of measurement accuracy. Other problems such as accumulation of contaminants in the inner solution that interfere with the spectrochemical reaction, irreversible reactions with the analyte, leaching and photobleaching of the indicator reagent also lead to unstable signals. Notwithstanding these limitations, FOCS have been successfully developed for certain applications. Blood gas CO_2, O_2, and pH FOCS are now commercially available (Miller *et al.*, 1987; Collison and Meyerhoff, 1990). Although these FOCS still drift, the drift is well-characterized and is rectified by relatively frequent (4–5 h; Collison and Meyerhoff, 1990) recalibration.

Fixed-reagent CO_2 FOCS have also been extensively studied for marine applications (e.g. Goswami *et al.*, 1989; Goyet *et al.*, 1992; Lefèvre *et al.*, 1993). Some of these sensors appear to have adequate seawater pCO_2 resolution ($\sim$1 μatm) but no long-term evaluations of stability were reported. These sensors undoubtedly suffer from drift similar to the CO_2 FOCS developed for other applications.

From these various research and development efforts it became apparent that FOCS internal wavelength referencing is not the expected panacea for chemical sensor drift. Researchers soon began to explore methods for flushing out the internal membrane solution and renewing the reagent to enhance the long-term stability and response time (e.g. Berman *et al.*, 1990; Inman *et al.*, 1989; Luo and Walt, 1989). Termed renewable-reagent FOCS, they utilize pressurized or reagent-impregnated membranes or active pumping of the reagent to the fiber optic tips. Researchers found that the advantages typically outweigh the additional complexity introduced by renewing the reagent. The renewable-reagent FOCS are ideal for applications that require superior sensitivity and stability, often the case in oceanography. Soon after the publication of the initial papers demonstrating the advantages of renewing the reagent, we began exploring the possibility of developing a similar sensor for seawater pCO_2 (DeGrandpre, 1991).

We were interested in developing an autonomous CO_2 system that could be deployed on ocean moorings for extended periods. Measurement of seawater pCO_2 presents several unique challenges, however, that would not be easy to implement in an autonomous mode. Measurement of seawater pCO_2 requires a resolution of 1 μatm or less over a range of approximately 100–800 μatm (i.e. $\sim$0.3% precision at 360 μatm). This sensitivity is over 2 orders of magnitude greater than blood gas FOCS. Accuracy of at least $\pm$1 μatm is also essential for air-sea flux calculations used in climate models (Siegenthaler and Sarmiento, 1993) and for calculation of other carbonate parameters (Millero *et al.*, 1993). Moreover, the difficulty of making reliable pCO_2 standards imposes that the instrument have exceptional long-term stability so that an automated calibration is not necessary. Shipboard pCO_2 methods, which are based on infrared or gas chromatographic analysis of air equilibrated with seawater, did not seem readily adaptable to autonomous measurements (although it has been done: see Friederich *et al.*, 1995). A renewable-reagent FOCS, however, was appealing because it would not require gas phase equilibration and could, in principle, be based on the same simple colorimetric method developed for fixed-reagent FOCS. These practical advantages made it feasible to design a simple, rugged, reliable and low power instrument.

6.3 SAMI-CO_2 SENSOR DESIGN

Our original FOCS design was modeled after an ammonia sensor reported by Berman and Burgess (1990). The CO_2 FOCS delivered fresh pH indicator through a small bore capillary tube to a gas-permeable silicone rubber membrane fixed at the fiber tips (Figure 6.1: DeGrandpre, 1991; 1993). The design required a compromise between minimizing the sensor flush volume and maximizing light throughput by using larger diameter fiber optics. In the final design we used a 0.40 mm diameter light delivery fiber, a 0.30 mm return fiber and 0.10 mm reagent capillary tubes housed within a 0.8 mm bore capillary tube. A 0.51 mm ID, 0.94 mm OD silicone tube was pulled over the capillary housing to form a 1.6 µL reservoir at the fiber tips.

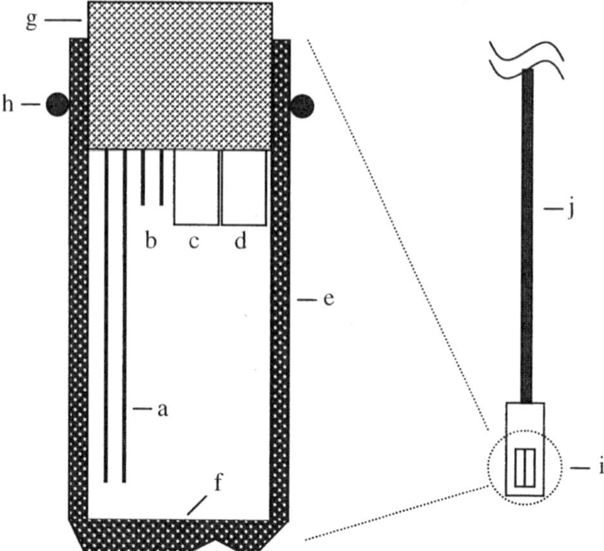

Figure 6.1 The first renewable-reagent pCO_2 sensor design: (a) reagent delivery capillary; (b) reagent exit capillary; (c) fiber optic from source; (d) fiber optic to detection system; (e) white silicon rubber membrane; (f) white silicone sealant; (g) epoxy; (h) o-ring; (i) sensor housing; (j) fiber optic cable. Individual components are not drawn to scale. Sensor outer diameter is ~1 mm.

Reagent delivery requires an appropriate pumping method. Power consumption worries led us to initially try pneumatic pumping using a N_2-pressurized reagent reservoir. This method was soon abandoned due to unstable flow and the impracticality of using gases under hydrostatic pressure. A syringe-type pump (syringe driven by a precision stepper-motor drive) was subsequently used for most of the initial evaluations.

During these initial tests, it was found that the sensor could be operated in either a diffusion-dependent or equilibrium mode depending upon the indicator flow rate (DeGrandpre, 1993). Although the diffusion-dependent (high flow rate) mode offered greater sensitivity, the equilibrium (low flow rate) response was found to be much less sensitive to flow rate and temperature, resulting in improved measurement precision.

The performance of the Figure 6.1 sensor was tested at sea and its precision was found to be comparable to shipboard methods (DeGrandpre, 1993). Although the sensor performance was encouraging, the design was difficult to fabricate and the sensor-to-sensor response was not consistent, due in part to different flushing characteristics for each sensor. A continuous tubular membrane from the reagent-delivery capillary to the optical cell simplified the fabrication and greatly improved flushing of the indicator (Figure 6.2). A syringe pump was (and still is)

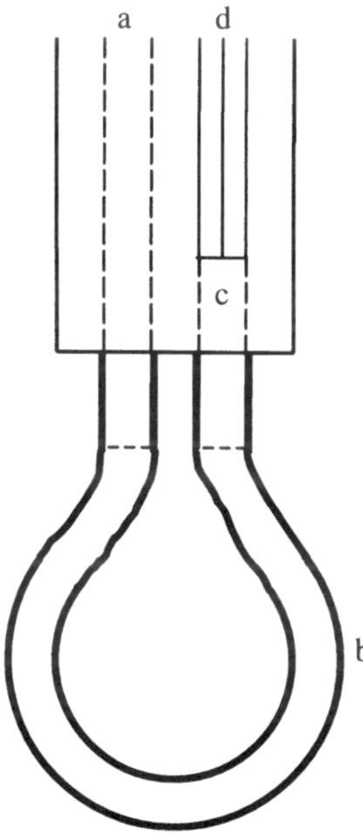

Figure 6.2 Schematic of the second sensor design: (a) capillary tube inlet from syringe pump; (b) tubular silicone rubber membrane; (c) optical cell; (d) fiber optics and indicator capillary tube outlet. Individual components are not drawn to scale. Sensor outer diameter is ~3 mm. Total membrane length is ~5 cm.

used with this sensor. The loop design proved to be very reliable and has been extensively used in laboratory experiments (DeGrandpre *et al.*, 1995b). Unfortunately, the ultimate limitation of both the Figure 6.1 and 6.2 designs is that only a small fraction of the light delivered to the fiber tips is backscattered into the return fiber optic. Hence, they require a high intensity 5 volt, 2 amp tungsten lamp to obtain an adequate signal-to-noise (S/N) for resolving 1 μatm CO_2. This lamp is an unacceptable power drain for an autonomous, battery-powered instrument and forced a major redesign in the sensor configuration. After about a year of trial-and-error, that was probably prolonged by our reluctance to abandon the single-ended sensor design, we settled upon a more conventional flow cell design with the fiber optics oriented in a transmission geometry (Figure 6.3: See Colour Plate section). Light throughput increased by over an order of magnitude making it possible to use a miniature 5 volt, 0.115 amp tungsten lamp. The combined flow cell and membrane equilibrator (Figure 6.3) makes the SAMI

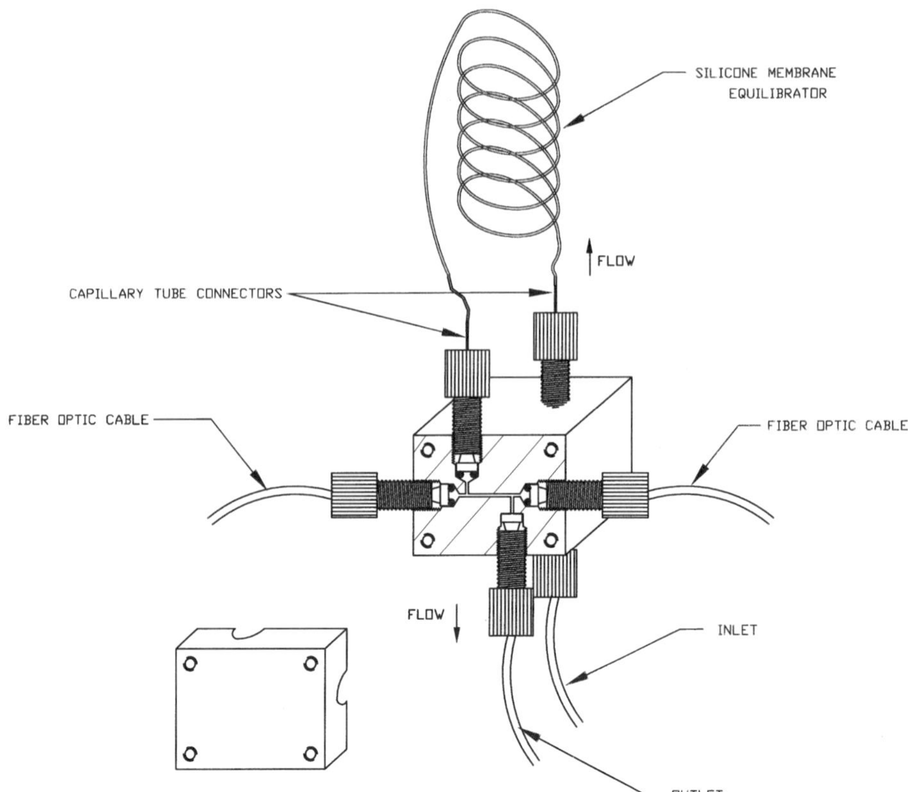

Figure 6.3 A cutaway view of the current SAMI sensor design. A fiber optic flow cell is combined with a tubular gas-permeable membrane equilibrator. Fiber optics are positioned in a transmission geometry to optimize throughput.

sensor more of a hybrid between a conventional analyzer and a FOCS; the distinction from an analyzer being that the SAMI does not actively pump the sample (seawater, in this case) into the instrument, but relies on passive diffusion of the analyte across a membrane into the reagent solution. A photograph of the SAMI is presented in Figure 6.4.

During the development of the new SAMI cell we started using miniature, low dead-volume solenoid pumps (The Lee Co.) to renew the reagent (DeGrandpre

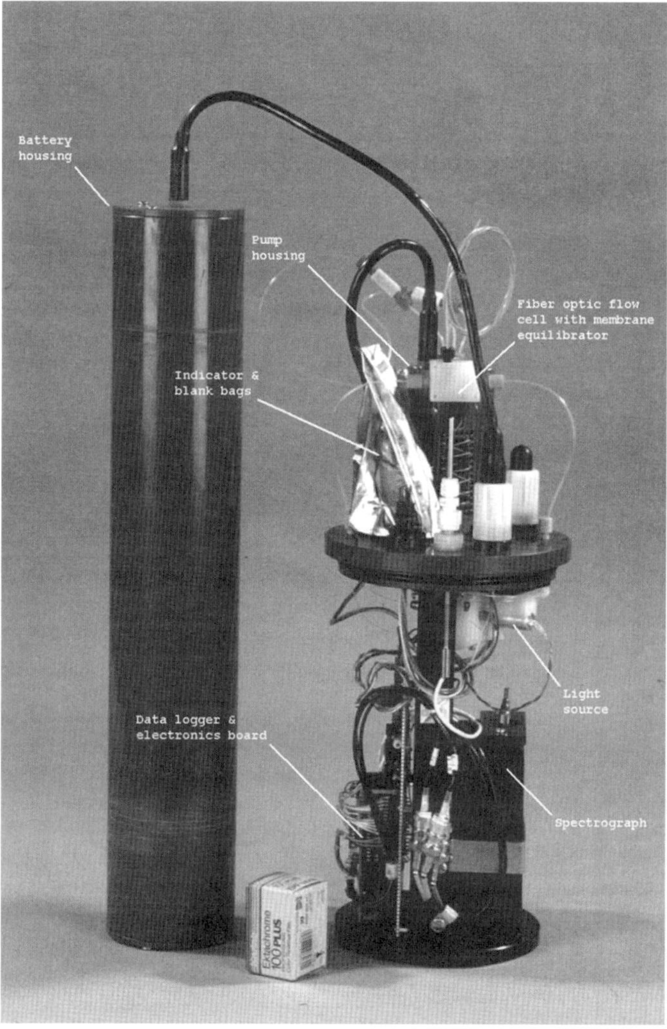

Figure 6.4 The Submersible Autonomous Moored Instrument for CO_2 (SAMI-CO_2). Note the size of the instrument relative to the film box. The main outer PVC pressure housing is not shown. The instrument is designed for depths up to 100 m.

et al., 1995a). These pumps are also used in the nitrate analyzer developed by Hans Jannasch and his colleagues (Jannasch *et al.*, 1994). Solenoid pumps are ideal for low power applications because pumping only requires a brief (< 1s) voltage pulse to activate the solenoid. The SAMI is operated in a stopped-flow mode so that the seawater CO_2 can equilibrate with the indicator solution (requiring about 5 minutes). During each measurement cycle, the pump is activated, pushing the CO_2-equilibrated solution into the fiber optic flow cell. The SAMI solenoid pump delivers ~50 µl of solution per pulse, enough to flush out the cell and membrane. The flow rate is not constant but, because the inner solution is in equilibrium with the seawater CO_2, the SAMI response is insensitive to the flow variations.

In the current SAMI design (Figure 6.4), light transmitted through the optical cell travels via a return fiber-optic to a miniature spectrograph (American Holographic) with three photodiode detectors (Hamamatsu Corp.). The photodiodes are positioned so that the center wavelengths are at 434, 620, and 740 nm. The photodiode currents are converted to an amplified voltage and recorded by a compact, low-power data logger (Onset Computer Corp.). Improvements such as a reduction in lamp and pump power and the utilization of low-power electronics have made it possible to deploy the SAMI for extended periods. Each measurement cycle lasts for ~1 minute with the average power consumed equal to ~7 milliamp hours (mAH). With one measurement every 1/2 hr the instrument should be capable of operating for nearly 6 months with the 30 AH battery pack (18 alkaline D cells) (Figure 6.4). In fact, the SAMI operated in an ice-covered lake for over 5 months at 2 °C and had enough reserve battery power to operate longer.

6.4 SAMI-CO_2 OPERATING PRINCIPLE AND PERFORMANCE

Although the SAMI response could be based solely upon the intensity transmitted at the absorbing wavelength of the indicator, a more sophisticated methodology is required to achieve the desired long-term stability. This methodology, described in detail in the following paragraphs, may be useful to others who are developing FOCS for oceanographic applications. As described above, CO_2 from the seawater diffuses across a gas-permeable membrane (Figure 6.3) into a sulfonephthalein acid-base indicator (bromothymol blue) solution. Sufficient time is allowed so that the seawater CO_2 reaches equilibrium with the indicator solution and, therefore, an equilibrium model can be used to describe the sensor response. The diffusion of CO_2 into the indicator solution leads to the formation of carbonic acid and bicarbonate which lowers the solution pH and establishes an equilibrium between the acid (HL^-) and base (L^{2-}) forms of the indicator:

$$HL^- \xleftrightarrow{K_a} L^{2-} + H^+ \tag{6.1}$$

The sulfonephthalein indicators are diprotic and are typically formulated as the sodium salt. The fully protonated form (H_2I) has a pK_a near 2 and does not exist in a significant amount at the indicator solution pH. The absorbance associated with the indicator solution-pCO_2 equilibrium is measured at the wavelengths corresponding to the molar absorptivity maxima of the two indicator forms. For the indicator bromothymol blue (BTB), these wavelengths are at 434 nm (acid or protonated HI^- form) and 620 nm (base or unprotonated I^{2-} form). The absorbance spectra for the acid and base form overlap and therefore both indicator forms contribute to the absorbance at the acid and base absorbance maxima. Either the acid or the base form absorbance could be used, both will change with varying levels of CO_2, but there are certain advantages to using ratios of the absorbances (A_{620}/A_{434}) rather than the absolute absorbances. One advantage is that the ratio will, in principle, be independent of cell pathlength. Furthermore, the temperature dependence of the absorbance ratio is dampened due to the similarity of the temperature dependence of the individual molar absorptivities (Byrne and Breland, 1989).

Rather than using the absorbance ratio for the SAMI response, we use a ratio that is directly related to the indicator solution pH. The pH can be described using a combination of the equilibrium equation (from Reaction 1) with Beer's law, identical to the method used to determine seawater pH by Robert-Baldo *et al.* (1985), where

$$pH = pK_a + \log([L^{2-}]/[HL^-]) \tag{6.2}$$

and the ratio $[L^{2-}]/[HL^-]$ is expressed as

$$\frac{[L^{2-}]}{[HL^-]} = \frac{R - {}_{620}\varepsilon_a/{}_{434}\varepsilon_a}{({}_{620}\varepsilon_b/{}_{434}\varepsilon_a) - R({}_{434}\varepsilon_b/{}_{434}\varepsilon_a)} \tag{6.3}$$

where $R = A_{620}/A_{434}$ and the subscripted ε's represent the molar absorptivities for BTB, with "a" the acid form of the indicator, and "b" the base form of the indicator. The SAMI response ratio, designated R_{CO_2}, is just a rearrangement of equation 6.2.:

$$R_{CO_2} = -\log([L^{2-}]/[HL^-]) = pK_a - pH. \tag{6.4}$$

A distinct advantage of defining the instrumental response this way is that direct comparisons of R_{CO_2} can be made to the theoretical response of the instrument (see below). Since R_{CO_2} is only dependent upon the absorbances and molar absorptivities of the indicator, every SAMI should give the same pCO_2 response, given that they have correct wavelength calibration and carefully controlled indicator chemistry.

This approach requires that true absorbances be determined, that is $A = -\log(I/I_0)$. The measurement of I_0, or the light intensity transmitted with no absorber present, is problematic because it requires a system blank. I_0 could be measured once prior to deployment with the hope that it remains constant, but experience has shown that it is susceptible to drift. As a compromise to running distilled water blanks between every pCO_2 measurement, we decided that a blank could be run infrequently if we were to employ a "blank constant" K to determine the absolute absorbances between the blank measurements. Accordingly, the SAMI indicator absorbance is calculated using the equation

$$A = -\log[I/(I_{ref}K)], \tag{6.5}$$

where I is the transmitted light intensity at the absorbing wavelength (434 or 620 nm), I_{ref} is the transmitted light intensity measured at a wavelength where the indicator does not appreciably absorb (740 nm for BTB), and $K = I_0/I_{ref}$, which is determined from the most recent blank measurement. By using I_{ref} for every measurement the instrument can correct for drift due to light intensity fluctuations from the light source, fiber bending, cell changes, etc.

The SAMI response ratio (R_{CO_2}) is empirically related to the seawater pCO_2 through calibration. The SAMI is calibrated in a thermostated water-filled chamber that is equilibrated with varying CO_2 gas concentrations. To obtain multiple calibration points, mass flow controllers are used to dilute a high concentration CO_2 standard. The equilibrated headspace CO_2 concentrations are continuously monitored with a nondispersive infrared CO_2 analyzer (NDIR) (Li-COR) calibrated with N.I.S.T. traceable CO_2 gas standards. After calibration, the R_{CO_2} from the SAMI and the NDIR data are combined to give a second-order polynomial calibration curve of R_{CO_2} versus pCO_2. A typical calibration plot along with the absorbances for the acid and base form of the indicator is shown in Figure 6.5.

It is important to note that the SAMI has a temperature response due to the temperature dependency of the indicator pK_a and the carbonate equilibria. Consequently, SAMI field data must be corrected for the difference between the *in-situ* and calibration temperatures. The temperature response of the instrument is determined by changing the calibration temperature and keeping the pCO_2 constant. Figure 6.6 shows the SAMI temperature response for two different pCO_2 concentrations. The slope does not vary significantly within the typical seawater pCO_2 range (300–400 μatm) so a constant temperature coefficient (0.0060 R_{CO_2} units $°C^{-1}$) is used to correct field measurements to the calibration temperature.

6.5 SAMI THEORETICAL RESPONSE

As shown in equation 6.4, the theoretical response for the SAMI is the difference between the indicator solution pH and the indicator pK_a. The theoretical pH can be calculated using the CO_2 equilibria equations combined with the charge balance (DeGrandpre, 1993). The indicator pK_a is determined experimentally using low ionic strength buffers (Covington *et al.*, 1983). We can use this theoretical response as a guide for determining optimal conditions and to determine if

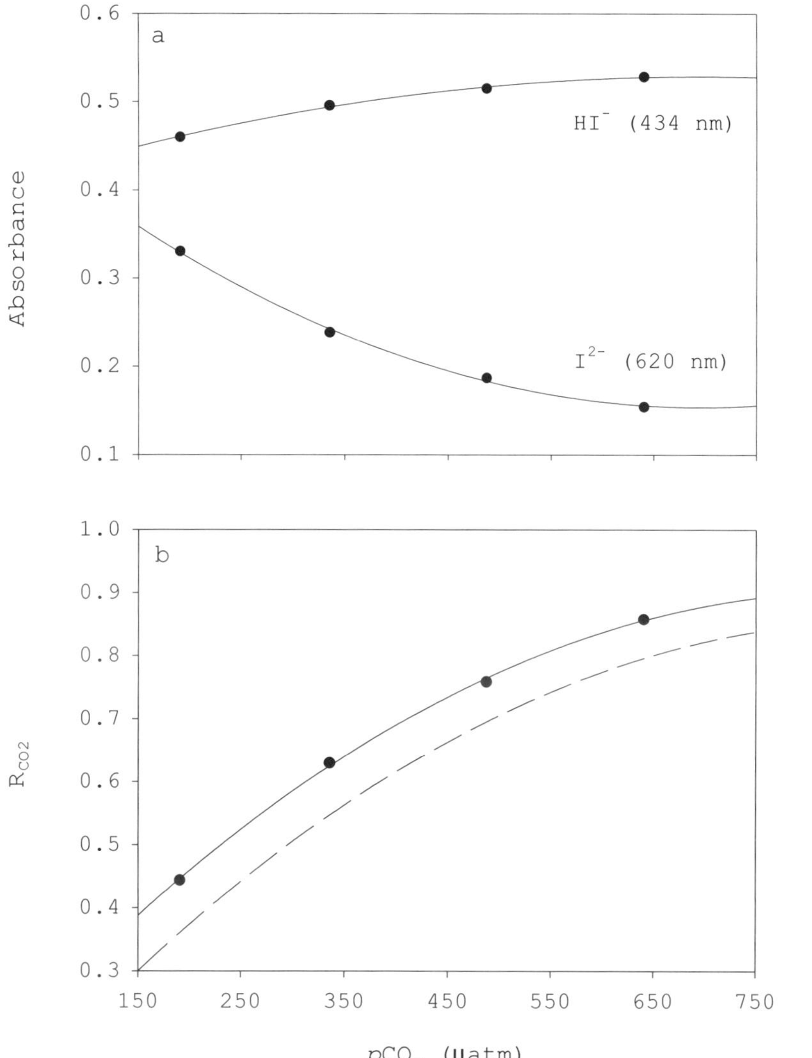

Figure 6.5 A typical SAMI calibration ($T = 20.5\,°C$). Plot (a) shows the individual absorbances of the acid and base forms of the indicator as a function of pCO_2. Plot (b) shows the instrumental response (R_{CO_2}) fit to a second-order polynomial (solid line). The dashed line represents the theoretical response curve at 20.5 °C over the same pCO_2 interval.

sensors are following expected behavior. Deviations of the SAMI response from the theoretical response can be used to direct troubleshooting efforts and lend insight into better sensor designs. A theoretical response curve for a SAMI calibration is shown in Figure 6.5(b). The actual response has the same shape as the

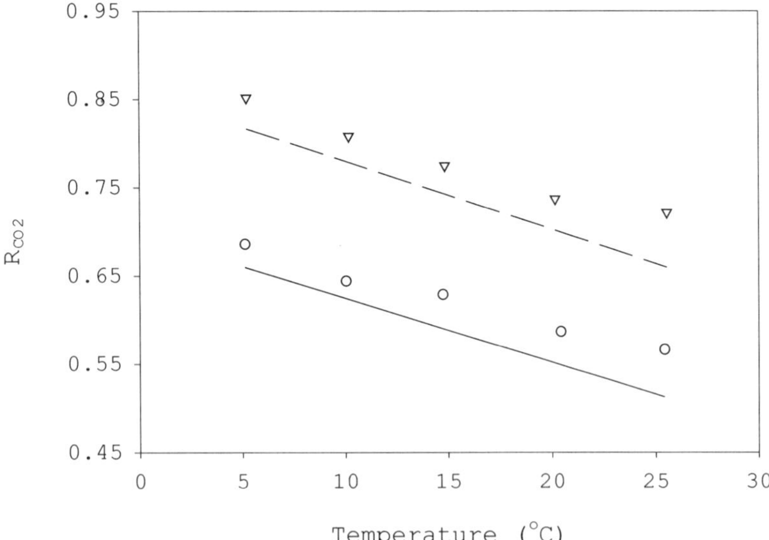

Figure 6.6 SAMI temperature response. The triangles and dashed line represent the actual and theoretical SAMI temperature response respectively, at 506 μatm pCO_2. The circles and solid line depict the actual and theoretical response respectively, at 325 μatm pCO_2. The experimental response (ΔR_{CO_2}/°C) at 506 μatm pCO_2 is 0.0076 and at 325 μatm pCO_2 it is 0.0060. Since the temperature response does not change significantly with pCO_2, a temperature coefficient of 0.0060 is used to correct typical seawater pCO_2 measurements (300–400 μatm) to the calibration temperature.

theoretical response curve but is offset. We believe that the disagreement between the actual and theoretical responses is due to the differences between the ionic strength of the buffer used to determine the indicator pK_a and that of the indicator solution itself. Although we have not yet quantified the magnitude of this pK_a activity coefficient correction, preliminary work shows that this correction will move the theoretical response in the direction of the actual response curve.

Theoretical plots proved useful in the optimization of the indicator solution used in this application. Before theory could be used however, a number of important factors needed to be considered. One important consideration is that the indicator pK_a should lie in the middle of the expected pH range (Peterson *et al.*, 1980). The indicator should also have high molar absorptivities at the acid-base absorbance maxima. In addition, the indicator solution should exhibit high sensitivity (large $\Delta pH/\Delta pCO_2$) but still maintain a large dynamic range (DeGrandpre, 1993). The pK_a of BTB makes it the indicator of choice for this application. In preparing the indicator solution it is necessary to optimize the indicator pH response range by the addition of a strong base (NaOH). Base (alkalinity) is added to the indicator solution to adjust the indicator absorbances so that they fall within a range of 0.3–1.0 absorbance units, where photometric precision is greatest (Willard *et al.*, 1988). This range corresponds to an optimal instrument response (R_{CO_2}) of 0.36–0.85. Theoretical calculations were used to demonstrate the effect of changing the indicator solution alkalinity on the instrumental response (Figure 6.7). Using Figure 6.7 as a guide, an indicator solution of 55 μM BTB with an addition of 42 μeq of alkalinity has been used with good results in field measurements for pCO_2 in the range of 200–700 μatm. Notice, however, that the base form (I^{2-}) absorbance (Figure 6.5a) resulting from this indicator formulation is below the range for optimal precision. Since we already had the response from a number of different instruments using this formulation, we did not change the indicator solution so that intercomparisons could be made. Now that this test has concluded, the low absorbance range for the base form will be adjusted in future indicator solutions by increasing the indicator concentration or the alkalinity. When higher pCO_2 (e.g. >700 μatm) is expected, such as in more productive inland and estuarine environments the alkalinity can be increased to keep the resulting absorbances in the optimal range.

Theory can also be used to validate empirical temperature responses. As can be seen in Figure 6.6, there is a slight disagreement between the SAMI temperature response and theory for the two pCO_2 concentrations shown. This offset was also demonstrated in Figure 6.5(b). However, the slopes of the empirical and theoretical temperature responses are nearly the same. This agreement supports the validity of the temperature corrections used for field pCO_2 data.

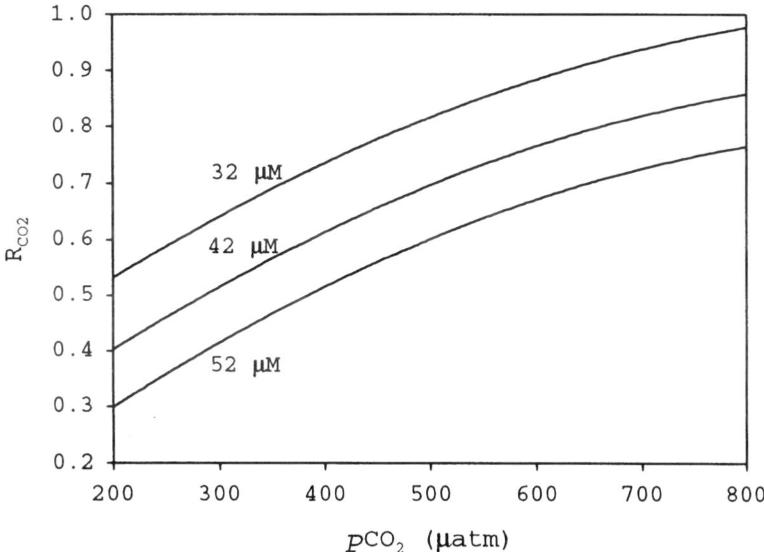

Figure 6.7 The effect of varying the amount of alkalinity added to the indicator solution (55 μM BTB) on the theoretical SAMI response ($T = 20.5\,°C$).

6.6 EVALUATION OF LONG-TERM STABILITY

Figure 6.8 illustrates the importance of the blank constant *K*, which is measured during each blank cycle (every two days). This figure shows the variability of the blank constant over a long-term deployment off Cape Hatteras, USA. If, for example, *K* was not used to account for instrumental drift in this deployment, an experimental error of ~0.060 R_{CO_2} units (or ~54 μatm pCO_2) would result by the end of the time period in Figure 6.8. The implementation of I_{ref} and *K* into the indicator absorbance measurements has considerably improved the long-term stability of the instrument. Initial tests of long-term stability in Woods Hole

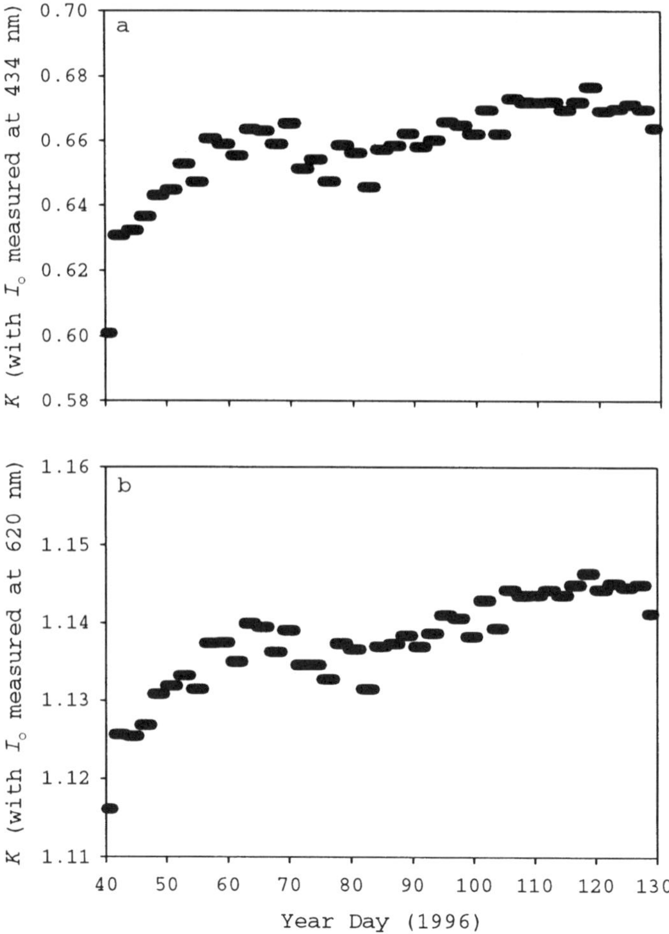

Figure 6.8 Blank constants measured during a long-term deployment. The acid form *K* is shown in (a). The base form *K* is shown in (b). Note that the *K* only changes every two days when it is measured in the blank cycle.

harbor showed no detectable drift over a thirty-day period (DeGrandpre *et al.*, 1995a). Verification of long-term accuracy during mooring deployments has been more difficult due to infrequent visits by supporting research vessels and spatial variability between the ship and mooring. However, a good example of the instrument long-term stability is given by comparing two instruments that were deployed at 10 and 22 m depths during 1996 (Figure 6.9). After nearly

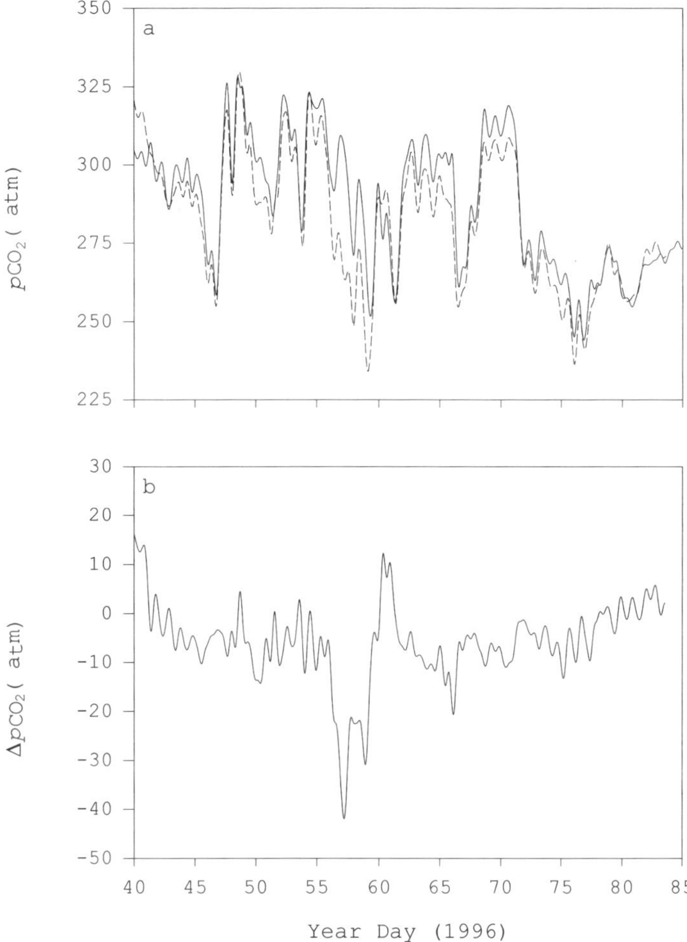

Figure 6.9 SAMI pCO_2 data obtained over long-term deployment off Cape Hatteras, USA. The data in (a) were obtained on the same mooring at 10 (dashed line) and 22 m (solid line) depths during 1996. The two highly variable signals were difficult to distinguish on the line plot so they were filtered using a low-pass (16 h) filter. The difference between the two SAMIs is shown in (b). The difference shows no systematic deviation over time supporting that neither instrument had significant measurement drift. Occasional large differences are due to advection of different water masses and periodic stratification.

50 days the two pCO_2 signals still overlay each other. The occasional large differences between the two instruments correspond to the presence of slight thermal stratification (DeGrandpre, unpublished results 1998).

6.7 EXPERIENCES DEPLOYING THE SAMI-CO_2

The SAMIs have primarily been deployed on surface and subsurface moorings in coastal waters. Unique risks and problems are found in mooring-based biogeochemical research and the reader may benefit from a brief description of the problems that we have encountered in deploying the SAMIs. Initial tests of the instrument took place in Woods Hole Harbor during the winter of 1994. Water temperature was below 2 °C and the pump housing oil became too viscous for effective pumping. We switched to low viscosity fluorinated hydrocarbon oil but this caused the internal valve tubing to swell, permanently damaging the valve. Since then a low viscosity (1 centistoke) silicone oil has been used without any problems. During the first mooring deployment, which took place off Cape Hatteras in 1994, the reagent bags were not sufficiently protected and began leaking after abrasion from wave action (DeGrandpre *et al.*, 1997). We have lost three instruments due to fishing-related incidents. In March 1996 an entire mooring was lost with two SAMIs and other instrumentation. A research vessel found the surface buoy adrift and it was suspected that a fishing trawler had dislodged the mooring. Through a communication snafu, the scientific crew onboard thought that the mooring was a guard buoy and, rather than lift the anchor on deck in heavy seas, they cut the chain. We later received a call from a New England scalloper who had drug up one of the SAMIs from this mooring. Unfortunately one of the fittings had snapped off and the SAMI had filled with water so all data were lost. In 1996 six instruments were deployed off Cape Hatteras, USA on subsurface moorings. One of these moorings was hauled up in a fisherman's net. In order to free the net, the upper portion of the mooring was cut free which then drifted away with a SAMI and other instrumentation. During this field program we also had 3 instruments disabled by fish bites on exposed Teflon tubing. Although fish bites are a well-known problem, we had never encountered it. Now all exposed tubing is covered with metal sleeves. Also during 1996 one battery pack leaked stopping the instrument. In spite of these many problems, most mooring deployments have been successful and have provided a wealth of novel and interesting data (DeGrandpre *et al.*, 1997; 1998). The problems you might expect from such an instrument, such as electronic or pump failure, have been infrequent.

Although most of our work has been in productive coastal waters, the SAMIs have rarely shown evidence of fouling artifacts. A copper mesh that is used around the membrane equilibrator appears to be an effective deterrent to algal growth in most cases. One exception was during a summer deployment in 15 m of water off the New Jersey, USA coast. Upon recovery, we found the membrane to be completely enclosed by a thick algal growth and the data showed a sharp and rapid decline in CO_2, which was probably caused by localized uptake of CO_2 around the fouled membrane.

6.8 FIELD RESULTS AND FUTURE DIRECTIONS

The SAMI-CO_2 was developed to improve our ability to study marine CO_2 variability, the premise being that better temporal coverage will improve quantitation and modeling of CO_2 sources and sinks. Progress has been made in this regard through deployment of the SAMI on moorings and other platforms such as the manned spar buoy, R/P FLIP (see DeGrandpre *et al.*, 1997; 1998). The SAMIs are usually deployed with dissolved O_2 sensors (YSI Inc.). The combined pCO_2/O_2 data sets provide a more in-depth understanding of the gas variability than would be possible with pCO_2 alone. However, rigorous interpretation and modeling of these data sets is often difficult because of advection of different water masses, a common complicating factor in all oceanographic research. In the unpublished data in Figure 6.9, the large rapid changes in pCO_2 and temperature (not shown) suggest that pCO_2 variability is dominated by advection during much of this period. Without knowledge of the spatial CO_2 distribution, interpretation of these results will be very difficult. However, long, continuous, time-series can capture extended periods when advection is not important, making it possible to test simple model predictions (DeGrandpre *et al.*, 1998). Future SAMI-CO_2 field projects will focus on improving how and where the instruments will be deployed to help simplify the data interpretation; for example, placing SAMIs on drifters that track a single water mass.

Significant progress has been made in the development and application of autonomous chemical sensors in marine research over the past 10 years. Because many of the most challenging and important questions in marine chemistry (such as, how do oceans regulate the uptake of CO_2?) require improvements in measurement technology, research on autonomous sensors will continue to grow. These new instruments will enable more and more research to be performed on unmanned platforms. It is exciting to think that with further miniaturization and reduction in sensor cost, it may someday be possible to obtain distributed chemical information over large ocean areas.

Acknowledgements

We would like to thank Steve Smith for his technical expertise in designing the SAMI electronics. This work has been funded by the U.S. DOE-Ocean Margins Program and U.S. NSF-Ocean Sciences.

References

Berman, R.J. and Burgess, L.W. (1990). Renewable-reagent fiber-optic-based ammonia sensor. In *SPIE vol. 1172 Chemical, Biochemical and Environmental Fiber Sensors*, pp. 206–214.

Berman, R.J., Christian, G.D. and Burgess, L.W. (1990). The measurement of sodium hydroxide concentration with a renewable-reagent-based fiber-optic sensor, *Anal. Chem.*, 62, 2066–2071.

Byrne, R.H. and Breland, J.A. (1989). High precision multiwavelength pH determinations in seawater using cresol red, *Deep-Sea Res. I*, 36, 803–810.

Collison, M.E. and Meyerhoff, M.E. (1990). Chemical sensors for bedside monitoring of critically ill patients, *Anal. Chem.*, 62, 425A-437A.

Covington, A.K., Whalley, P.D. and Davison, W. (1983). Procedures for the measurement of pH in low ionic strength solutions including freshwater, *Analyst*, 108, 1528–1532.

Cowles, T.J., Moum, J.N., Desiderio, R.A. and Angel, S.M. (1989). *In situ* monitoring of ocean chlorophyll via laser-induced fluorescence backscattering through an optical fiber, *App. Opt.*, 28, 595–600.

Dandonneau, Y. (1995). Sea-surface partial pressure of carbon dioxide in the eastern equatorial Pacific (August 1991 to October 1992): a multivariate analysis of physical and biological factors. *Deep-Sea Res. II*, 42, 349–364.

DeGrandpre, M.D., Hammar, T.R. and Wirick, C.D. (1998). Short term pCO_2 and O_2 dynamics in California coastal waters. *Deep-Sea Res. II*, 45, 1557–1575.

DeGrandpre, M.D., Hammar, T.R., Wallace, D.W.R. and Wirick, C.D. (1997). Simultaneous mooring-based measurements of seawater CO_2 and O_2 off Cape Hatleras, North Carolina, *Limnol. Oceanog.*, 42, 21–28.

DeGrandpre, M.D. and Bellerby, R.G.J. (1995). Chemical sensors in marine chemistry, *Oceanus*, 38, 30–32.

DeGrandpre, M.D., Hammar, T.R., Smith, S.P. and Sayles, F.L. (1995a). *In situ* measurements of seawater pCO_2, *Limnol. and Oceanog.*, 40, 969–975.

DeGrandpre, M.D., McGillis, W.R., Frew, N.M. and Bock, E.J. (1995b). Laboratory measurements of seawater CO_2 gas fluxes, In *Proceedings of the Third International Symposium on Air-Water Gas Transfer*, edited by Jähne, B. and Monahan, E.C. pp. 375–383. AEON publ.

DeGrandpre, M.D. (1993). Measurement of seawater pCO_2 using a renewable-reagent fiber optic sensor with colorimetric detection, *Anal. Chem.*, 65, 331–337.

DeGrandpre, M.D. (1991). A renewable reagent fiber optic sensor for ocean pCO_2, In *Proceedings SPIE vol. 1587 Chemical, Biochemical and Environmental Fiber Sensors III*, pp. 60–66.

Friederich, G.E., Brewer, P.G., Herlien, R. and Chavez, F.P. (1995). Measurement of sea surface partial pressure of CO_2 from a moored buoy, *Deep-Sea Res. I*, 42, 1175–1186.

Goswami, K., Kennedy, J.A., Dandge, D.K., Klainer, S.M. and Tokar, J.M. (1990). A fiber optic chemical sensor for carbon dioxide dissolved in sea water, In *Proceedings SPIE vol. 1172 Chemical, Biochemical and Environmental Sensors*, pp. 225–232.

Goyet, C., Walt, D.R. and Brewer, P.G. (1992). Development of a fiber optic sensor for measurement of pCO_2 in sea water: design criteria and sea trials, *Deep-Sea Res. I*, 39, 1015–1026.

Inman, S.M., Stromvall, E.J. and Lieberman, S.H. (1989). Pressurized membrane indicator system for fluorogenic-based fiber-optic chemical sensors, *Anal. Chim. Acta*, 217, 249–262.

Jannasch, H., Johnson, K.S. and Sakamoto, C.M. (1994). Submersible, osmotically-pumped analyzers for continuous determination of nitrate *in situ*, *Anal. Chem.*, 66, 3352–3361.

Jenkins, W.J. (1998). Studying subtropical thermocline ventilation and circulation using tritium and 3He, *J. Geophys. Res.*, 103, 15,817.

Jones, G.A., Gagnon, A.R., Schneider, R.J., von Reden, K.F. and McNichol, A.P. (1994). High-precision AMS radiocarbon measurements of central Arctic Ocean seawaters. Nuclear Instruments and Methods B92, pp. 426–430.

Kawabata, Y., Kamichika, T., Imasaka, T. and Ishibashi, N. (1989). Fiber-optic sensor for carbon dioxide with a pH indicator dispersed in a poly(ethylene glycol) membrane, *Anal. Chim. Acta*, 219, 223–229.

Lefèvre, N., Ciabrini, J.P., Michard, G., Brient, B., DuChaffaut, M. and Merlivat, L. (1993). A new optical sensor for pCO_2 measurements in seawater, *Mar. Chem.*, 42, 189–198.

Luo, S. and Walt, D.R. (1989). Fiber-optic sensors based on reagent delivery with controlled-release polymers, *Anal. Chem.*, 61, 174–177.

Melling, P.J., Truett, W.L. and Propster, G.D. (1991). Remote FTIR spectroscopy via optical fibers, *Amer. Lab.*, 32C–32G.

Miller, W.W., Yafuso, M., Yan, C.F., Hui, H.K. and Arick, S. (1987). Performance of an *in-vivo*, continuous blood-gas monitor with disposable probe, *Clin. Chem.*, 33, 1538–1542.

Millero, F.J., Byrne, R.H., Wanninkhof, R., Feely, R., Clayton, T., Murphy, P. and Lamb, M.F. (1993). The internal consistency of CO_2 measurements in the equatorial Pacific, *Mar. Chem.*, 44, 269–280.

Munkholm, C., Walt, D.R. and Milanovich, F.P. (1988). A fiber-optic sensor for CO_2 measurement, *Talanta*, 35, 109–112.

Parker, J.W., Laksin, O., Yu, C., Lau, M., Klima, S., Fisher, R., Scott, I. and Atwater, B.W. (1993). Fiber-optic sensors for pH and carbon dioxide using a self-referencing dye, *Anal. Chem.*, 65, 2329–2334.

Peterson, J.I., Goldstein, S.R., Fitzgerald, R.V. and Buckhold, D.K. (1980). Fiber optic pH probe for physiological use, *Anal. Chem.*, 52, 864–869.

Robert-Baldo, G.L., Morris, M.J. and Byrne, R.H. (1985). Spectrophotometric determination of seawater pH using phenol red. *Anal. Chem.*, 57, 2564–2567.

Seitz, W.R. (1988). Chemical sensors based on immobilized indicators and fiber optics, *CRC Crit. Rev. Anal. Chem.*, 19, 135–173.

Severinghaus, J.W. and Bradley, A.F. (1958). Electrodes for blood pO_2 and pCO_2 determination, *J. Appl. Physiol.*, 13, 515–520.

Siegenthaler, U. and Sarmiento, G. (1993). Atmospheric carbon dioxide and the ocean, *Nature*, 365, 119–125.

Vodacek, A., Hoge, F.E., Swift, R.N., Yungel, J.K., Peltzer, E.T. and Blough, N.V. (1995). The use of *in situ* and airborne fluorescence measurements to determine uv absorption and DOC concentrations in surface waters, *Limnol. Oceanog.*, 40, 411–415.

Vurek, G.G., Fuestel, P.J. and Severinghaus, J.W. (1983). A fiber optic pCO_2 sensor, *Ann. Biomed. Eng.*, 11, 499–510.

Watson, A.J., Law, C.S., Van Scoy, K.A., Millero, F.J., Yao, W., Friederich, G.E., Liddicoat, M.I., Wanninkhof, R.H., Barber, R.T. and Coale K.H. (1994). Minimal effect of iron fertilization on sea-surface carbon dioxide concentrations, *Nature*, 371, 143–145.

Weiss, R.F. (1974). Carbon dioxide in water and seawater: the solubility of a non-ideal gas, *Mar. Chem.*, 2, 203–215.

Willard, H.H., Merritt, L.L., Dean, J.A. and Settle, F.A. (1988). *Instrumental Methods of Analysis*, 7th edn., Wadsworth Publishing Co.

Wolfbeis, O.S., Weis, L.J., Leiner, M.J.P. and Ziegler, W.E. (1988). Fiber-optic fluorosensor for oxygen and carbon dioxide, *Anal. Chem.*, 60, 2028–2030.

Zhujun, Z. and Seitz, W.R. (1984). A carbon dioxide sensor based on fluorescence, *Anal. Chim. Acta*, 160, 305–309.

7. OPTICAL MICROSENSORS AND MICROPROBES

GERHARD HOLST, INGO KLIMANT, MICHAEL KÜHL
and OLIVER KOHLS

7.1 INTRODUCTION

Biogeochemical processes in the ocean are (to a large extent) regulated by the physico-chemical characteristics of the microenvironment where the processes occur. In the pelagic, phytoplankton, bacteria and small grazers interact and regulate the productivity in response to environmental variables like temperature, salinity or availability of nutrients and trace elements. Hot spots of metabolic activities are found in aggregates of microalgae or e.g. in planktonic foraminifera or radiolaria harbouring microalgal symbionts. Also, during the continuous export of biomass (e.g. dead or dying phytoplankton, faecal material and other organic debris) from the euphotic zone of the ocean, ca. 0.5–5 mm large aggregates (marine snow) are formed that are rapidly mineralised during their journey to the sea floor. In the open ocean, recycling of carbon and other essential elements thus mainly takes place in the water column, while only refractory material reaches the seafloor, where it is slowly degraded and buried. The deep sea sediment is thus a major sink for carbon on a global scale.

In coastal waters with shallower depths, an increasing amount of the pelagic production reaches the sea floor and causes more intense remineralisation in the sediment. This pelagic-benthic coupling is replaced in waters where sufficient light reaches the sea floor, by photosynthetic production in the upper sediment layers. Production and remineralisation become closely coupled in these photosynthetic sediments, and occur within a few mm thick overlapping reaction zones in the top layers of the sediment (Figure 7.1). In the extreme case of biofilms and microbial mats, this leads to extremely productive communities fuelled by microbenthic phototrophs, which partly rely on an efficient recycling of carbon and nutrients via heterotrophic microbes in their immediate neighbourhood. Here the complete range of production and remineralisation processes, usually found over *a* > 1 km deep water column and the sediment in the open ocean, are compressed to a few millimetres.

In marine aggregates and photosynthetic sediments, production and mineralisation processes are closely coupled and take place in a densely populated structured microenvironment of cells and exopolymers. The high volumetric conversion rates in these systems, in combination with the fact that hydrodynamic and diffusive boundary layers form at the interface between the system and the turbulent surrounding water phase, result in the formation of steep gradients of physico-chemical variables over spatial distances of < 0.1–1 mm. There can be steep gradients of light intensity and spectral composition due to intense

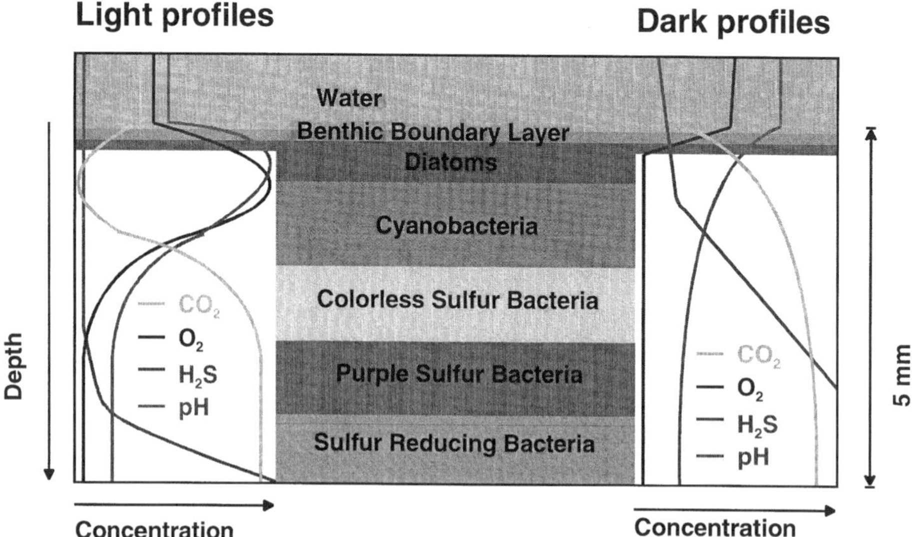

Figure 7.1 Vertical zonation of microbial mats and shallow water sediments. Typical concentration profiles of O_2, pH, CO_2 and H_2S vs. penetration depth are given for day (light profiles) and night (dark profiles) reflecting the photsynthetic activity of the upper layers.

scattering and absorption in the matrix of pigmented cells, exopolymers, and mineral grains. Furthermore, there can be gradients of metabolic substrates/products like oxygen, carbon dioxide, hydrogen sulphide, methane, pH, nutrients and dissolved organic compounds.

Gradients are also present in deep-sea sediments, but due to the refractory quality of available organic material the gradients over the diffusive boundary layer and within the sediment are not as steep as in shallower waters. While oxygen penetration is less than 0.5 mm in the most active sediments, oxygen penetrates to greater than 50 cm in cold deep-sea sediments. Important exceptions from this general observation are the hydrothermal vents and seeps found in tectonically active areas of the deep sea. In these areas, dense and active benthic communities develop on the basis of geothermally fed chemolithotrophic microbes.

In all systems, many of the gradients result from mass transfer and metabolic activity. Provided that (non- or minimally-invasive) measurements at high spatial and temporal resolution can be performed, it is, therefore, possible to obtain detailed information about relevant transport processes, their interaction with the consumption/production of substrates/products, and the zonation and magnitude of the various conversion processes.

Such information is very important to understand the dynamics and regulation of biogeochemical processes in the ocean. Most traditional techniques in biogeochemistry are essentially "black box" approaches, where measured conversion rates rely on assumptions about the actual microenvironment, where the processes occur. Using microsensors and microprobes, these assumptions can be investigated and, if necessary, modified according to experimental information determined within the microenvironment.

The theme of this chapter is a description of one class of miniature measuring devices, fibre-optic microsensors, that allow such high resolution measurements to be performed either in the laboratory or directly in the ocean via use of special remote operating vehicles. We focus on technical aspects of instrumentation, sensor construction and performance. Only a few applications are described. Information about electrochemical microsensors and further examples of microsensor application can found in other chapters of this book and in some recent reviews (e.g. Revsbech and Jørgensen, 1986; Revsbech, 1994; Kühl *et al.*, 1994a; Kühl and Revsbech, 1998; Amann and Kühl, 1998).

7.2 OPTICAL MICROSENSORS AND MICROPROBES

7.2.1 Optical Microsensors, Microprobes

Fibre optical devices and instrumentation for fine scale measurements of physicochemical variables in the aquatic environment typically operate to a spatial resolution of < 0.1 mm. It is important to discriminate between fibre optic microprobes that collect optical information from the medium surrounding the measuring fibre tip and direct it to a detector system, and fibre optic microsensors, microoptodes, that measure a chemical or physical variable via a reversible change in the optical properties of an indicator which is immobilized at the measuring fibre tip. Both types of devices are based on the use of single strand optical fibres for directing light information to and from the site of measurement. Optical fibres and optical components like fibre couplers and switches have been tremendously improved by the advancements in telecommunication industry and they display: low weight, robustness, immunity to electromagnetic radiation, high temperature stability and many other advantageous characteristics for use in measuring systems (Dakin and Culshaw, 1988). Due to the rapid progress of optoelectronics, many classical spectroscopic techniques can now be applied to measurements with small and easy to handle instruments. Especially, the current availability of robust fibre optical components and compatible light sources and detectors presents a huge advancement for the design of optical measuring systems. This has resulted in a large number of fibre optical sensors and sensor systems in recent years, and this development seems to be accelerating (Dakin and Culshaw, 1988; Wolfbeis, 1991). Most practical applications of such sensors have been realised in medicine and biotechnology, while the development and use of fibre optic sensors in environmental applications has been more limited. Nevertheless, fibre optic microprobes and microoptodes have seen numerous applications in oceanography, limnology, and biogeochemistry. (Kühl and Revsbech, 1998; Klimant *et al.*, 1997).

7.2.1.1 Spectroscopic Principles of Optical Sensing

Principally every interaction between light and matter which exhibits a reproducible and reversible sensitivity to a chemical or physical variable can be used to detect this variable. In this part we focus on the basics of spectroscopic principles used in microoptodes that use appropriate dye materials, which interact with the analyte to be measured. Most of the experiences with such interactions exist in the visible part of the spectrum – which ranges from approximately 390 to 730 nm, and the processes and principles used in microoptodes mainly exploit this part of the electromagnetic spectrum.

The energy of the light can interact with the outer electrons of dye molecules with large electron systems. The π-electron system plays a key role. On the atomic scale, the molecular structure (and therefore the optical properties) of the dye are changed by either a chemical reaction or a change to its molecular environment. This can be due to protonation/deprotonation, oxidation/reduction or the presence of special kind of species. In any case, energy of incident light has to be absorbed by the dye, which only can happen if the energy levels within the molecular system exactly correspond to the incident light energy level.

In the Jablonski diagram (energy level scheme, Figure 7.2) the dye molecule is promoted from the electronic ground state S_0 to an excited level S_1 respectively S_2 depending on the light energy. The energy levels can be split into different sub-levels due to various possible rotational and vibrational energy levels of the

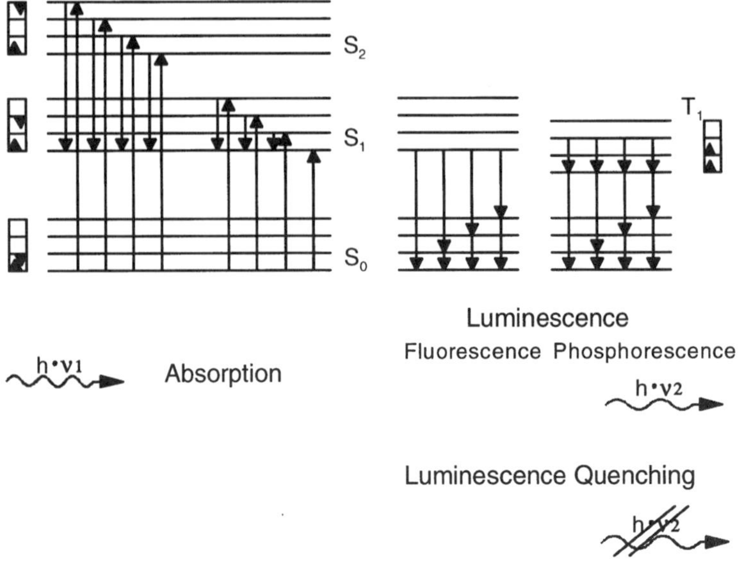

Figure 7.2 Jablonski diagram (energy level scheme) with the energetic ground level S_1, the excited singlet levels S_1 and S_2, and an excited triplet level T_1 ($h\nu$ = energy of light radiation).

molecule. The dye excitation/absorption spectrum thus consists of a huge amount of single energy transition processes.

Many dyes exhibit relaxation to the ground state S_0 via repetitive energy transfer to rotational and vibrational levels (as we can see in the Jablonski diagram between the excited levels S_1 and S_2 in Figure 7.2). These dyes are called absorption dyes. Fibre optical sensors, which have the dye immobilised at the fibre tip measure absorption indirectly, via the reflected and backscattered light from the fibre tip. In section 2.2.4 such an absorption dye based pH microsensor is described, where the protonated and the deprotonated form of the dye exhibit different absorption spectra.

Other dyes return to the ground level by emission of a photon. The photon usually has less energy than the absorbed photon, due to vibrational relaxation prior to photon emission. Therefore, the emission spectra of these dyes are shifted towards the red part of the spectrum as compared to the excitation spectrum, this effect is called the Stokes shift. If there is no intersystem crossing between the excited levels, the time between absorption and emission of a photon is very short (in the range of nanoseconds). In this case we talk about fluorescence. If there is intersystem crossing between triplet (T_1) and singlet (S_1) levels in the excited state, the relaxation process takes much longer, milliseconds to minutes. Here we talk about phosphorescence. Phosphorence only occurs when a strong internal interaction in the dye molecule is present. This is found, for instance, in some large metallo-organic molecules like platinum- or palladium-porphyrins. Both processes, fluorescence and phosphorescence, are generally called photo luminescence.

Luminescence dyes enable measurements at wavelengths different from absorption allowing an efficient separation from background signal. In many applications this improves the efficiency of measurement. Some molecules, oxygen for example, are able to absorb energy from the excited dye by collision. The dye returns to the ground state without emitting a photon. The luminescence is quenched as a function of the amount of quenching molecules present. The quenching phenomenon changes both the intensity of emitted radiation and the lifetime of the luminescence. This principle is used for the measurement of e.g. oxygen by many optical measuring techniques.

7.2.1.2 Construction of Microoptodes

Optical fibres guide light through total internal reflection, which is caused by the difference in refraction indices between the fibre core and the cladding material. The described fibre optical sensors are made from quartz "gradient index" fibres. Gradient index describes the fact, that the refraction index profile perpendicular to the fibre axis has a curved ("gradient") shape instead of the partially constant refraction index of step index fibres. The core of the presented fibre sensors has a diameter of 100 µm with a 20 µm thick layer of cladding material (100/140 µm fibre). For mechanical protection, the fibre itself carries an additional polymer jacket and is inserted in a PVC tube with carbon fibres. This robust fibre cable finally has an outer diameter of 3 mm. At one end, the cable is terminated with a standard ST-plug. This allows connection to instruments as well as to other fibre cables with high precision and stability. For high spatial resolution measurements an optical fibre sensor diameter of 140 µm is too high. However, thinner fibre material has concomitant handling problems. To achieve optimum spatial resolution the fibre diameter is reduced to a diameter which is necessary for minimum invasive applications (typically < 50 µm, see part 1). This can be done either by etching of the fibre tip or by tapering in the heat of a flame. The size of the flame, the pulling force and the timing all influence the final taper geometry (Figure 7.3a). This geometry of the taper is highly important for the sensor performance and has to be optimised for each application. (A short, steep taper can compress the probe during profiling measurement. On the other hand, light passing down to a long taper may be lost due to interaction with the sample, resulting in low signals.) After tapering the fibre down to a diameter of approximately 10 µm the taper is typically cut further back at a diameter of

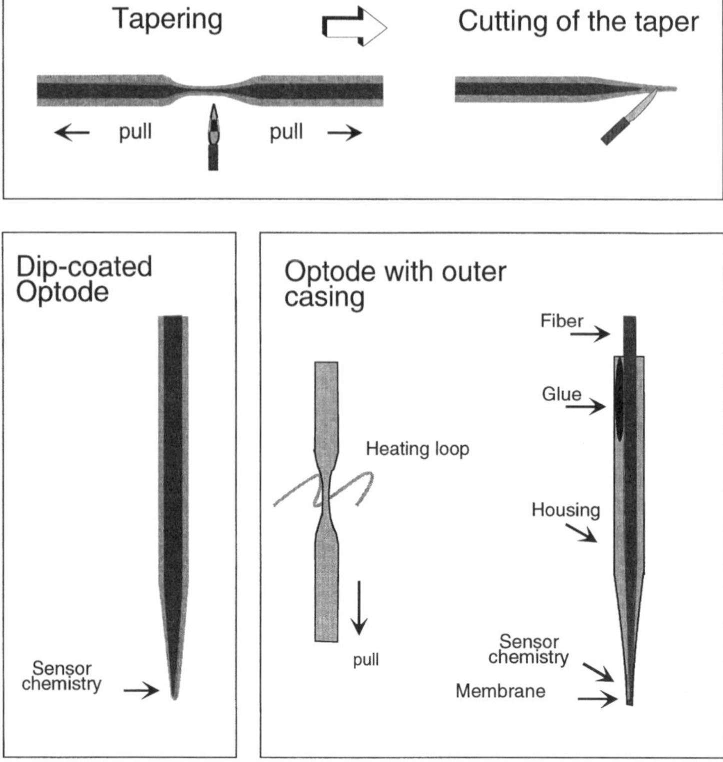

Figure 7.3 Preparation of optical microsensors: (a) tapering and cutting of optical fibre; (b) dip-coated microoptode; and (c) optode with outer casing.

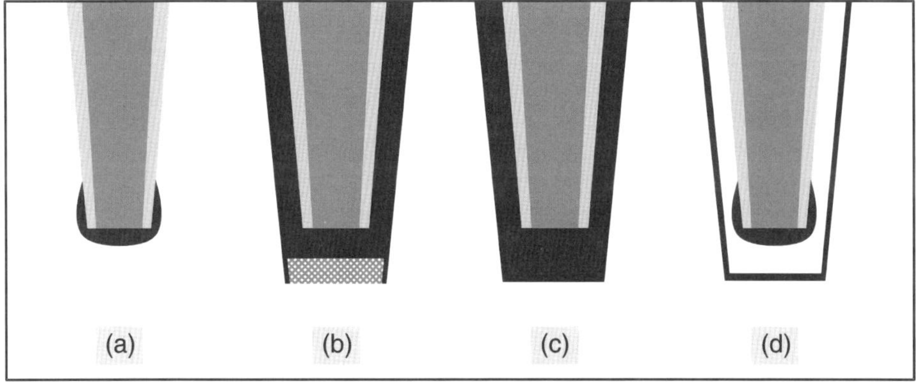

Figure 7.4 Structural overview (schematic cross sections) of microoptodes: (a) dip-coated fibre taper, various layers may be added; (b) plain fibre taper in an open microcapillary housing that is filled with indicator solution and closed with a membrane; (c) plain fibre taper in a closed microcapillary housing that is filled with indicator solution; and (d) dip-coated fibre taper in a closed microcapillary housing that is filled with a gas.

about 30 μm. The next step in fabrication is defined by the structure and the type of sensor (Figures 7.4 and 7.18). For example, simple optical microprobes consist of tapered fibres that are aluminium coated and which have, depending on the light parameter that should be measured, scattering spheres or discs attached to the fibre tip (Figure 7.18). For microoptodes the sensing chemistry is immobilised on or around the fibre tip. From the practical point of view the easiest set-up for an optical chemical sensor is the immobilisation of the dye in or on a matrix material directly attached to the fibre tip by dip-coating. A matrix material for the immobilisation of the dye is needed which is permeable to the analyte and prevents leaching of the dye. Furthermore, the material needs to adhere well to the quartz surface and should not change the spectral characteristics of the dye. If necessary, the tip of the sensor can be over-coated by an additional black layer as an optical insulation or a semipermeable membrane for increasing the selectivity of the sensor. For some applications the dip-coated single optical fibre is not suitable. Another approach is to surround the optical fibre by a glass casing (Figure 7.3c). The casing can be manufactured from a glass capillary by tapering it in the heat of a heating loop. For some sensors the geometry of the fibre and the casing is of high importance and has to fit well together. The control parameters for the preparation of the casing are similar to the taper process of the fibre, the size of the heated part and the pulling force.

Depending on the type of sensor the housing is closed by melting e.g. for special kind of temperature optodes (Figure 7.4c) or closed with an analyte permeable membrane e.g. silicone for carbon dioxide optodes (Figure 7.4b). Then the tip of the housing is filled with the sensor chemistry in a liquid phase in which the tapered fibre is placed. Alternatively, the dip-coated taper can be placed in a closed casing which is a possible solution for temperature micro-optodes (Figure 7.4d).

The fixation of the optical fibre in the casing simplifies the handling of the microoptodes in experimental set-ups, which can be micromanipulators in laboratory flow chambers as well as benthic landers for *in situ* applications.

7.2.1.3 Oxygen Microoptode

The established type of microsensor for oxygen measurements is the Clark-type microelectrode (Revsbech, 1989). The manufacturing of this sensor is a complex and time consuming process. Normally, the Clark-type electrode exhibits a stirring sensitivity based on the electrochemical reduction of oxygen, which can be assumed as negligible for microsensors. Nevertheless, for high resolution oxygen measurements at low oxygen levels, problems can occur according to the measuring resolution of the electrodes near to their zero current.

Most of the oxygen optodes are based on luminescence quenching (Kautsky, 1939; Lübbers and Opitz, 1975). Therefore, the oxygen optode exhibits the highest signal at low oxygen concentrations. Furthermore the oxygen is not consumed and therefore stirring has no influence on the detected signal. The design of the oxygen optode is carried out as described above as sensor type A. The optode consists of a tapered optical fibre which is dip-coated with the appropriate sensing chemistry. Compared to the micro electrode production the micro optode can be prepared fast and at low costs.

As oxygen sensitive material two groups of dyes are established in literature: Ruthenium-dyes which consists of Ruthenium(II) as the central atom in the metallo-organic complex (Demas, 1976; Wolfbeis *et al.*, 1984; Lippitsch *et al.*, 1988; Klimant *et al.*, 1992; MacCraith *et al.*, 1994; Hartmann and Leiner, 1995) and Porphyrine-dyes which commonly have Platinum or Palladium as central atom in a porphyrine molecule (Wilson *et al.*, 1987; Khalil *et al.*, 1988; Papkovsky *et al.*, 1993). These dyes have excitation maxima in the blue respectively blue/green part of the spectrum and exhibit a strong orange respectively red phosphorescence. Figure 7.5 shows the spectral characteristics of the molecule Tris(4,7-diphenyl-1,10-phenanthroline)-ruthenium(II) (RuDPP) the most frequently used oxygen indicator.

The compatibility of the spectral characteristics of the dyes enables the use of light emitting diodes (LED) as excitation light sources, and therefore the measurement of the luminescence intensity (or the luminescence life-time) with minimal opto-electronics (see section 3.3). The relation between the luminescence intensity I and the oxygen concentration c was first described by Stern and Volmer (1919)

$$\frac{I}{I_0} = \frac{\tau}{\tau_0} = \frac{1}{1 + K_{sv} \cdot c}$$

where I_0 is the luminescence in the absence of oxygen and K_{SV} the Stern-Volmer or quenching coefficient. The same equation also is valid for life-time measurements, only the intensity has to be replaced by the respective life-time parameters

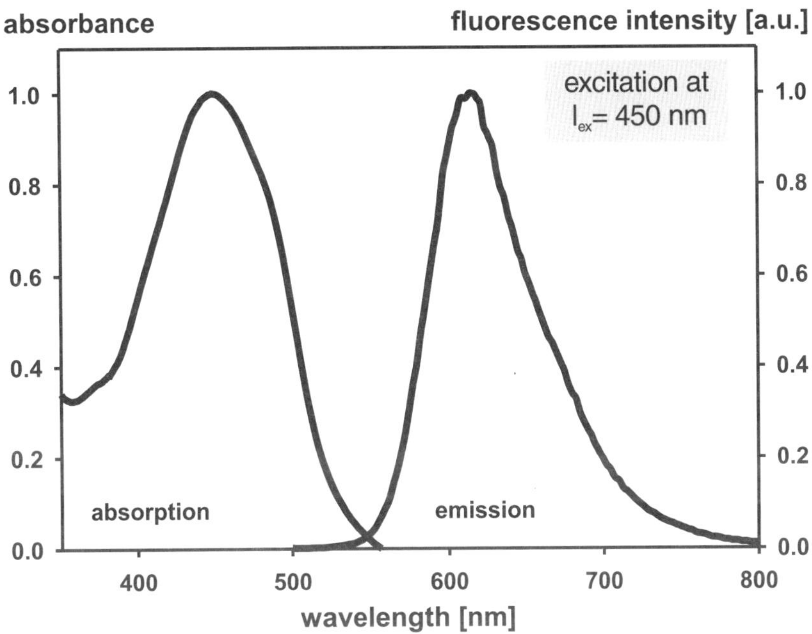

Figure 7.5 Absorption/emission spectra of Tris(4,7-diphenyl-1,10-phenanthroline)-ruthenium(II) solved in chloroform.

τ_0 and τ. In most of the real systems this relation does not sufficiently describe the measured data. Based on a two component model proposed by Carraway *et al.* (1991) the equation can be extended by a second quenchable component:

$$\frac{I}{I_0} = \frac{\tau}{\tau_0} = \frac{A}{1 + K_{sv1} \cdot c} + \frac{B}{1 + K_{sv2} \cdot c}.$$

For many reasons it is possible to simplify this model (Klimant *et al.*, 1995) by assuming one component to be non quenchable:

$$\frac{I}{I_0} = \frac{\tau}{\tau_0} = \frac{f}{1 + K_{sv} \cdot c} + (1 - f),$$

f describes the fraction of the indicator that is quenchable. The usual procedure in characterising an indicator-matrix combination is to measure calibration curves with a sufficient amount of microoptodes, made with this sensor chemistry. Then, a best fit is calculated with three variables: τ_0, K_{SV} and f. Usually we found the fraction f to be constant. Therefore a two point calibration procedure describes the calibration curve sufficiently. Figure 7.6 demonstrates the fit of the

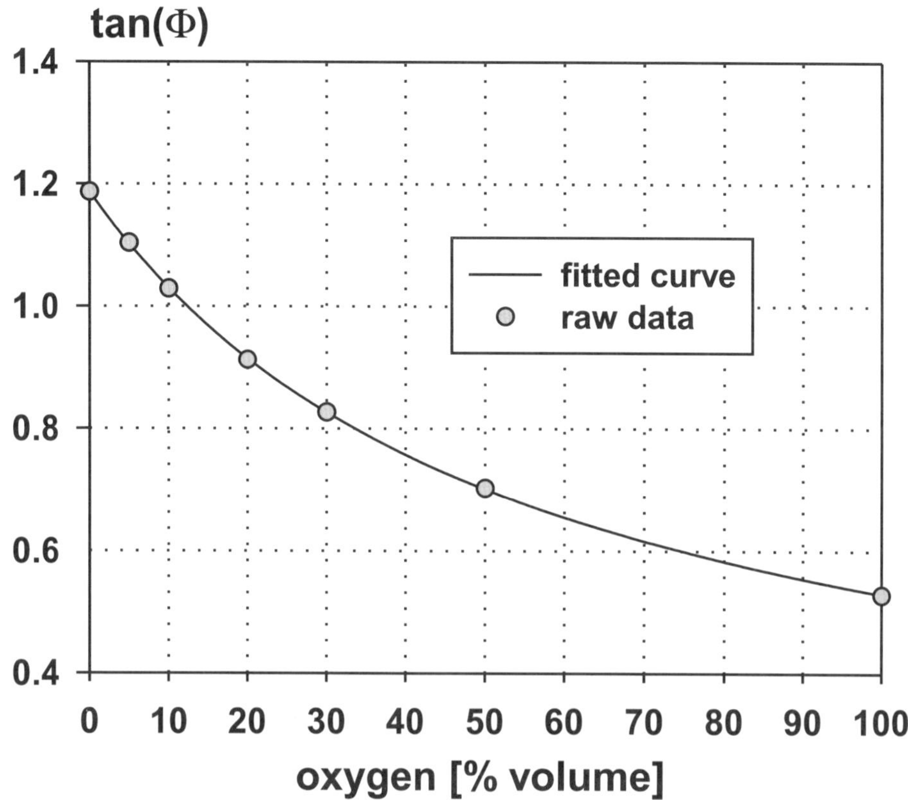

Figure 7.6 Comparison of raw measured data and the corresponding model fit for a RuDiPhe – sol-gel oxygen optode: $\Phi_0 = 49.9°$ ($\tau = 3.78\,\mu s$), $K_{SV} = 0.0184\%$, frac $= 0.854$, $f = 50\,kHz$.

model with the calibration data and the corresponding curve of a RuDPP indicator in an organically modified sol-gel matrix, which gave a correlation factor of $r = 0.999984$.

The choice of the dye and its immobilisation govern the sensitivity of the oxygen optodes. A high oxygen solubility leads to absolute high amounts of quenching substance at a given oxygen partial pressure. Therefore, the sensor will be very sensitive at low oxygen concentrations. A matrix material with a high oxygen solubility is silicone. Compared to silicone the oxygen solubility of polystyrene is much lower. In the last few years, sol-gel materials have became popular matrices in optical sensing, because by appropriate choice of the preparation procedure and of the precursors the permeability can be adjusted (McDonagh *et al.*, 1998). All these materials have good adhesion properties to silica (glass), which is an important prerequisite for their application as micro-optodes. The oxygen sensitivity of the indicators is a complex phenomenon.

Generally due to the significant lower luminescence decay-time, the Ruthenium-complexes are less sensitive then the Porphyrine-compounds. In Figure 7.7 normalised calibration curves for three different indicator-matrix combinations are shown. The sensitivity of each combination is optimum for a different oxygen concentration range (Figure 7.7, see legend box), which demonstrates the ability to tune the oxygen microoptode for the corresponding application. Optodes with very high sensitivity in the low oxygen range (Figure 7.7, 0–10%) as well as sensors with nearly homogeneous sensitivity over the whole range (Figure 7.7, 0–800%) can be prepared.

Typically, the oxygen microoptode is a type A sensor (Figure 7.4a) (tapered fibre and dip-coated sensor chemistry). The thin sensing layer usually does not absorb all the excitation light. Therefore, a certain amount of light may illuminate the near environment of the fibre tip – which can cause problems in photosynthetic environments, because the excitation light for the dye changes the

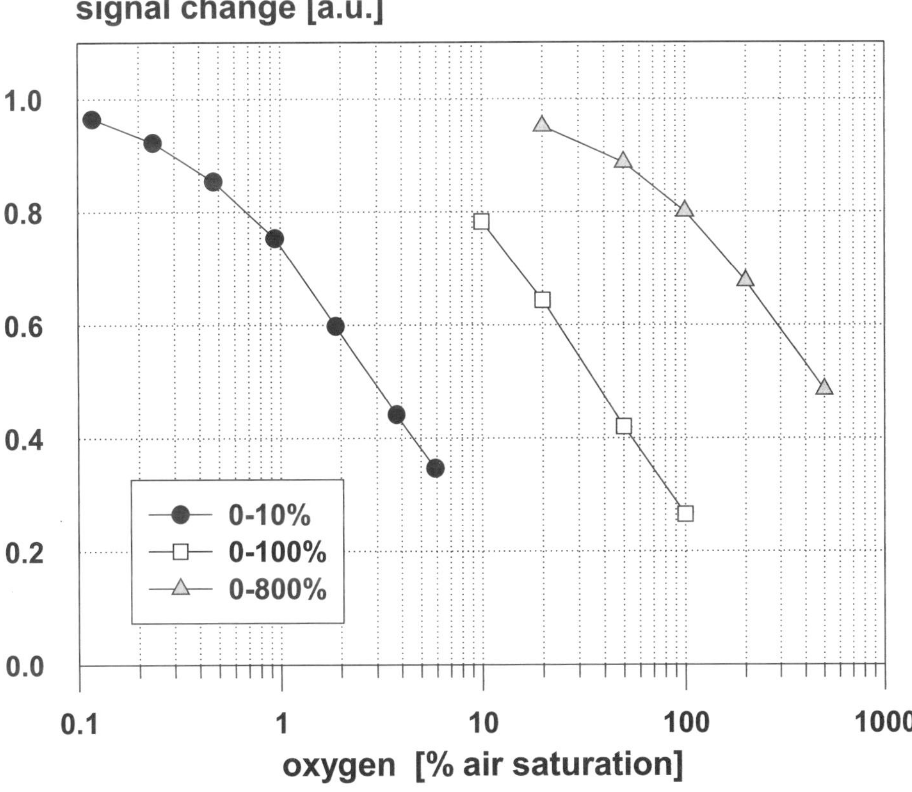

Figure 7.7 Comparison of calibration curves of oxygen microoptode optimised for different oxygen measuring ranges (see legend box). The oxygen zero point of all curves is suppressed because of the logarithmic scale.

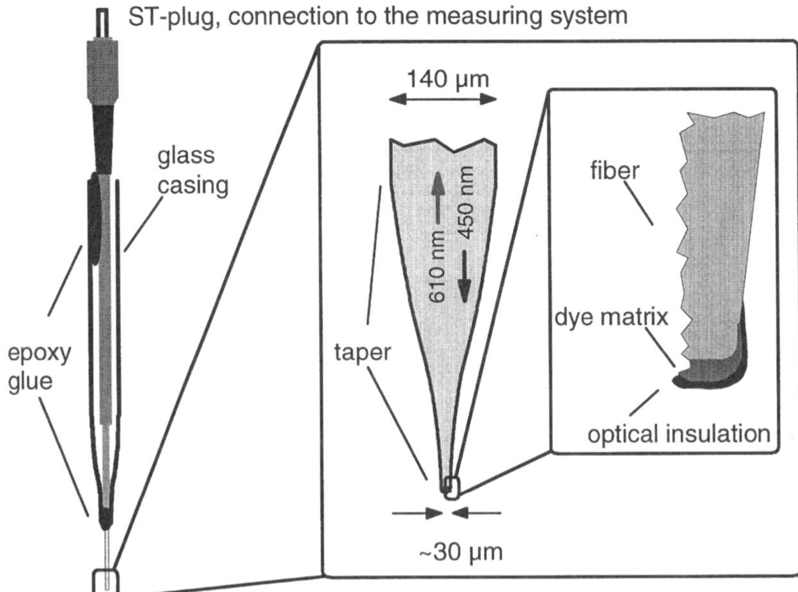

Figure 7.8 Schematic drawing of an oxygen microoptode for application in benthic landers.

local light conditions for the microorganisms and might induce photosynthesis. This problem can be solved by an additional layer on the sensor tip which prevents the light from leaving the sensor. This layer is called optical insulation and usually is made of a highly oxygen permeable material like carbon black silicone. In Figure 7.8 an oxygen microoptode consisting of an additional casing for better handling and optical insulation is shown in detail.

The oxygen microoptode was successfully applied in laboratory set-ups, as can be seen in Figure 7.9. An oxygen microoptode and an oxygen microelectrode were combined to form a double tip sensor with an approximate distance of 100 μm between the tips. With the combined sensor depth profiles of oxygen were measured in a dense microbial mat from a hypersaline lake (Solar Lake, Egypt). The mat was kept in a steady flow, at constant salinity and temperature in a flow chamber. The results obtained with illumination and darkness illustrate nicely the equivalence of both microsensors. The negligible differences in the curves are due to the natural heterogeneity of the mat and the spatial distance between the tips. Furthermore, oxygen microoptodes have been applied on lander systems (Glud *et al.*, 1997). Test and characterisation measurements exhibited a small (for practical work, negligible) influence of hydrostatic pressure on the sensor response down to a water depth of 6000 m (Kohls *et al.*, 1996–1998, unpublished results). Therefore, oxygen measurements in the deep sea on lander systems are entirely feasible, as demonstrated by Wenzhöfer *et al.* (1998).

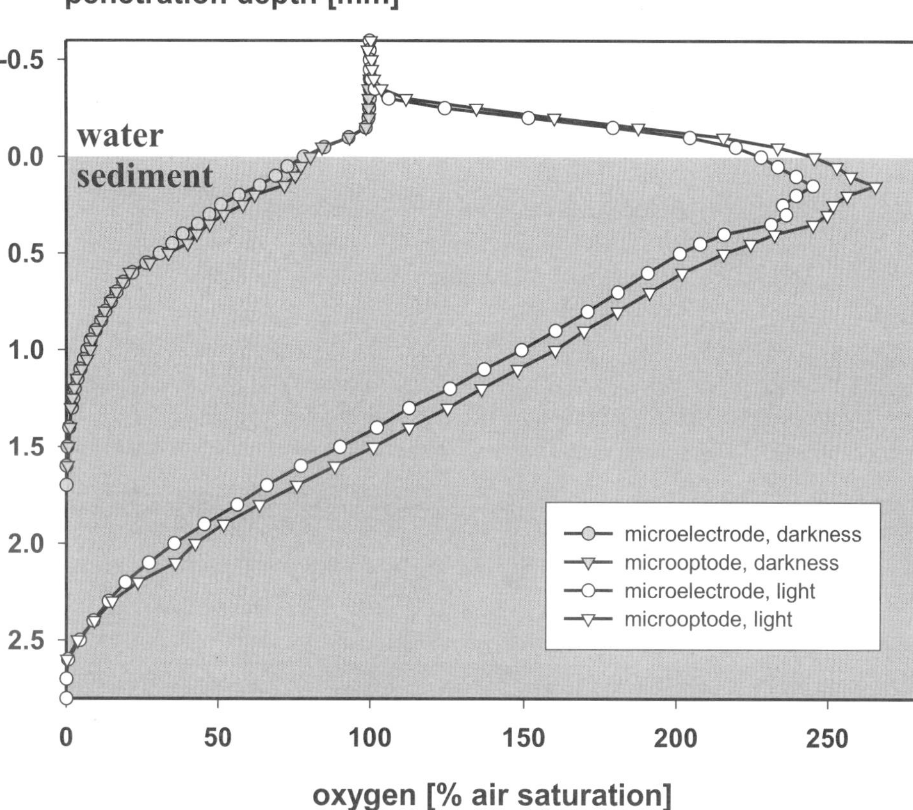

Figure 7.9 Oxygen depth profiles of a microsensor combination made of an oxygen microoptode and an oxygen microelectrode measured in a microbial mat from a hypersaline lake (Solar Lake, Egypt) in light and darkness conditions (see legend box). The distance between the two microsensor tips was approximately 100 μm.

7.2.1.4 pH Microoptode

A broad range of pH sensitive fluorescence and absorption based indicators is described in the literature (Bishop, 1972; Munkholm *et al.*, 1986; Lübbers *et al.*, 1977; Leiner, 1991, and reviewed in Wolfbeis, 1991). For the analytical pH range of seawater, pH 7 to 9 the number of useful dyes is limited. An optical pH measurement usually can determine pH to a very high accuracy over a dynamic range of ± 2 pH units of the pK value of the dye (The pK value is the pH value when 50% of the indicator is protonated). Therefore the pK of the immobilised dye has to be in the seawater range. An optimum pH indicator has distinct different spectral characteristics in its protonated and deprotonated form.

Generally many indicators have more than one corresponding pK value but usually only one pK is analytically useful.

$$H_nI \xleftrightarrow{K_1} H^+_{n-1} + I^{1-} \xleftrightarrow{K_2} 2 \cdot H^+_{n-2} + I^{2-} \xleftrightarrow{K_3} \wedge \ldots$$

Over the relatively small pH range of the marine environment usually just one pK is used for measurement. The constant K for a single equilibrium is expressed by

$$K = \frac{[H^+] \cdot [I^-]}{[HI]}$$

where it has to kept in mind that the equilibrium constant refers to the activity of the species. If the activity coefficient f is taken into account, the relation between the dye forms and the pH value can be expressed by the Henderson-Hasselbalch equation:

$$pH = pK + \log\left(\frac{[I^-]}{[HI]}\right) + \log\left(\frac{f_{I-}}{f_{HI}}\right)$$

Absorption dyes useful for application at marine pH values are e.g. sulphophthalein indicators like thymol blue, m-cresol purple, cresol red or phenol red (Robert-Baldo *et al.*, 1986). Furthermore a number of luminescence dyes are known, such as hydroxypyrene trisulfonic acid (Leiner, 1991) or different rhodamin and fluorescein derivates like SNARF and SNAFL dyes (Molecular Probes Inc.). The most common technique in optical pH determination is based on intensity measurement because the lifetime of pH sensitive luminescence dyes is rather short (range of ns). A set-up for optical pH measurements using lifetime evaluations would be much more complex and expensive than electrochemical methods and, therefore, offer no real alternative to existing microelectrodes at the moment.

The most critical point for all kinds of optical pH measurement is the immobilisation of the dye. In order to reach a high solubility for protons, it is necessary to immobilise the dye in or at a hydrophilic material such as cellulose derivatives or sol-gels. Due to the fact, that pH dyes are polar substances they are usually soluble in water. This enables the leaching of the dyes from their matrix. Therefore the dyes have to be bound covalently or at least absorbed at a material which has a higher affinity to the dyes than water. The absorption dye N9 (Merck) can bound covalently on a prepared cellulose acetate material and is therefore not leachable. In Figure 7.10 the pH dependent absorption of the dye is shown. Detailed characterisation of the pH dependent absorption of the dye identifies the optimum pH sensitive regions (450 nm and 590 nm). Wavelengths at 480 nm and >800 nm do not exhibit a distinct pH depending absorption and can thus be used for referencing purposes (e.g. correction of fibre bending effects and other pH-independent changes in light intensity).

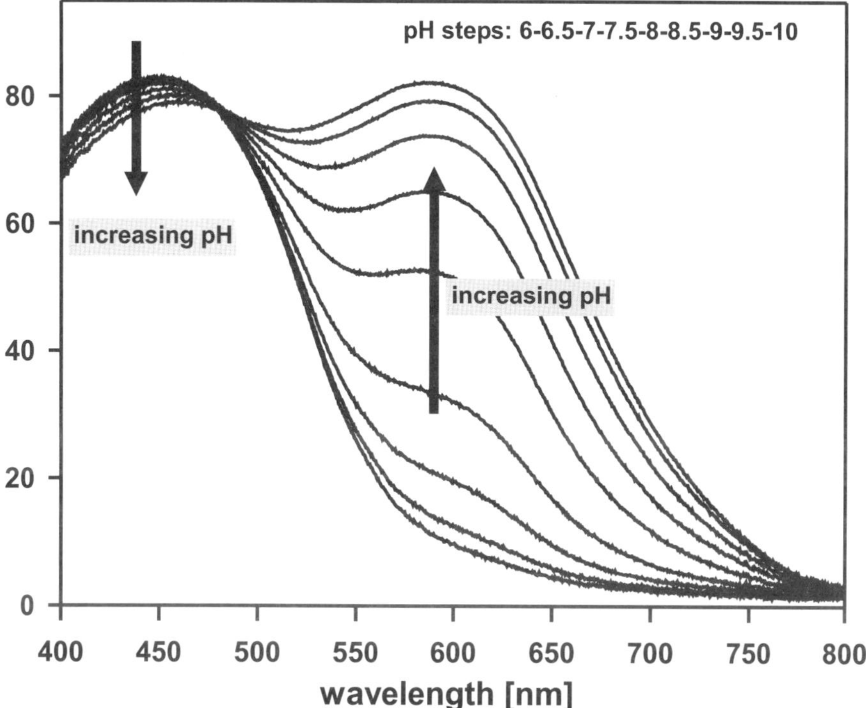

Figure 7.10 pH dependent absorption spectra of the indicator N9. The arrows indicate the direction of change of each spectrum when the pH is increased. The isosbestic point is at 480 nm.

In practice, the amount of reflected and backscattered light from the indicator at the fibre tip is monitored via an optoelectronical set-up, where blue and yellow LEDs are used as light sources and a photodiode for detection. By using yellow light for monitoring the 590 nm region this dye exhibits an apparent pK around the natural pH of seawater, i.e. at pH 8.2 (Figure 7.11). The pH microoptode with a tip diameter smaller then 40 μm allows pH measurements at an approximate accuracy of ± 0.05 pH units over a range of pH 7–9. First pH depth profiles (Kohls *et al.*, 1997, Figure 7.12) in a coastal sediment resulted in pH microprofiles similar to the depth profiles that have been observed with pH microelectrodes (Revsbech *et al.*, 1983).

7.2.1.5 Carbon dioxide Microoptode

CO_2 in the gas phase can be measured directly by infrared absorption. This is the most exact way to determine CO_2 with negligible cross sensitivity. For practical microsensor applications this technique is not useful due to the low IR absorption

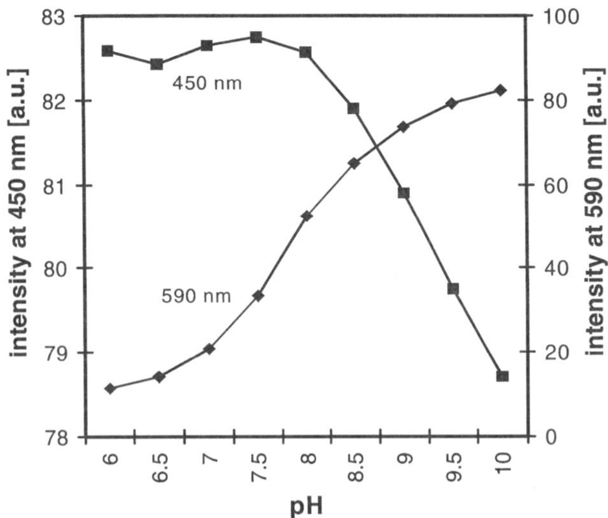

Figure 7.11 Calibration curves of the pH microoptode (N9), measured light intensity vs. pH units, with (squares) and (diamonds) illumination light.

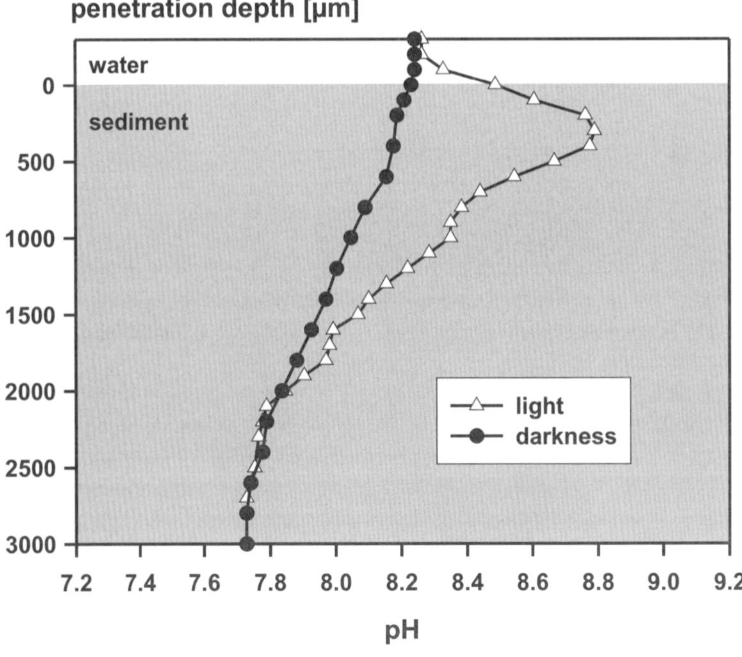

Figure 7.12 pH depth profiles, obtained with an absorption based (N9) pH microoptode, measured in a North Sea sediment in light and darkness conditions (see legend box).

coefficient of CO_2 and therefore a poor sensitivity. Other techniques are based on the Severinghaus-principle (Severinghaus and Bradley, 1958; Lübbers and Opitz, 1975; Munkholm *et al.*, 1988; Leiner, 1991). Here CO_2 diffuses through a gas-permeable membrane into a buffer reservoir, where the induced shift in pH is monitored with a pH sensor. This can be easily realised with optical microsensors. The sensor, type B, consists of an outer glass-microcapillary, which is sealed with a thin silicone membrane and filled with a solution of an appropriate dye. The dye must have its pK value at the pK of the equilibrium between hydrogencarbonate and carbonate,

$$CO_2 + H_2O \xleftrightarrow{K_1} HCO_3^- + H^+ \xleftrightarrow{K_2} CO_3^{2-} + 2 \cdot H^+$$

A useful dye for this purpose is e.g. hydroxypyrene trisulfonic acid (HPTS, Leiner, 1991). HPTS is a pH sensitive luminescence dye with a high quantum yield and an apparent pK value of 7.5. The absorption/emission spectrum and molecule structure of HPTS is shown in Figure 7.13. HPTS exhibits an excitation maximum at 455 nm (compatible to blue LEDs) and emitted green fluorescence

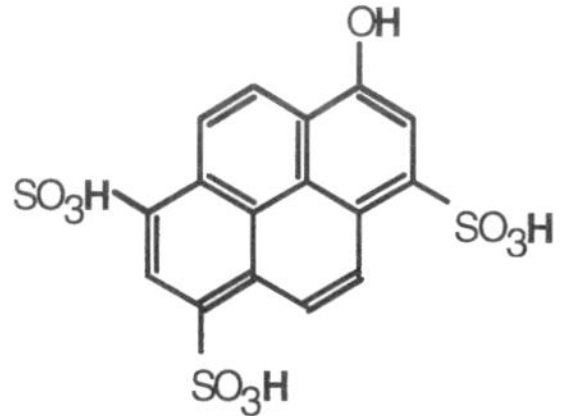

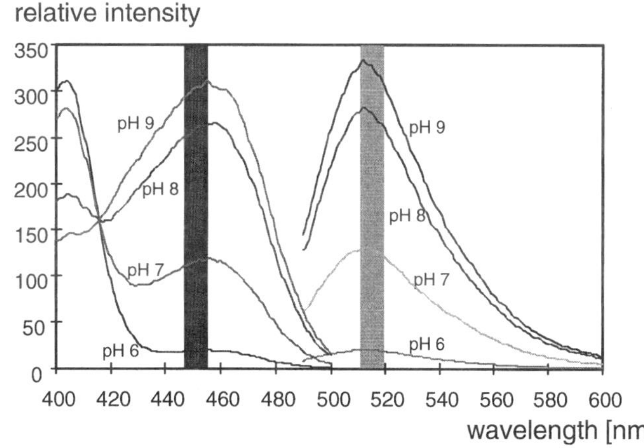

Figure 7.13 Molecular structure and excitation/emission spectra of hydroxypyrenetrisulfonicacid (HPTS) at different pH values. The marked areas indicate the optimum wavelengths for excitation (left) and emission (right).

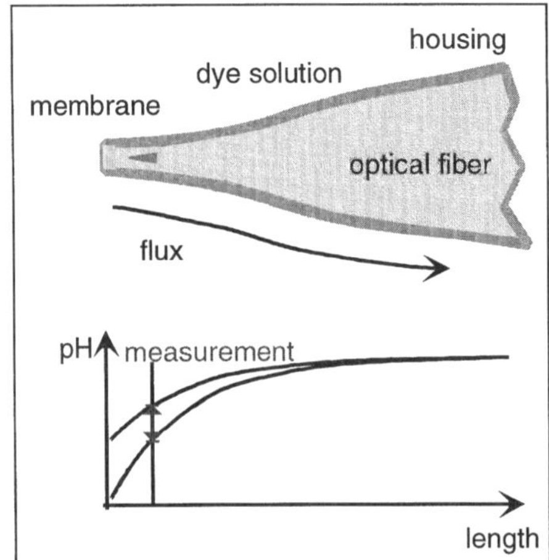

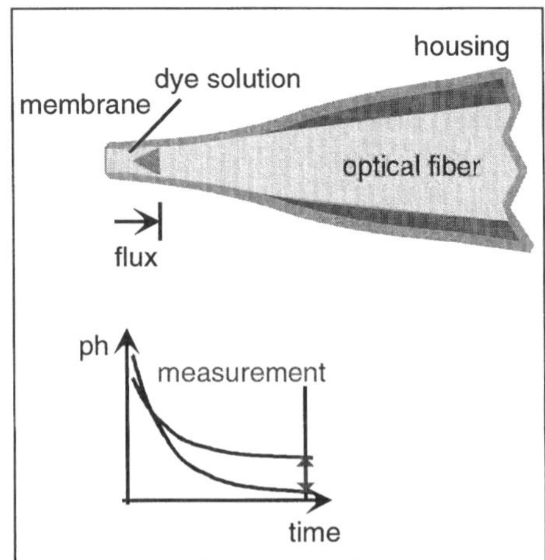

Figure 7.14 Structural pathways for CO_2 microoptodes with (a) a dynamic variation and (b) a static variation of the Severinghaus-principle.

at 515 nm. With increasing pCO_2 HPTS becomes protonated resulting in a decreased fluorescence intensity.

Two pathways using the Severinghaus principle are possible in microscale: Figure 7.14(a) shows a dynamic measuring principle, where an efficient diffusive exchange from the tip environment to a larger buffer reservoir occurs, and Figure 7.14(b) shows a static measuring principle with a semi-closed buffer reservoir in the sensor tip (Kohls and Epping, 1997, unpublished results). The performance of the optode can be "tuned" by the internal dye concentration and the buffering capacity. High dye and buffer concentrations give a high sensor signal and high detection limits. For measurements in the marine environment high CO_2 sensitivity is required (seawater contains only 10–15 μM dissolved CO_2). By using the static type CO_2 optode a sufficient sensitivity with a detection limit < 5 μM CO_2 has been demonstrated (Figure 7.15). The response time of a CO_2 microoptode with a tip diameter below 40 μm is less than 2 minutes – which is fast for a CO_2 sensor but somehow slower than other microoptodes.

7.2.1.6 Temperature Microoptode

Both respiration and photosynthesis of microorganisms are strongly effected by changes in temperature. The determination of the temperature distribution in sediments and microbial mats is therefore of importance to better understand the regulation of these metabolic processes. Furthermore, if chemical parameters are measured with microsensors, the presence of temperature gradients would cause errors in the measurement due to the lack of temperature compensation of the measuring signal. Obviously, there is interest in microsensors which allow the measurement of temperature to the same spatial resolution as chemical micro-

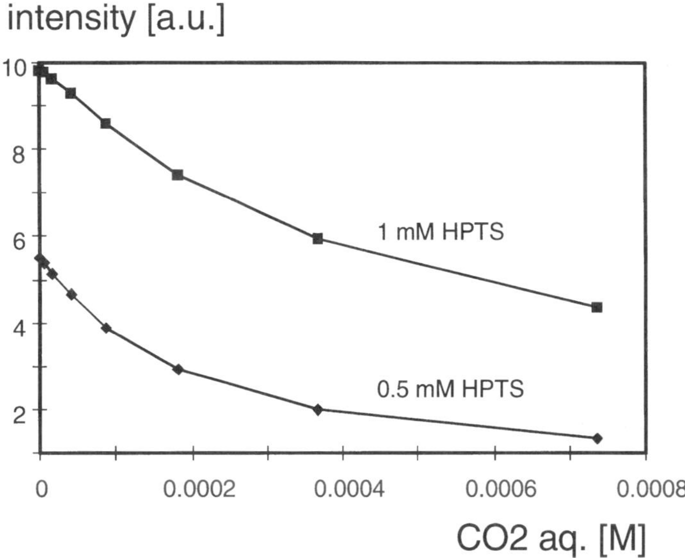

Figure 7.15 Calibration curves for the CO_2 microoptode with high (squares) and low (diamonds) dye-concentration/buffering capacity.

sensors. A potential strategy for the design of temperature microoptodes is the measurement of the temperature dependent luminescence of luminophores with long decay times. The excited state of certain luminophores is thermally deactivated, resulting in a correlation between temperature and luminescence lifetime. Certain solid materials (such as Cr^{3+}-doped crystalline materials) or transition metal complexes are potentially useful for optical temperature sensing (Zhang *et al.*, 1993; Alcala *et al.*, 1995; Grattan and Zhang, 1995). Demas and De Graff (1992) suggested to use luminescent ruthenium(II)-complexes as temperature indicators because they exhibit a strong temperature dependence of the quantum yield and the luminescent lifetime. Recently new temperature microoptodes based on this concept were presented (Holst *et al.*, 1997; Klimant *et al.*, 1997). The ruthenium(II)-tris-1,10-phenanthroline complex ($Ru[phen]_3$) was selected as most promising indicator for designing new temperature microoptodes, since its luminescence lifetime is strongly effected by temperature in the relevant range between 0 and 50 °C (appr. 2.5%/°C). Three different concepts to design temperature microoptodes were realised, which correspond to the microoptode types that are shown in Figure 7.4(a), (c) and (d).

The design of sensor type A (Figure 7.4) is identical to the described oxygen microoptodes and is based on tapered optical fibres which are dip-coated with a thin temperature sensitive film. The sensing film contains the $Ru[phen]_3$-complex (Figure 7.16, ind 1) embedded in a matrix where oxygen quenching is minimised. The most favourable matrices for immobilisation are PVC and sol-gel

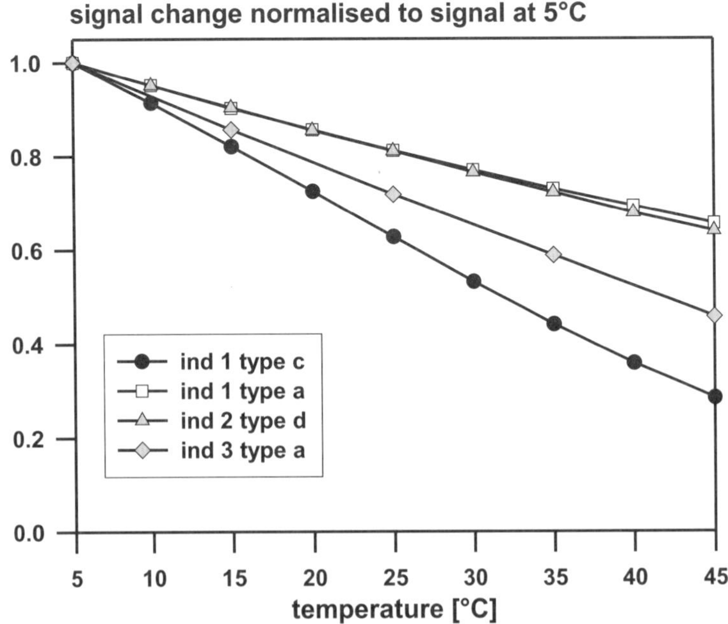

Figure 7.16 Temperature calibration curves of 4 different temperature microoptodes (see legend box and text for explanation) in the range of 5–45 °C. The signal of each sensor has been normalised to the value at 5 °C.

glasses with a very low oxygen permeability. Microoptodes of type [a] (Figure 7.4) allow the miniaturisation down to 10 μm by using the preparation process for dip-coated microoptodes (see section 7.3.2). Unfortunately, the ruthenium complex embedded in a solid matrix (Figure 7.16, and ind 1 type [a]), shows a temperature coefficient that is a factor of 2–3x lower compared to the aqueous solution (Figure 7.16, ind 1 type [c]). The advantage of type [a] sensors (Figure 7.4) are their very simple preparation.

The type [c] sensor (Figure 7.4) consists of a tapered and sealed glass capillary, which is filled with a deaerated solution of the temperature indicator (Figure 7.16, ind 1 type [c]). A tapered optical fibre is then introduced into the tip of the glass capillary to measure the luminescence lifetime (Figure 7.4c) in the small volume that is illuminated by the fibre tip. The internal solution contains $Ru[phen]_3$ (Figure 7.16, ind 1) and an excess of sodium dodecylsulfate or polystyrene sulfonic acid to enhance the luminescence lifetime and the signal intensity. Sodium sulphite was added as oxygen scavenger because oxygen is (as described in section 7.2.1.3) an efficient quencher of luminescence. The outstanding feature of this microoptode is its very high temperature coefficient which makes it suitable for high resolution temperature sensing (Figure 7.16, ind 1 type [c], 7.17).

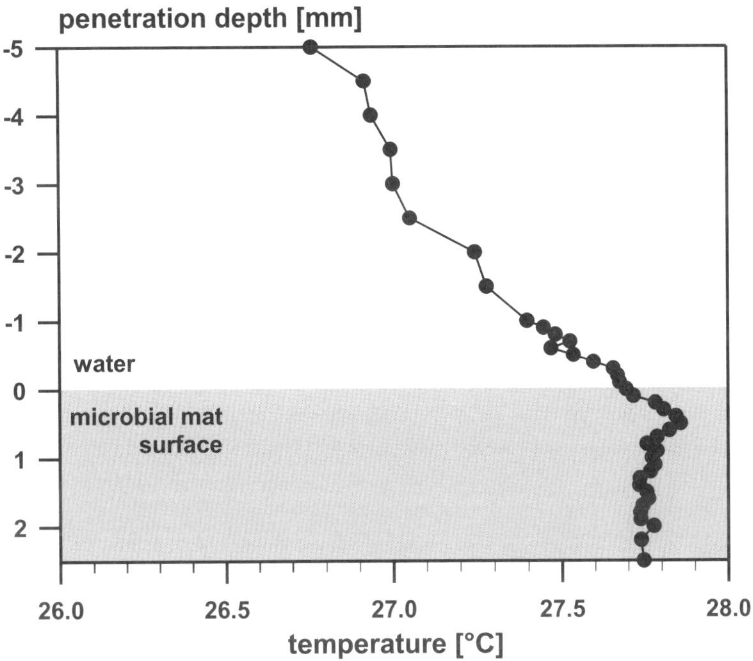

Figure 7.17 Temperature depth profile measured with a temperature microoptode type [c] in a microbial mat from a hypersaline lake (Solar Lake, Egypt) under light condition.

Type [d] sensors (Figure 7.4) use the cross sensitivity of temperature on the response of an oxygen microoptode. The sensor consists of an oxygen microoptode which is located in a sealed glass capillary which is filled with air. Therefore, the oxygen partial pressure is kept constant during the measurement. Temperature has two well-known effects on the sensor response. First, the excited state of the luminescence indicator is thermally deactivated, resulting in a decrease of the luminescence lifetime with increasing temperature. Second, an increase in temperature results in an increase of the quenching efficiency and therefore decreases the luminescence lifetime. This is caused by the increase in oxygen permeability of the polymer with increasing temperature. Type [d] sensors (Figure 7.4) allow the use of luminescence indicators having longer lifetimes since, due to the constant oxygen concentration, the oxygen does not need to be excluded. Therefore, phosphorescent platinum(II)-porphyrins (Figure 7.16, ind 3 type [d]) which have luminescence lifetimes in range up to 100 μs have been selected as suitable.

Sensors of type [a] (Figure 7.4) show the fastest temperature response (a few milliseconds) allowing real time measurements. Type [d] sensors (Figure 7.4) exhibit the slowest response since there is a kinetic limitation to the exchange of oxygen between the sensing layer and the gas phase with changing temperature.

Therefore, the response is limited by the response towards oxygen, which is in the order of 100 ms up to a few seconds, depending on the immobilisation matrix used and the thickness of the sensing layer. Since the luminescence lifetimes of the temperature microsensors are in the same order of magnitude than these of oxygen microoptodes, the same instrumentation may be used. An example of a measured temperature gradient in a dense light absorbing microbial mat with a temperature microsensor of type [c] is given in Figure 7.17. A significant increase of temperature in the biological system was observed during illumination with the visible light of a halogen lamp.

7.2.2 Optical microprobes

While fibre-microsensors for physico-chemical variables (microoptodes) rely on reversible changes of an optical indicator immobilised at the tapered tip of an optical fibre, optical microprobes transmit the optical information from the surroundings of the optical fibre tip direct to the detector. Spectra can be acquired by coupling the microprobe fibre to a spectrum analyser, i.e. a spectrograph coupled to diode arrays and CCD's (e.g. Kühl and Jørgensen, 1992) or, alternatively, a scanning monochromator coupled to a photomultiplier or photodiode (e.g. Vogelmann and Björn, 1984; Garcia-Pichel, 1995; Jørgensen and Des Marais, 1986). Specific spectral information within a given wavelength region, e.g. the light available for photosynthesis (400–700 nm), can be obtained by connecting the microprobe to a photomultiplier or a photodiode, where the detector response has been modified with band pass and/or colour compensation filters (Kühl *et al.*, 1997). The actual spectral range depends on the characteristics of the light detector system and on the optical properties of the fibre material. Currently, microprobes and measuring systems are available for light measurements from UV to NIR.

The following section describes fibre-optic microprobes for light measurements in studies of photosynthesis and sediments. More details about definitions, construction of microprobes and their application can be found elsewhere (Kühl and Jørgensen, 1994; Kühl *et al.*, 1994a).

7.2.2.1 Field Radiance Microprobe

The first fibre-optic microprobes were developed for measuring the radiation flux incident from a specific direction and solid angle in biofilms and sediments (Jørgensen and Des Marais, 1986) as well as in leafs (Vogelman and Björn, 1984). This parameter is the field radiance, which is the fundamental optical parameter for describing light fields. Field radiance microprobes can easily be made from a single strand optical fibre with a flat cut light collecting end, which can be tapered down to even sub-micron diameters (Figure 7.18[a]). In most practical applications in aquatic environments, microprobes with 10–50 μm tip diameter are used. For microscale light measurements, it is important to coat the tapered sides of the

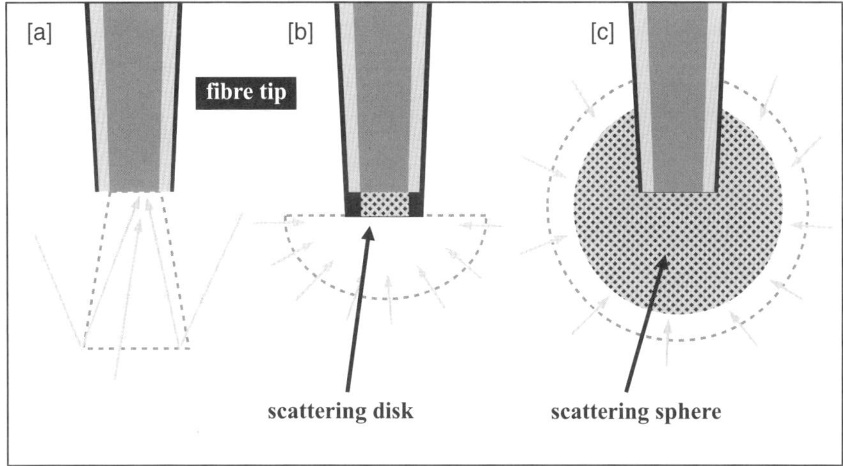

Figure 7.18 Schematic cross sectional overview on different microprobes: [A] radiance microprobe and sensor for surface detection; [B] irradiance with attached scattering microdisk; and [C] scalar irradiance microprobe with attached scattering microsphere. For radiance and irradiance measurements the fibre tips are Aluminium coated.

fibre tip with a metal film or some enamel paint in order to avoid light leakage into the fibre via the taper.

Field radiance microprobes can be used for high spatial resolution measurements of radiance spectra in sediments and biofilms (Kühl and Jørgensen, 1992). This enables studies of the zonations of differently pigmented phototrophs often found in these systems. It is also possible to monitor the migration of phototrophs by such measurements, e.g. by following the change in spectral composition of backscattered and reflected radiance (Kühl *et al.*, 1994a). Furthermore, by measuring depth profiles of field radiance at different angles relative to the incident light, it is possible to separate collimated and scattered light fluxes in sediments and biofilms (Figure 7.20). From such measurements of angular radiance distribution it is possible, after some mathematical treatment of the measured data (Kühl and Jørgensen, 1994; Fukshansky-Kazarinova *et al.*, 1997, 1998), to extract information of the inherent optical properties of the investigated sediment or biofilm. Thus, detailed information about light fields and optical properties of sediments and biofilms can be gained from the extensive application of field radiance microprobes.

The most recent application of field radiance microprobes involves microscale measurements of fluorescence. For such measurements, the microprobe is connected to the excitation source and the detector via a similar set-up as used for microoptodes (see Figure 7.26). By appropriate choice of excitation wavelength, excitation and emission filters, it is possible to monitor the microscale distribution of photopigments, such as chlorophyll *a*, in sediments and biofilms (M. Kühl

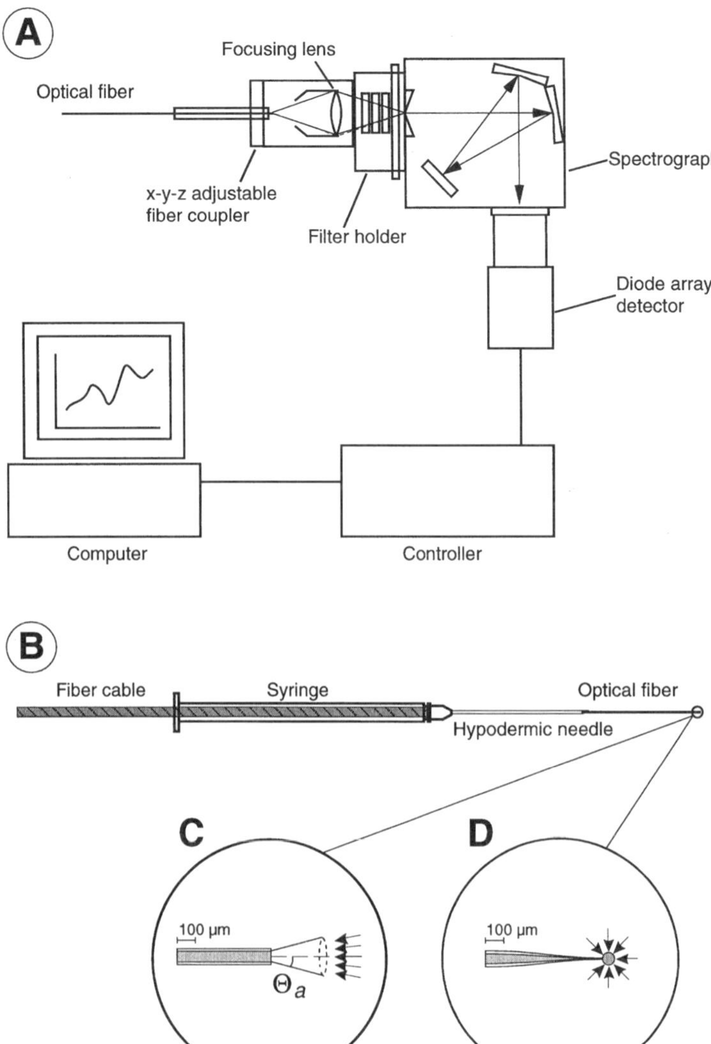

Figure 7.19 Example of optoelectronical set-up for fibre-optic microprobes and types of microprobes: A. Intensified diode array detector system; B. Basic design of a fibre-optic microprobe; C. Tip of microprobe for measuring field radiance. The fibre tip can also be tapered to smaller diameters; and D. Spherical tip of a microprobe for measuring scalar irradiance (from Kühl and Jørgensen, 1992).

unpublished results). Furthermore, by applying a more sophisticated measuring system, it is possible to measure photosynthetic electron transport via fluorescence measurements with a radiance microprobe (Schreiber *et al.*, 1996).

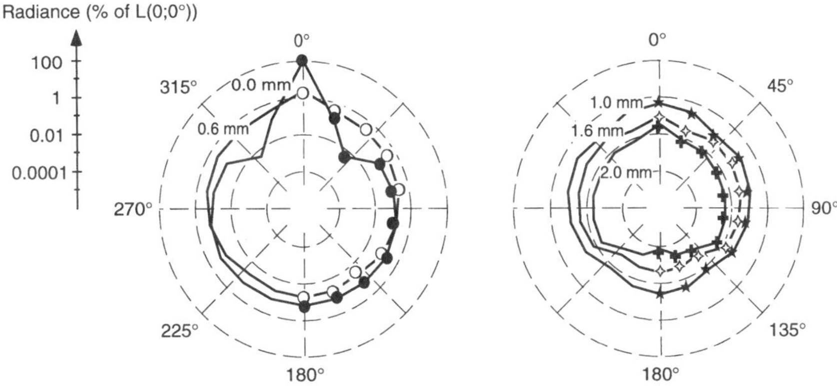

Figure 7.20 Angular radiance distributions of 650 nm light measured with a fibre-optic microprobe in a coastal sediment. Numbers on curves indicate depth below the sediment surface. Symbols indicate measured values expressed on a log-scale in % of the vertically incident radiance (0°) at the sediment surface. From such data sets, the optical properties of sediments can be calculated (from Kühl *et al.*, 1994a).

7.2.2.2 Irradiance and Scalar Irradiance Microprobe

While detailed information about the light field in sediments and biofilms can be gained from numerous measurements with field radiance microprobes, they are not applicable for routine measurements of total available light for e.g. photosynthesis. Sediments and biofilms are strongly scattering and absorbing systems, where the photosynthetic microbes live in a diffuse light field (Figure 7.20) and receive significant amounts of light from all directions around them. Measurements with radiance microprobes in a certain direction, therefore, only collect a fraction of the available light. The total light flux from all directions can be estimated by integration of angular radiance distribution data. This is very time consuming and not easily applicable to photosynthesis studies of sediments and biofilms. Alternatively, the light collecting properties of the optical fibre tip can be modified to integrate the radiant flux over certain directions (Figure 7.18).

The integral of field radiance incident on a horizontal plane from an upper or lower hemisphere is called the downwelling and upwelling irradiance, respectively. A microprobe for irradiance can be manufactured by fixing a small disk of light scattering material at the end of a tapered fibre, and such a microprobe acts as an almost ideal cosine collector, i.e. an ideal irradiance meter (Lassen *et al.*, 1994; Kühl *et al.*, 1994b, Figure 7.18[b]).

Irradiance is a common measure of light intensity and is often used for estimating light availability in the aquatic sciences. However, in turbid waters and especially in sediments, biofilms, plant and animal tissue and other optically dense media irradiance measurements are not appropriate for quantifying the total available light. A better measure is the integral of radiance from all directions about a point, i.e. the scalar irradiance. The scalar irradiance is the most relevant light field parameter for measuring total light availability in scattering.

Scalar irradiance microprobes can be made by immobilising a small light scattering sphere (70–150 μm diameter) at the tapered end of an optical fibre (Figure 7.18[c]). The sphere efficiently scatters incident light, and this leads to the same probability for incident photons (irrespective of their angle of incidence) to be channelled into the fibre tip and, subsequently, to the detector. A microprobe with such an isotropic light collection thus measures scalar irradiance. The sphere can either be formed by dip coating of the fibre tip in a TiO_2-doped methacrylate (Lassen *et al.*, 1992) or by forming a vitro-ceramic sphere by careful heating of a fibre tip, which is drawn to a thin filament and dipped in MgO (Garcia-Pichel, 1995).

Scalar irradiance microprobes can be used to obtain information about the microscale light availability in photosynthetic sediments and biofilms, e.g. by measurements of attenuation spectra (Figure 7.21) and depth profiles of light penetration (Figure 7.22b). Such measurements can then be combined with

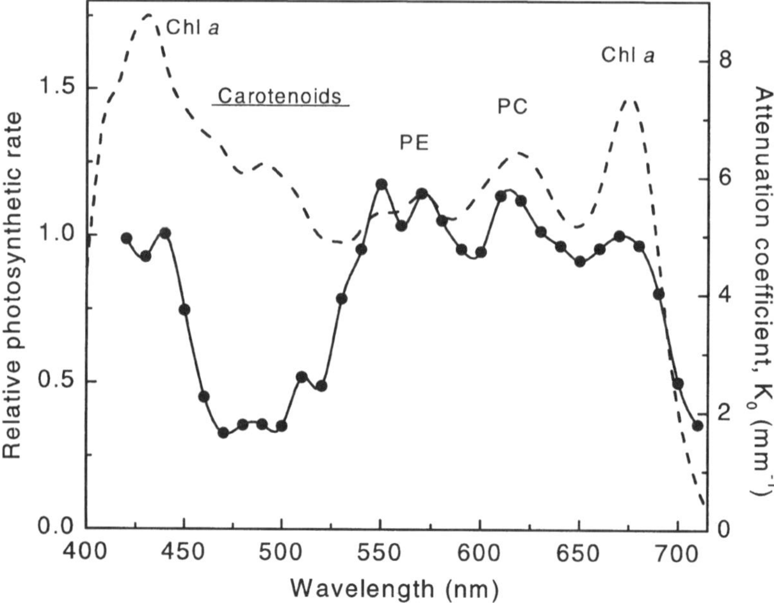

Figure 7.21 Attenuation spectrum of scalar irradiance (dashed curve) measured with a fibre-optic microprobe over the the upper 0.1 mm of a photosynthetic biofilm. Scalar irradiance measurements were also combined with measurements of photosynthesis with an oxygen microsensor to measure the action spectrum of photosynthesis. Absorption regions of Chorophyll *a*, carotenoids, and phycobiliproteins (phycocyanin, PC, and phycoerythrin, PE) are indicated on the Figure (from Kühl *et al.*, 1996).

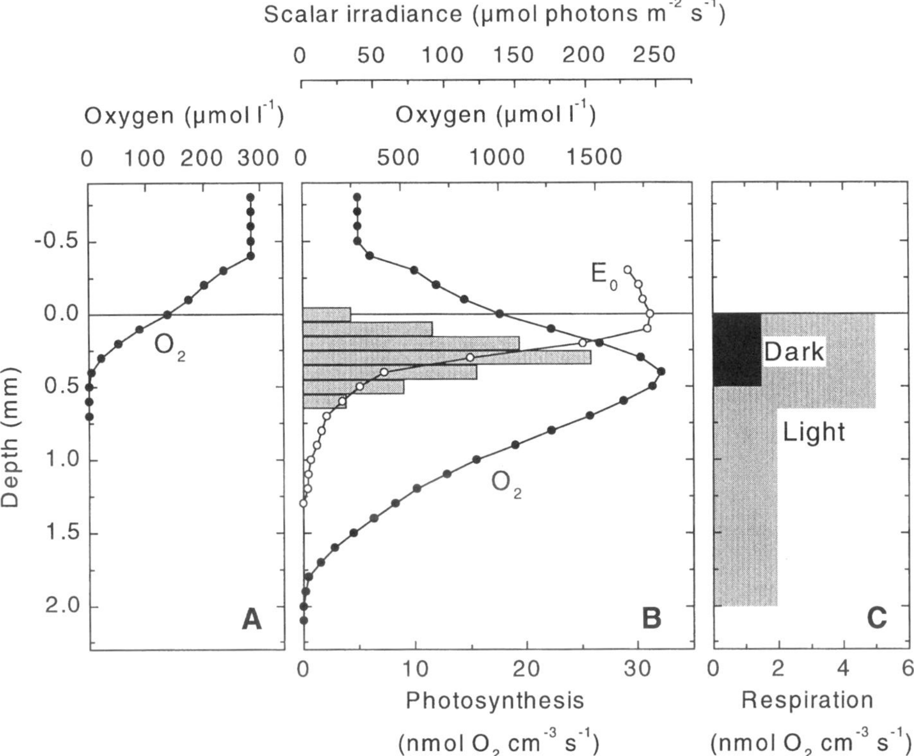

Figure 7.22 Oxygen, photosynthesis and 400–700 nm light penetration measured with microsensors for oxygen and scalar irradiance in a photosynthetic biofilm: A. In the dark, oxygen only penetrates 0.5 mm; B. In the light, photosynthesis (indicated with bars in the graph) leads to supersaturation of the biofilm with oxygen. Photosynthesis is confined to the upper 0.5 mm of the biofilm due to strong attenuation of scalar irradiance, E_0, by the densely populated biofilm; and C. Calculated respiration rates within the biofilm in dark and light. The respiration is stimulated in the light.

microsensor measurements of photosynthesis and chemical variables to obtain information about the light regulation of photosynthesis and respiration in sediments and biofilms (Figures 7.21, 7.22; Kühl *et al.*, 1996; 1997; Garcia-Pichel and Bebout, 1996).

7.2.2.3 Microprobe for Surface Detection

An important prerequisite for obtaining useful information from microsensor measurements (e.g. calculation of diffusive boundary layer thickness, reaction rates and fluxes), is the ability to determine the exact position of the sensor tip relative to the interface between the biogeochemical system and the overlying water (i.e. the water-sediment interface). In the laboratory, penetration of the

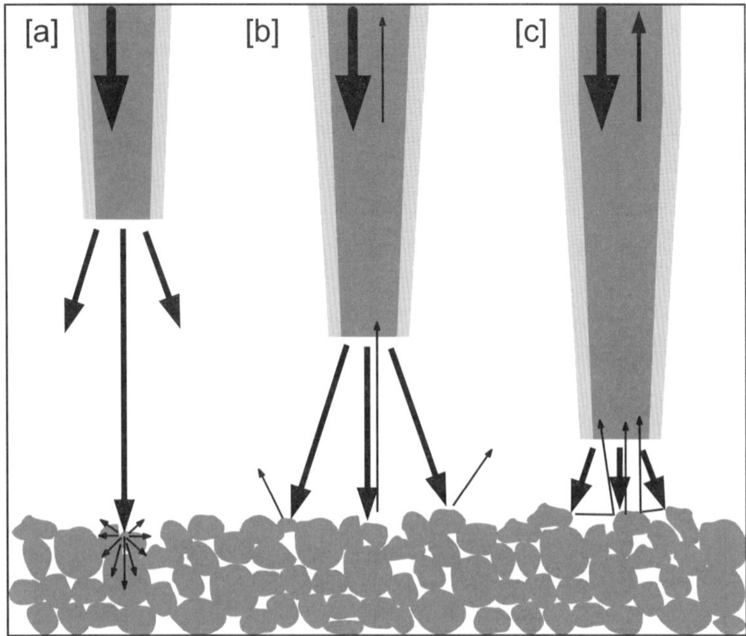

Figure 7.23 Principle of sediment surface detection (schematic drawing): (a) The NIR light (downward pointing black arrows) emitting fibre taper is far away from the surface, the light that reaches the surface is reflected and scattered (biggest part is forward scattered) by the sand grains of the sediment surface, but no light is captured by the microsensor; (b) The fibre tip approaches the surface and first amounts of the reflected and backscattered light from the surface are captured by the fibre (upward pointing black arrows); and (c) The fibre is near to touch the surface, before the amount of reflected and backscattered light depends on single grains and the pore space more light is captured by the sensor tip (more upward pointing black arrows, compared to b).

sensor tip into a sediment or biofilm is often determined visually using for instance a dissection microscope. However, visual inspection can be imprecise or even false – especially if the surface is characterised by a high level of heterogeneity or the overlying water layer is turbid. During *in-situ* application of microsensors visual determination of the surface is generally impossible and is usually estimated from measurements with relatively large resistivity probes (Andrews and Bennett, 1981; Tengberg *et al.*, 1995).

Reflection based optical methods for surface detection and positioning control have found many industrial applications (Dakin and Culshaw, 1988). Such sensors illuminate an object surface and measure the level of reflected and backscattered light via the same optical fibre.

Recently, this method was adapted in the design of a very simple optical measuring system for detecting the sediment-water interface with a high spatial

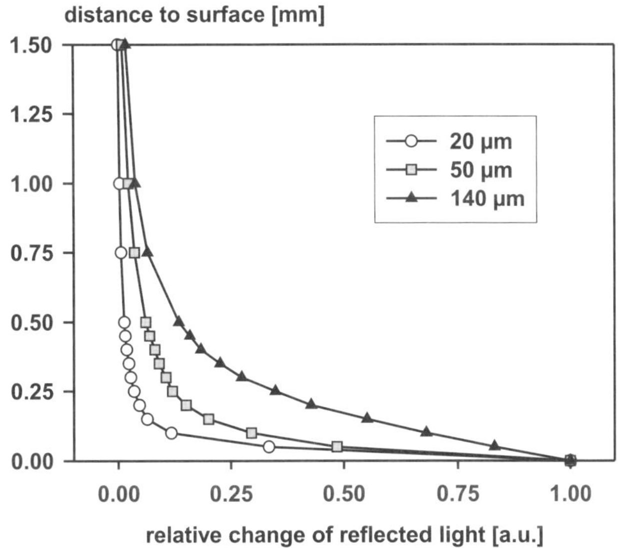

Figure 7.24 Results of surface detection measurements obtained by surface detection microprobes with different tip diameters (see legend box).

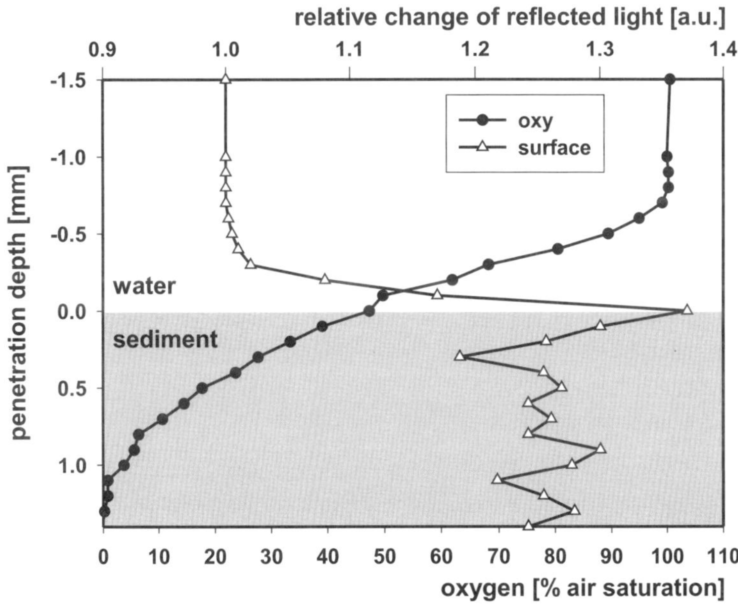

Figure 7.25 Oxygen depth profile and surface detection measurement of a microsensor combination made of an oxygen microoptode (oxy) and a surface detection microprobe (surface).

resolution by use of a simple fiber-optic microprobe (Klimant *et al.*, 1998). The microprobe is a tapered optical fibre with a tip diameter of 20–40 µm, terminated with a standard ST-fibre connector. The optical system is a simplified version of a set-up developed for optical oxygen microsensors (Figure 7.28, Figure 7.26). A NIR laserdiode was selected as light source ($\lambda = 780$ nm), since at this wavelength the intrinsic light absorption of sediments is low and no stimulation of photosynthesis occurs. The laser light was amplitude-modulated to separate the signal from ambient light. Due to the efficient coupling of the laser diode light into the fibre, a simple photodiode was used to measure the reflected light with a sufficient signal-to-noise-ratio. A 2×2 fibre coupler (see section 7.3.1.2) was used to minimise contributions from reflected light generated in the optical system.

The operating principle of the surface detection system is simple. When the microprobe tip is moved towards the water-sediment interface (Figure 7.23, [a]–[c]), an increasing amount of backscattered and reflected light from the sediment surface is captured by the fibre (Figure 7.23, smaller black arrows). The highest relative signal change occurs, when the fibre tip reaches the water-sediment interface (Figures 7.24, 7.25) causing a significant shift in the scattering properties.

The spatial resolution of the surface detection is related to the diameter of the fibre tip. The precision of the measurement increases with decreasing sensor size. With a 20 µm sensor the surface of a highly scattering surface may be determined with a resolution better than 50 µm (Figure 7.24). It is significant that miniaturisation not only improves the spatial resolution of the measurement but also results in a strong increase of the signal intensity. The shape of the sensor response curve not only depends on the sensor geometry but also on the optical properties of the sediment surface itself, i.e. refractive index, size and shape of the sediment particles. Best results were obtained in fine or silty sediments, whereas the use of the surface detection sensors in sandy sediments with larger grains (< 100 µm) was complicated (Klimant *et al.*, 1998).

The surface detection system can be combined with other microsensors, simply by fixing the tapered fibre to a microelectrode or microoptode. Figure 7.25 shows the respective depth profiles measured with such a combined sensor in a sandy North Sea sediment. The results show a good correspondence of the optically determined surface position and the surface position one would infer from the oxygen measurements.

An important field of application, which can be envisioned for the surface detection system in combination with microsensors, is for *in situ* measurements with benthic profiling landers, where visual inspection of surface penetration is not possible.

7.2.3 Submicron or Nano Sensors

The smallest fibre optical chemical sensors (so-called sub-micron or nanooptodes) were developed by Tan *et al.* (1992) and Rosenzweig and Kopelman (1995) by adapting a technique which was already successfully used in near field microscopy. Silica single mode fibres with a core diameter of a few micrometers and a cladding diameter of 125 μm were tapered in a CO_2 laser beam to extremely thin tips. The diameter of the fibre tip and the form of the taper is controlled by the pulling velocity during the heating process and the heat itself via the intensity of the laser beam. Fibre tips down to 100 nm were produced with this technique. Then, the taper was coated with a reflective aluminium layer to ensure that the excitation light leaves the nanosensor only via its tip. The appropriate indicator was immobilised using a procedure described by Munkholm *et al.* (1986), where the fluorescent indicator is covalently cross linked in an UV curing polyacryl amid matrix using intense UV-illumination via the optical fibre itself. In this way, Kopelman *et al.* have developed fibre optic microsensors for the measurement of various parameters such as pH, O_2, glucose and ionic species like Ca^{2+}, potassium or nitrate. The pH and Ca^{2+} nanosensors, in particular, are very powerful tools for intracellular measurements (Barker *et al.*, 1998; Rosenzweig and Kopelman, 1995; Shortreed *et al.*, 1996a, b; Tan *et al.*, 1992).

The experimental set-up for working with sub-micron optodes consists of the following key components: a laser light source (for efficient coupling of excitation light into the single mode fibre), the coated tapered fibre with the sensing layer, a microscope and a photomultiplier tube. The optical fibre only acts as carrier for the excitation light (see Figure 7.26c). The fluorescence signal is collected by the lens optic of a microscope and guided to the detector. Therefore, applications are limited to samples which are almost clear and transparent and can only be performed on a microscope table. Significant losses may be caused by scattering or intrinsic absorption in the sample itself. The major potential of submicron optodes are intracellular measurements (Shortreed *et al.*, 1996b), where nano optodes are a real alternative to fluorescence microscopy techniques. Due to their small size, nano optodes can penetrate the cell wall without destroying the cell. Furthermore, injection of a luminophore, which may interfere with the cellular processes, can be avoided.

7.3 MEASURING SYSTEMS FOR OPTICAL MICROSENSORS AND MICROPROBES

In general, light of a specific light wavelength has to be guided to the fibre tip and interact in the sensing layer within an analyte. The modified or generated light has to be returned up the same fibre to be detected and further processed. Optical microsensors and microprobes are based on a single optical fibre, which serves as light guide in one or both directions. Therefore, the essential elements of the optical set-up are the optical coupling, the light source, and the detector. Once the electrooptical set-up is defined, the measuring method with appropriate evaluation schemes determines the measuring system with its resolution and precision.

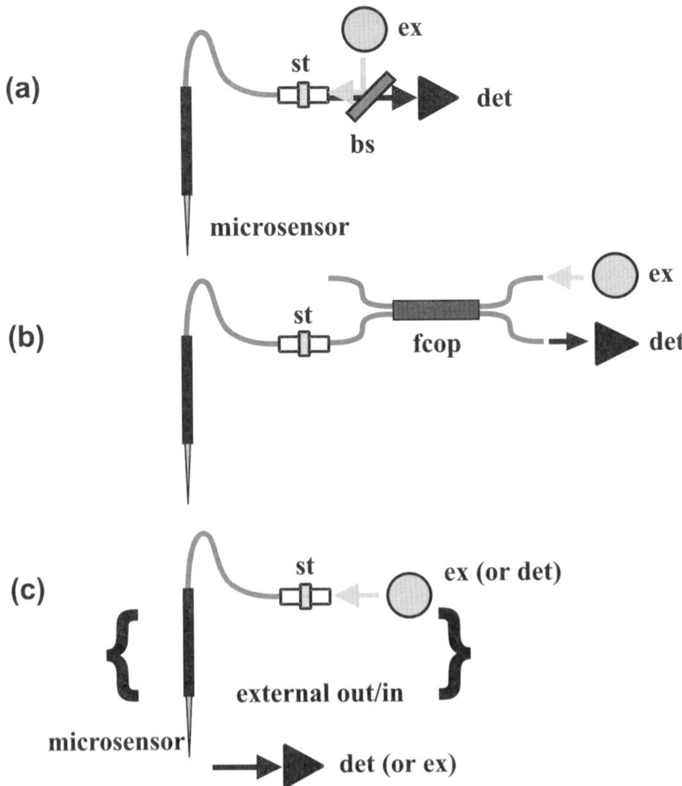

Figure 7.26 Schematic comparison of different electrooptical set-ups for optical microsensors: (a) bs – beamsplitter, det – photo detector, ex – excitation light source, st – standard fibre connectors; (b) fcop – 2 × 2 fibre coupler, det – photo detector, ex – excitation light source, st – standard fibre connectors; and (c) external in or out-coupling of light signal, means either (as shown) external detection, det – photo detector, ex – excitation light source, st – standard fibre connectors, or (not shown) external in-coupling.

7.3.1 Electrooptical Set-up

7.3.1.1 Light Sources

Light sources for optical microsensors range from natural sources, like sunlight (for spectrographic investigations), bio- and chemo-luminescence through to artificial sources: flash lamps, arc lamps, halogen lamps, lasers, laser diodes (LD) and light emitting diodes (LED). Because of their broad spectra and high light output, lamps are generally used in spectrographic measuring instruments where monocromators or variable optical filters select specific wavelengths that should be used for the measurement. In contrast, in most microsensor systems lasers, LD's and LED's are used. This is due to either the high light output power and the inherent wavelength selectivity of lasers and LD's, or the low price and simplicity of LED's. Developments within the last few years have further improved the quality of the LED's especially concerning light output – even towards the blue end of the spectrum. Furthermore, LED's are long-term stable and have smaller power requirements than lasers or laser diodes. Because all semiconductor light sources are primarily used in the (optical) telecommunications systems, there is a vast range of standard fixings and connectors for coupling light in optical fibres. So, adaption or modification for specific measuring system is easy.

In optical microprobe applications, sunlight (or a lamp with a broad wavelength range) is used if whole spectra need to be measured, while a near infra-red (NIR) LD is used for surface detection. The measuring systems for microoptodes use LED's and appropriate optical filters as light sources. Both LD's and LED's are mounted using standard receptacles to facilitate interconnection. If the coupling efficiency needs to be increased additional lenses, such as ball or gradient refraction index lenses, can be used to image the light generating chip of the LD or LED to the fibre core. Alternatively the chips can be placed as near as possible to the fibre or they can be purchased with a fibre already attached, so-called "pig-tail" solutions.

7.3.1.2 Couplers

The first coupling elements were standard devices from the telecommunication industry: melting and grinded couplers (Figure 7.26b). In the first, the cladding of two fibres is partly removed, the fibres are twisted together, softly melted, stabilised with a polymer and, for mechanical robustness, fixed in a (steel tube or plastic) casing. Most of these are 2 × 2 couplers. In the second, the fibres are cut, ground to a fixed angle and then glued together, resulting in two half fibres sharing a common diameter. This glued assembly is fixed to a single fibre, which results in a 1 × 2 coupler. In newer couplers, light is transmitted more efficiently using gradient refraction index lenses (GRIN). All fibre-based couplers (Figure 7.26b) have a high directivity and the light distribution of the output branches can be manipulated to a certain extent (splits of 10:90 or 50:50, for example). Commercially available fibre couplers are supplied with connectors, do not need additional adjustments, they are are rugged, and they do not exhibit any change in performance over time. Despite their high quality and low price, fibre optics sometimes generate certain problems in sensing applications

(1) Availability. If a special type of silica fibre with a non-standard diameter is required, it might either be impossible (or very expensive) to get an appropriate coupler.
(2) Background noise. As the telecommunications only works with three wavelengths, 850 nm, 1.3 μm and 1.55 μm, all fibre optics are usually optimised for use at these wavelengths. When applied to luminescence detection, the blue excitation light source also excites many of the epoxy glues that are used for

fixing the coupler in the casing and the white dyes used in ceramics of the connector ferrule. This causes additional and unwanted luminescence, that can be larger than the signal. Therefore, couplers have to be carefully chosen.

The second possible coupler is a beam-splitter (Figure 7.26a). This coupler needs a carefully adjusted mechanical set-up to couple the light efficiently. The directivity can be improved by optical surface filters, e.g. the replacement of the common prism by a dichroic mirror. But, this is only possible if input and output wavelengths are sufficiently distinct for these mirrors, which usually have weak optical blocking properties. Additionally, any necessary optical filter must have a high blocking efficiency because they are the only means to prevent the intense excitation wavelengths from reaching the detector. Background luminescence caused by epoxy or white dyes in ceramics is like mentioned above an issue to take care of.

Finally, external coupling (Figure 7.26c) can be treated as an set-up bound "coupler". Here the fibre of the sensor is used as a light guide in one direction only. This either means the transmission of the exciting light (Tan *et al.*, 1992; Rosenzweig and Kopelman, 1995) to the microsensor or the return of captured light to the detector (radiance and irradiance microprobes). The former needs an external collection of light e.g. by a microscope lens and couples the light via a beam-splitter to a detector, while the latter has a broad illumination of the sample, e.g. by sun light or a halogen lamp, to measure the spectral information coming from the specific microsensor tip (Figure 7.18).

Various microoptode measuring systems (Klimant *et al.*, 1995; Holst *et al.*, 1995; Holst *et al.*, 1997; Kohls *et al.*, 1997) have used 2 × 2 melting couplers with silica fibres, similar in size and refraction index to the microoptodes themselves, while recent investigations also give promising results with a beamsplitter. All microprobes with 2 × 2 fibre couplers (except for surface detection and local luminescence measurements) utilise external in-coupling (Figure 7.27).

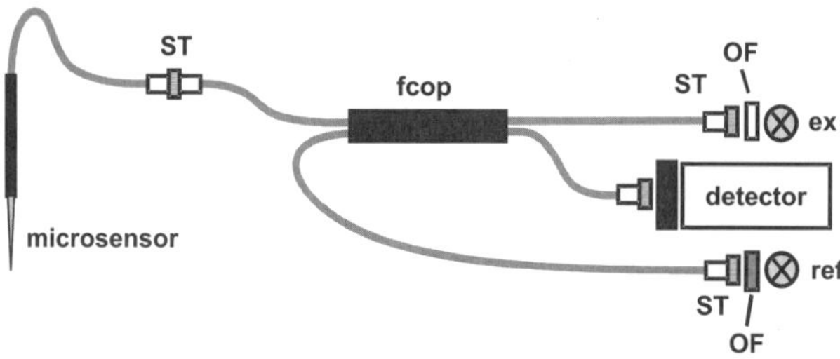

Figure 7.27 Schematic drawing of the electrooptical set-up of the measuring system for oxygen and temperature microoptodes: fcop – 2 × 2 fibre coupler; detector – photomultiplier tube; ex – excitation light source (LED, $\lambda p = 505$ nm); ST – standard fibre connectors; OF – optical filters for excitation, emission and attenuation; ref – reference light source (LED, $\lambda p = 620$ nm).

7.3.1.3 Detectors

The detectors for light captured by the tip of an optical microsensor have to be very sensitive, because the amount of light is very small. The most sensitive detectors are photomultiplier tubes (PMT sensitivity in terms of electrical output signal related to optical input signal). Based on photo-induced electron amplification, they are highly sensitive in the ultraviolet and visible spectral range but can only be used up to a maximum of 800 nm. Although currently available PMT-assemblies contain on-board high voltage power supply and current-to-voltage con- version circuits, they are expensive, can be destroyed too much light, highly temperature dependent, and have a distinctive noise behaviour due to the ampli- fication process. The second commonly used detectors are semiconductor avalanche photodiodes (APD), which are usefully sensitive from 500–900 nm. They are fast, need to be thermostated and have a noise behaviour like a PMT but with a lower sensitivity in the visible range. Finally, there are photodiodes, which are low in noise and temperature dependency, sensitive from 500–900 nm, slower in response time compared to an APD and cheap. APD's have better sensitivities than photodiodes due to the avalanche amplification process. PD-arrays or charge coupled devices, that are often used for spectrometers, principally show a behaviour comparable to PDs.

In nearly all optical microsensor applications PMTs have been used as detectors because the light power usually is in the range of tenth of pW. Nevertheless, improvements in sensor chemistry, optical set-ups and light sources will see PDs being used more in the future. This is preferable because of their better noise behaviour, higher stability and (not least) their low cost. The exception is given by the described surface detection, that already uses PDs as detectors.

7.3.2 Light Intensity Based Measuring Methods and Systems

In spectrographic measurements the collected light is measured in relation to its wavelength. The light captured by the microsensor tip is guided through the fibre, passes either a variable optical filter or a monochromator and reaches a sensitive detector like a PMT, or reaches an optical grating where the light is diffracted across an array of photodiodes or a line CCD-chip. In all these applications the measured spectra are compared to a reference spectrum obtained before or after from the excitation light source. This light source has to be very stable because any spectrally varying influence along the light path (extreme bending of the fibre, for instance) can not be referenced. In addition, external light "noise" would have to be measured separately with an identical detection path to enable sample beam referencing, or the light source and the detection would have to be modulated to separate the spectra from ambient light. For these reasons, most of these measurements are performed under defined and exact light conditions.

Another class of sensing applications are where a change in absorption spectra (or luminescence spectra of an immobilised indicator molecule) characterises the sensing event. To employ these chemical sensors as microsensors it is necessary to make them insensitive towards indirect environmental changes e.g. change in ambient light, change of external absorption or refraction index, bending of the fibre etc.

There are two possible strategies to reduce problems due to ambient or stray light: time-multiplex beam referencing (measurement with and without the light signal and subtracting to resolve the intensity information of interest) and amplitude modulation of the signal (sinusoidal or squarewave, and appropriate electronic filtering). The latter can be of advantage because it reduces the necessary bandwidth, and therefore the noise, of the measuring system. Extra care is needed to ensure the stability of the amplitude of the light signal that excites (or is absorbed by) the indicator at the microsensor tip.

For absolute intensity measurements only ratioing can reference changes in the optical environment that may potentially influence the light guiding property of the microsensor fibre and the sensor tip. Either light at two different wavelengths is used for excitation and detection (time multiplex with one detector, or two different detectors) or signals of two wavelengths are separately measured with two detectors. Then by ratioing all influences on the transmitted light the absolute intensity can be calculated (assuming that the influences are the same for the two wavelengths). Another solution is the optical separation of the sensing event from the environment, e.g. by an optical insulation of the sensor tip (Klimant *et al.*, 1995) and a fixation of the fibre (lander set-up).

One advantage of intensity-based measuring systems is their relative simplicity (Gruber *et al.*, 1993). The first measuring system for oxygen microoptodes (Klimant *et al.*, 1995) was intensity based with an amplitude modulation (rectangular signal). The light output of the excitation source, a blue LED ($\lambda = 450$ nm), was measured and controlled, and, to be independent of all optical properties of the environment, the microoptodes were optically insulated. Based on this first prototype, a small unit was developed for benthic landers which resulted in the first lander applications in shallow water (Glud *et al.*, 1998) and deep sea (Wenzhöfer *et al.*, 1998).

Surface detection systems (Klimant *et al.*, 1998) also detect light intensity. Here the influence of bending of the fibre is of minor importance as the biggest change in signal is from the surface being detected, unless the fibre is near to fracture. As shown in Figure 7.28 the system consists of a modulated ($f_{mod} = 10$ kHz) excitation light source, a laser diode ($\lambda = 780$ nm), a 2×2 silica fibre coupler (that is used as 2×1 coupler) and a PIN photodiode as detector. The unused branch of the coupler is immersed in oil to reduce unwanted reflection from the fibre end face. The analogue signal, which is displayed and present at the output (Figure 7.28), is directly proportional to the amount of reflected and backscattered light.

The measuring system for pH microoptodes is also intensity based. As the measuring scheme uses absorption based indicators, the measurement relies on the amount of reflected and backscattered excitation light. The excitation light is amplitude modulated ($f_{mod} = 1$ kHz) and, with a fibre optical 1×2 switch, two different LEDs were alternately used for the measurement, at wavelengths

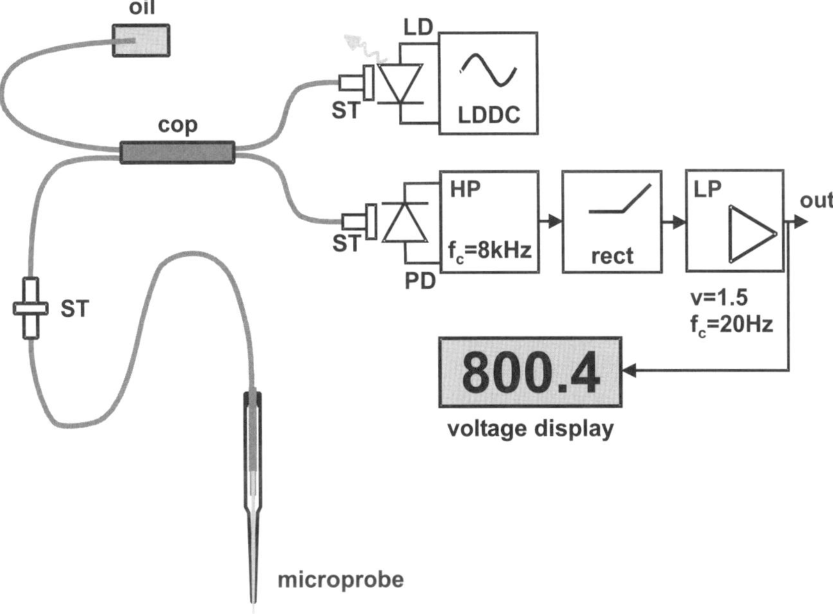

Figure 7.28 Schematic overview of light intensity based measuring system for surface detection: oil – one branch of the fibre coupler is immersed in oil to prevent fibre end face reflection; cop – 2 × 2 fibre coupler; ST – standard fibre connectors; LD – laser diode; LDDC – laser diode driving circuit, controls light output and modulates at 10 kHz; PD – PIN photodiode; HP – electronical highpass filter; fc – cut-off frequency; rect – precision rectifier; v – amplification; out – signal output in Volts.

$\lambda_1 = 470\,\text{nm}$ and $\lambda_2 = 595\,\text{nm}$. By a ratioing method the pH can be evaluated independent of changes in optical pathlength and the optical environment.

7.3.3 Light-time Based Measuring Methods and Systems

Many interactions between light and molecules not only influence absorption or emission properties, but also influence the rates – a time parameter that describes the transition probability between different energy levels. A well known class of molecules include the luminescent O_2 indicators, which by dynamic quenching of their luminescence change both the emitted intensity as well as the time parameter that characterises the decay curve, the luminescence decay or lifetime. This parameter does not depend on the indicator concentration, dye leaching or photo-bleaching (if the effect does not generate compounds that interact with the indicator), and it does not depend on the optical environment or the light path. Therefore, luminescence lifetimes are independent of influences such as: fibre bending, refraction index change, absorption change and so on.

There are two pathways to evaluate the time parameter, a measurement in the time domain: a pulse decay technique, or a measurement in the frequency domain: a phase modulation technique. They are equivalent but each has technical and practical advantages and disadvantages. Figure 7.29 compares the timing of the corresponding excitation (Figure 7.29, ex) and the luminescence emission (Figure 7.29, lum) signal. A luminophore with a single exponential decay curve is assumed.

In the pulse-decay method (Figure 7.29a) the shape of the excitation light signal is a rectangular pulse. The luminescence rises at a constant rate until it reaches a steady state. When the excitation is switched off, the time measurement starts. When the intensity falls below 1/e times the intensity I_0 (Figure 7.29a, the intensity when the excitation reaches zero), the resulting pulse width equals the decay time. The electronic circuits for this method need to be broadband and to avoid additional impedance contributing to the decay curve – which can be a disadvantage concerning noise reduction. Furthermore, the separation of constant light compounds, that can be very high due to incoupled ambient light, is very difficult. Ambient effects need to be subtracted from the decay time and very early in the amplification stages of a measuring system. There are other techniques (Lippitsch *et al.*, 1988; Zhang *et al.*, 1993) to evaluate the luminescence lifetime in the time-domain, but pulse excitation and level crossing (or integration) is common, for instance.

The phase modulation technique (Figure 7.29b) measures the luminescence lifetime in the frequency domain (Lakowicz, 1983; Gratton and Limkeman, 1984; Berndt and Lakowicz, 1992; Alcala *et al.*, 1995; Gruber *et al.*, 1995; Holst *et al.*, 1995). For that purpose the excitation light is sinusoidally modulated at a modulation frequency f_{mod}. The reaction is an emission of sinusoidal light, that is weaker in intensity and delayed compared to the excitation. This delay is called

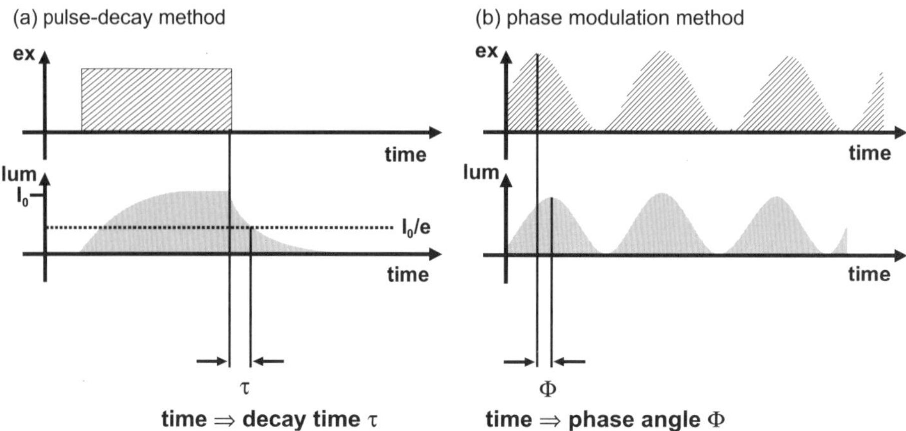

Figure 7.29 Timing relations of excitation and luminescence light signals for pulse-decay and phase modulation measurements.

phase angle Φ between two sinusoidal signals and the relation to the luminescence lifetime τ is given (Lakowicz, 1983):

$$\tan(\Phi) = 2\pi \cdot f_{mod} \cdot \tau$$

If the two lifetimes τ_1 and τ_2, that describe the boundaries of the measuring range for the analyte, are known, the optimum modulation frequency f_{opt} can be derived to (Holst *et al.*, 1997):

$$f_{\text{opt}} = \frac{1}{2\pi \cdot \sqrt{\tau_1 \cdot \tau_2}}.$$

As the phase modulation technique uses a single frequency, the signal can be electronically filtered (with a narrow bandwidth) which simplifies the suppression of noise and constant/ambient light. Nevertheless, in both methods care has to be taken when referencing zero time, because all electronic components additionally influence the signal.

If there are more luminescent decay products, they can be time-domain resolved by measuring more points along the decay curve and (least squares) curve analysis. Alternatively, they can be resolved in the frequency domain, by measuring at multiple frequencies and spectral analysis (Lakowicz, 1983; Gratton and Limkeman, 1984). Whether the time parameter is measured in the time or the frequency domain finally depends on the specific application and the most convenient technical solution for the measuring system, since the information content is equivalent.

The measuring system for oxygen microoptodes uses the phase modulation technique to evaluate the oxygen dependent lifetime change of the applied indicators. The excitation light source, a LED ($\lambda_p = 505$ nm, Figure 7.27, ex), is sinusoidally modulated at frequencies of $f_{\text{mod}} = 1.25$ kHz, 5 kHz or 37.5 kHz (for different lifetime ranges). The light is coupled through a shortpass filter ($\lambda_c = 540$ nm, Figure 7.27, OF) into a 2×2 fibre coupler (Figure 7.27, fcop). One branch of the coupler is connected via a standard fibre connector (Figure 7.27, ST) with the microoptode. The other branch is used for referencing purposes (Figure 7.27, ref). The luminescence is guided back through the coupler via a longpass emission filter ($\lambda_c = 590$ nm) to the detector, a photomultiplier tube (Figure 7.27, detector). Alternately, the excitation LED (Figure 7.27, ex) and the reference LED (Figure 7.27, ref) are switched on and their phase angle is measured. Finally, the difference of both give the luminescence lifetime induced phase angle whereas electronic influences are cancelled out (except any difference between light sources). For the phase angle evaluation there are two possible solutions: (1) a zero-crossing detector (Holst *et al.*, 1995), which basically compares the zero crossing event of a reference and the measuring signal and converts the corresponding pulse width into a phase angle, and (2) a dual phase lock-in detector (Gruber *et al.*, 1995; Holst *et al.*, 1998) (Figure 7.30) that is better suited for PMT detectors.

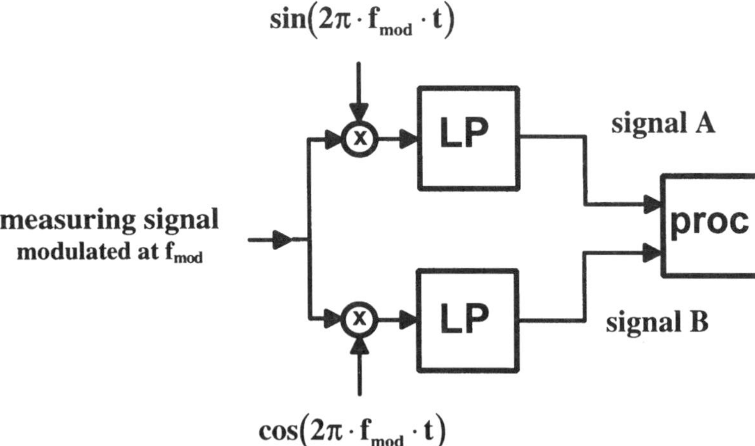

Figure 7.30 Principle of lock-in detection: fmod – modulation frequency; LP – electronical lowpass filter; proc – calculations that are explained in the text; signal A and B – output signals of the lowpass filters of the respective signal paths; x – multiplication.

As shown in Figure 7.30 the signal processing of a lock-in detector starts with the multiplication (in this case a demodulation) of the measuring signal that might been amplified and pre-filtered. The signal is multiplied both, with a reference sine- and cosine-signal of the same frequency f_{mod}. The products contain many different frequency contributions. If the product signals are now lowpass filtered with a cut-off frequency much lower than f_{mod}, the resulting signals A and B (Figure 7.30, signal A, signal B) consist of the amplitudes of the measuring and the reference signals, and of the in-phase and quadrature angle between the measurement and reference signals. Therefore, the amplitude and the phase angle Φ of the measuring signal can be simply calculated (Figure 7.30, proc) by:

$$\mathrm{amp} = \sqrt{(\mathrm{signal})^2 + (\mathrm{signal})^2}$$

$$\Phi = \arctan\left(\frac{\mathrm{signal\ A}}{\mathrm{signal\ B}}\right)$$

This method of phase angle evaluation is very precise and can resolve even very noisy signals. As this phase angle is measured for excitation light, Φ_{sig}, and reference light, Φ_{ref}, alternately, the real phase angle, caused by the luminescence lifetime, is the difference between: $\Phi_{\mathrm{lum}} = \Phi_{\mathrm{sig}} - \Phi_{\mathrm{ref}}$.

Although the phase modulation technique has many advantages compared to intensity measurements there are noise events that can have a severe impact on the quality of the measurement. If there are additional luminescence signals that interfere with the analytical signal, the phase angle measured at a single frequency is the sum of both signals. Therefore, it not only depends on the phase angle of the indicator based signal but also on the ratio of both amplitudes – an intensity parameter. If the noise signal is constant, it can be separately measured and mathematically corrected. If it is variable, the measurement has to be made at multiple frequencies to obtain the correct information.

Despite the fact that they require more complex electronics compared to intensity based systems, light-time measuring systems reveal better and reliable results in conjunction with optical microsensors. However, their sensing principles depend on a change in a time parameter rather than on the intensity of the optical signal. Therefore, they have been to-date limited to oxygen and temperature, but will in the future be extended to pH and carbon dioxide measurements.

7.4 FROM MICRO TO MACRO SCALE

Historically, principles for optical sensing were used on large scales. The driving force for the developments and investigations came from biomedical applications (Lübbers and Opitz, 1975; Gehrich *et al.*, 1986; Tusa *et al.*, 1986; Leiner, 1991; Wolfbeis, 1991). These larger sensors, when they are made as sensing layers on transparent supports called planar optodes, can as well be applied in oceanography. They exhibit many advantages: high pressure and long-term stability, immunity to stirring effects etc.. Therefore, applications in the water column have already been reported (DeGrandpre, 1993) and are still a matter of research (Holst *et al.*, 1998, unpublished results). Systems have become cheaper, more reliable and stable and the preparation of the sensing layers is much easier. In contrast, due to diffusion effects the response time of physically larger sensors is increased. As the sensing principle of all optical chemical sensors is diffusion based, spherical diffusion to a microsensor is reduced to a nearly one-dimensional diffusion for planar optodes. Consequently the diffusion boundary layer governs the response time of planar optodes.

Another potential application of large scale optodes has been presented (Glud *et al.*, 1996) and is continuously investigated (Holst *et al.*, 1998), mapping of two-dimensional oxygen distributions by planar oxygen optodes that are placed perpendicular to the sediment-water interface. This technique gives full access to the heterogeneity of the oxygen distribution in sediment and microbial communities, which previously could only be achieved by profiling measurements. Conclusively, optical chemical sensors have great potential for analytical measurements in oceanography.

References

Alcala, J.R., Liao, S.-C. and Zheng, J. (1995). Real Time Frequency Domain Fiberoptic Temperature Sensor, *IEEE Transactions on Biomedical Engineering*, 42, 471–476.

Amann, R. and Kühl, M. (1998). *In situ* methods for assessment of microorganisms and their activities, *Current Opinion in Microbiology*, in press.

Andrews, D. and Bennett, A. (1981). Measurements of Diffusivity Near the Sediment-Water Interface with a Fine-Scale Resistivity Probe, *Geochimica et Cosmochimica Acta*, 45, 2169–2175.

Bacon, J.R. and Demas, J.N. (1987). Determination of Oxygen Concentrations by Luminescence Quenching of a Polymer-Immobilized Transition-Metal Complex, *Analytical Chemistry*, 59, 2780–2785.

Baldini, F. and Del Bianco, A. (1992). Optical Fiber Sensor for Oxygen Detection Working on an Absorption Basis, *Fiber and Integrated Optics*, 11, 123–133.

Barker, S.L.R., Kopelman, R., Meyer, T.E. and Cusanovich, M.A. (1998). Fiber-Optic Nitric Oxide-Selective Biosensors and Nanosensors, *Analytical Chemistry*, 70, 971–976.

Bergman, I. (1986). Rapid-Response Atmospheric Oxygen Monitor Based on Fluorescence Quenching, *Nature*, 218, 396.

Berndt, K.W. and Lakowicz, J.R. (1992). Electroluminescent Lamp-Based Phase Fluorometer and Oxygen Sensor, *Analytical Biochemistry*, 201, 319–325.

Bishop, E. (1972). Indicators, New York: Pergamon Press.

Carraway, E.R., Demas, J.N., DeGraff, B.A. and Bacon, J.R. (1991). Photophysics and Photochemistry of Oxygen Sensors Based on Luminescent Transition-Metal Complexes, *Analytical Chemistry*, 63, 337–342.

Carraway, E.R., Demas, J.N. and DeGraff, B.A. (1991). Luminescence Quenching Mechanism for Microheterogenous Systems, *Analytical Chemistry*, 63, 332–336.

Carraway, E.R. and Demas, J.N. (1991). Photophysics and Oxygen Quenching of Transition-Metal Complexes on Fumed Silica, *Langmuir*, 7, 2991–2998.

Dakin, J. and Culshaw, B. (1988). Optical Fibre Sensors: I Principles and Components, II Systems and Applications, Artech House, Boston and London.

DeGrandpre, M.D. (1993). Measurement of Seawater pCO_2 Using a Renewable-Reagent Fiber Optic Sensor with Colorimetric Detection, *Analytical Chemistry*, 65, 331–337.

Demas, J.N. (1976). Luminescence Decay Times and Bimolecular Quenching, *Journal of Chemical Education*, 53, 657–663.

Demas, J.N. and DeGraff, B.A. (1992). On the Design of Luminescence Based Temperature Sensors, *Proceedings SPIE*, 1796, 71–75.

Fukshansky-Kazarinova, N., Fukshansky, L., Kühl, M. and Jørgensen, B.B. (1997). General theory of three-dimensional radiance measurements with optical microprobes, *Applied Optics*, 36, 6520–6528.

Fukshansky-Kazarinova, N., Fukshansky, L., Kühl, M. and Jørgensen, B.B. (1998). Solution of the inverse problem of radiative transfer on the basis of measured internal fluxes, *Journal of Quantitative Spectroscopy and Radiation Transfer*, 59, 77–89.

Garcia-Pichel, F. (1995). A Scalar Irradiance Fiber-Optic Microprobe for the Measurement of Ultraviolet Radiation at High Spatial Resolution, *Photochemistry and Photobiology*, 61(3), 248–254.

Garcia-Pichel, F. and Bebout, B. (1996). Penetration of Ultraviolet Radiation into Shallow Water Sediments: High Exposure for Photosynthetic Communities, *Marine Ecology Progress Series*, 131, 257–262.

Gehrich, J.L., Lübbers, D.W., Opitz, N., Hansmann, D.R., Miller, W.W., Tusa, J.K. and Yafuso, M. (1986). Optical Fluorescence and its Application to an Intravascular Blood Gas Monitoring System, *IEEE Transactions on Biomedical Engineering*, 33, 117–121.

Glud, R.N., Klimant, I., Holst, G., Kohls, O., Meyer, V., Kühl, M. and Gundersen, J.K. (1999). Adaptation, test and *in situ* measurements with O_2 microopt(r)odes on benthic landers, *Deep sea Research*, 46, 171–183.

Glud, R.N., Ramsing, N.B., Gundersen, J.K. and Klimant, I. (1996). Planar Optrodes: a New Tool for Fine Scale Measurements of Two-Dimensional O_2 Distribution in Benthic Communities, *Marine Ecology Progress Series*, 140, 217–226.

Grattan, K.T.V. and Zhang, Z.Y. (1995). Fiber Optic Fluorescence Thermometry, London Chapman & Hall.

Gratton, E. and Limkeman, M. (1984). Resolution of Mixtures of Fluorohores Using Variable-Frequency Phase and Modulation Data, *Biophysical Journal*, 46, 479–486.

Green, T.J., Wilson, D.F., Vanderkooi, J.M. and Defeo, S.P. (1988). Phosphorimeters for Analysis of Decay Profiles and Real Time Monitoring of Exponential Decay and Oxygen Concentrations, *Analytical Biochemistry*, 174, 73–79.

Gruber, W.R., Klimant, I. and Wolfbeis, O.S. (1993). Instrumentation for Optical Measurement of Dissolved Oxygen Based on Solid State Technology, *Proceedings SPIE*, 1885, 448–457.

Gruber, W.R., O'Leary, P. and Wolfbeis, O.S. (1995). Detection of Fluorescence Lifetime Based on Solid State Technology and its Application to Optical Oxygen Sensing, *Proceedings SPIE*, 1885, 448–457.

Hales, B., Burgess, L. and Emerson, S. (1997). An Absorbance-Based Fiber-Optic CO_2 (aq) Measurement in Porewaters of Sea Floor Sediments, *Marine Chemistry*, 59, 51–62.

Hartmann, P. and Leiner, M.J.P. (1995). Luminescence Quenching Behavior of an Oxygen Sensor Based on a Ru(II) Complex Dissolved in Polystyrene, *Analytical Chemistry*, 67, 88–93.

Holst, G., Kühl, M. and Klimant, I. (1995). A Novel Measuring System for Oxygen Microooptodes based on Phase Modulation Technique, *Proceedings SPIE*, 2508, 387–398.

Holst, G., Kühl, M., Klimant, I., Liebsch, G. and Kohls, O. (1997). Characterization and Application of Temperature Microooptodes for Use in Aquatic Biology, *Proceedings SPIE*, 2980, 164–170.

Holst, G., Glud, R. N., Kühl, M. and Klimant, I. (1997). A Microoptode Array for Fine-Scale Measurement of Oxygen Distribution, *Sensors and Actuators B*, 38–39, 122–129.

Holst, G., Kohls, O., Klimant, I., König, B., Kühl, M. and Richter, T. (1998). A Modular Luminescence Lifetime Imaging System for Mapping Oxygen Distribution in Biological Samples, *Sensors and Actuators B*, 51, 163–170.

Jørgensen, B.B. and Des Marais, D.J. (1986). A simple fiber-optic microprobe for high resolution light measurements: application in marine sediment, *Limnology and Oceanography*, 31, 1376–1383.

Kautsky, H. (1939). Quenching of Luminescence by Oxygen, *Transactions of Faraday Society*, 35, 216–219.

Karsten, U. and Kühl, M. (1996). Die Mikrobenmatte – das kleinste ökosystem der Welt, *Biologie in unserer Zeit*, 26, 16–26.

Khalil, G.-E., Gouterman, M. and Green, E. (1988). Method and Sensor for Measuring Oxygen Concentration, In United States Patent, Abbott Laboratories, USA, 23.

Klimant, I., Belser, P. and Wolfbeis, O.S. (1994). Novel Longwave Absorbing and Emitting Transition Metal Complexes for Use in Optical Oxygen Sensing, *Talanta*, 41, 985–991.

Klimant, I., Meyer, V. and Kühl, M. (1995). Fiber-Optic Oxygen Microsensors, a New Tool in Aquatic Biology, *Limnology & Oceanography*, 40, 1159–1165.

Klimant, I., Holst, G. and Kühl, M. (1995). Oxygen Microoptodes and their Application in Aquatic Environment, *Proceedings, SPIE*, 2508, 375–386.

Klimant, I., Kühl, M., Glud, R.N. and Holst, G. (1997). Optical Measurement of Oxygen and Temperature in Microscale: Strategies and Biological Applications, *Sensors and Actuators B*, 38, 29–37.

Klimant, I., Holst, G. and Kühl, M. (1997). A Simple Fiberoptic Sensor to Detect the Penetration of Microsensors into Sediments and other Biogeochemical Systems, *Limnology & Oceanography*, 42, 1638–1643.

Kohls, O., Klimant, I., Holst, G. and Kühl, M. (1997). Development and Comparison of pH Microoptodes for Use in Marine Systems, *Proceedings SPIE*, 2978, 82–93.

Kühl, M. and Jørgensen, B.B. (1992). Spectral Light Measurements in Microbenthic Phototrophic Communities with a Fiber-Optic Microprobe Coupled to a Sensitive Diode Array Detector, *Limnology & Oceanography*, 37, 1813–1823.

Kühl, M., Lassen, C. and Jørgensen, B.B. (1994a). Optical Properties of Microbial Mats: Light Measurements with Fiber-Optic Microprobes, In: Microbial Mats: Structure, Development, and Environmental Significance (Stal, L.J. and Caumette, P., eds), Springer, Berlin, 149–167.

Kühl, M., Lassen, C. and Jørgensen, B.B. (1994b). Light Penetration and Light Intensity in Sandy Marine Sediments Measured with Irradiance and Scalar Irradiance Fiber-Optic Microprobes, *Marine Ecology Progress Series*, 105, 139–148.

Kühl, M., Glud, R.N., Ploug, H. and Ramsing, N.B. (1996). Microenvironmental Control of Photosynthesis and Photosynthesis-Coupled Respiration in an Epilithic Cyanobacterial Biofilm, *Journal of Phycology*, 32, 799–812.

Kühl, M., Lassen, C. and Revsbech, N.P. (1997). A Simple Light Meter for Measurements of PAR (400 to 700 nm) with Fiber-Optic Microprobes: Application for P vs E_0 (PAR) Measurements in a Microbial Mat, *Aquatic Microbial Ecology*, 13, 197–207.

Kühl, M. and N.P. Revsbech (1998). Microsensor for the Study of Interfacial Biogeochemical Processes. In: Boudreau, B.P. and Jørgensen, B.B. (eds), The Benthic Boundary Layer, Oxford University Press, Oxford, (in press).

Lakowicz, J.R. and Cherek, H. (1981). Phase-Sensitive Fluorescence Spectroscopy: A New Method to Resolve Fluorescence Lifetimes or Emission Spectra of Components in a Mixture of Fluorophores, *Journal of Biochemical and Biophysical Methods*, 5, 19–35.

Lakowicz, J.R. and Balter, A. (1982). Analysis of Excited-State Processes by Phase-Modulation Fluorescence Spectroscopy, *Biophysical Chemistry*, 16, 117–132.

Lakowicz, J.R. (1983). Principles of Fluorescence Spectroscopy, New York: Plenum Press.

Lakowicz, J.R. and Maliwal, B.P. (1985). Construction and Performance of a Variable-Frequency Phase-Modulation Fluorometer, *Biophysical Chemistry*, 21, 61–78.

Lakowicz, J.R. and Szmacinski, H. (1993). Fluorescence Lifetime-Based Sensing of pH, Ca^{2+}, K^+ and Glucose, *Sensors and Actuators B*, 11, 133–143.

Lassen, C., Ploug, H. and Jørgensen, B.B. (1992). A Fibre-Optic Scalar Irradiance Microsensor: Application for Spectral Light Measurements in Sediments, *FEMS Microbiology Ecology*, 86, 247–254.

Lassen, C. and Jørgensen, B.B. (1994). A Fiber-Optic Irradiance Microsensor (Cosine Collector): Application for *In Situ* Measurements of Absorption Coefficients in Sediments and Microbial Mats, *FEMS Microbiology Ecology*, 15, 321–336.

Leiner, M.J.P. (1991). Luminescence Chemical Sensors for Biomedical Applications: Scope and Limitations, *Analytica Chimica Acta*, 225, 209–222.

Lippitsch, M.E., Pusterhofer, J., Leiner, M.J.P. and Wolfbeis, O.S. (1988). Fibre-Optic Oxygen Sensor with the Fluorescence Decay Time as the Information Carrier, *Analytica Chimica Acta*, 205, 1–6.

Lübbers, D.W. and Opitz, N. (1975). Die pCO_2/pO_2-Optode: eine neue pCO_2- bzw. pO_2-Messonde zur Messung des pCO_2 oder pO_2 von Gasen und Flüssigkeiten, *Zeitschrift für Naturforschung C*, 30, 532–533.

Lübbers, D.W., Opitz, N., Speiser, P.P. and Bisson, H.J. (1977). Nanoencapsulated Fluorescence Indicator Molecules Measuring pH and pO_2 Down to Submicroscopical Regions on the Basis of the Optode-Principle, *Zeitschrift für Naturforschung C*, 32, 512–1–512–2.

MacCraith, B.D., O'Keefe, G., McDonagh, C. and McEvoy, A. (1994). LED-based Fibre Optic Oxygen Sensor Using Sol-Gel Coating, *Electronics Letters*, 30 (11), 888–889.

McDonagh, C., MacCraith, B.D. and McEvoy, A.K. (1998). Tailoring of Sol-Gel Films for Optical Sensing of Oxygen in Gas and Aqueous Phase, *Sensors and Actuators B*, 70, 45–50.

Mills, A. and Chang, Q. (1993). Fluorescence Plastic Thin-Film Sensor for Carbon Dioxide, *Analyst*, 118, 839–843.

Munkholm, C., Walt, D., Milanovich, F. and Klainer, S. (1986). Polymer Modification of Fibre Optic Chemical Sensors as a Method for Enhancement of Signals for pH Measurement, *Analytical Chemistry*, 58, 1427–1430.

Munkholm, C., Walt, D.M. and Milanovich, F.P. (1988). A Fiber Optic Sensor for Carbon Dioxide, *Talanta*, 35, 109–114.

Papkovsky, D.P., Olah, J. and Kurochkin, H.N. (1993). Fibre-Optic Lifetime-Based Enzyme Biosensor, *Sensors and Actuators B*, 11, 525–530.

Peterson, I. and Fitzgerald, R.V. (1984). Fiber-Optic Probe for *In Vivo* Measurement of Oxygen Partial Pressure, *Analytical Chemistry*, 56, 62–67.

Revsbech, N.P., Jørgensen, B.B., Blackburn, T.H. and Cohen, Y. (1983). Microelectrode studies of photosynthesis and O_2, H_2S, and pH profiles of a microbial mat, *Limnology & Oceanography*, 28, 1062–1074.

Revsbech, N.P. and Jørgensen, B.B. (1986). Microelectrodes: their Use in Microbial Ecology, *In Advances in Microbial Ecology*, Plenum Publishing Corporation.

Revsbech, N.P. (1989). An Oxygen Microelectrode with a Guard Cathode, *Limnology & Oceanography*, 34, 474–478.

Revsbech, N.P. (1994). Analysis of Microbial Mats by Use of Electrochemical Microsensors: Recent Advances, In: Microbial mats: Structure, Development, and Environmental Significance (Stal, L.J. and Caumette, P., eds), Springer, Berlin, 135–147.

Robert-Baldo, G.L., Morris, M.J. and Byrne, R.H. (1987). Fibre-Optic pH Sensor for Seawater Monitoring, *Proceedings SPIE*, 798, 294–301.

Rosenzweig, Z. and Kopelman, R. (1995). Development of a Submicrometer Optical Fiber Oxygen Sensor, *Analytical Chemistry*, 67, 2650–2654.

Sacksteder, L.A., Demas, J.N. and DeGraff, B.A. (1993). Design of Oxygen Sensors based on Quenching of Luminescent Metal Complexes: Effect of Ligand Size on Heterogeneity, *Analytical Chemistry*, 65, 3480–3483.

Schreiber, U., Kühl, M., Klimant, I. and Reising, H. (1996). Measurement of Chlorophyll Fluorescence within Leaves Using a Modified PAM Fluorometer with a Fiber-Optic Microprobe, *Photosynthesis Research*, 47, 103–109.

Severinghaus, J.W., Bradley, A.F. (1958). Electrodes for blood P_{O_2} and P_{CO_2} Determination, *Journal of Applied Physiology*, 13, 515–520.

Sharma, A. and Wolfbeis, O.S. (1988). Fiberoptic Oxygen Sensor Based on Fluorescence Quenching and Energy Transfer, *Applied Spectroscopy*, 42, 1009–1011.

Sipior, J., Bambot, S., Romauld, M., Carter, G.M., Lakowicz, J.R. and Rao, G. (1995). A Lifetime-Based Optical CO_2 Gas Sensor with Blue or Red Excitation and Stokes or Anti-Stokes Detection, *Analytical Biochemistry*, 227, 309–318.

Shortreed, M., Bakker, E. and Kopelman, R. (1996). Miniature Sodium-Selective Ion-Exchange Optode with Fluorescent pH Chromoionophores and Tunable Dynamic Range, *Analytical Chemistry*, 68, 2656–2662.

Shortreed, M., Kopelman, R. Kuhn, M. and Hoyland, B. (1996). Fluorescent Fiber-Optic Calcium Sensor for Physiological Measurements, *Analytical Chemistry*, 68, 1414–1418.

Stern, O. and Volmer, M. (1919). Über die Abklingzeit der Fluoreszenz, *Physikalische Zeitschrift*, 20, 183–188.

Tan, W., Shi, Z.-Y., Smith, S., Birnbaum, D. and Kopelman, R. (1992). Submicrometer Intracellular Chemical Optical Fiber Sensors, *Science*, 258, 778–781.

Tengberg, A., De Bovee, F., Hall, P., Berelson, W., Chadwick, D., Ciceri, G. *et al.* (1995). Benthic Chamber and Profiling Landers in Oceanography – A Review of Design, Technical Solutions and Functioning, *Progress in Oceanography*, 35, 253–294.

Tusa, J.K., Hacker, T., Hansmann, D.R., Kaput, T.M. and Maxwell, T.P. (1986). Fiber Optic Microsensor for Continous *In-Vivo* Measurement of Blood Gases, *Proceedings SPIE*, 713, 137–143.

Vanderkooi, J.M., Maniara, G., Green, T.J. and Wilson, D.F. (1987). An Optical Method for Measurement of Dioxygen Concentration Based upon Quenching of Phosphorescence, *Journal of Biological Chemistry*, 262, 5476–5482.

Vogelmann, T.C. and Björn, L.O. (1984). Measurement of Light Gradients and Spectral Regime in Plant Tissue with a Fibre Optic Probe, *Physiologia Plantarum*, 60, 361–368.

Wenzhöfer, F., Kohls, O. and Holby, O. (1998). Deep Penetration of Oxygen Measured *in situ* by Oxygen Optodes, *Deep-Sea Research*, (subm. f. publ.).

Wilson, D.F., Vanderkooi, J.M., Green, T.J., Maniara, G., Defeo, S.P. and Bloomgarden, D.C. (1987). A Versatile and Sensitive Method for Measuring Oxygen, In Oxygen Transport To Tissue IX, New York: Plenum Press, 71–77.

Wolfbeis, O.S., Offenbacher, H., Kroneis, H. and Marsoner, (1984). A Fast Responding Fluorescence Sensor for Oxygen, *Mikrochimica Acta*, I, 153–158.

Wolfbeis, O.S., Leiner, M.J.P. and Posch, H.E. (1986). A New Sensing Material for Optical Oxygen Measurement, with the Indicator Embedded in an Aqueous Phase, *Mikrochimica Acta, III*, 359–366.

Wolfbeis, O.S., Weis, L.J., Leiner, M.J.P. and Ziegler, W. (1988). Fiber-Optic Fluorosensor for Oxygen and Carbon Dioxide, *Analytical Chemistry*, 60, 2028–2030.

Wolfbeis, O.S. (1991). Fiber Optic Chemical Sensors and Biosensors, CRC Press, Boca Raton.

Wolthuis, R.A., McCrae, D., Saaski, E., Hartl, J. and Mitchell, G. (1992). Development of a Medical Fiber-Optic pH Sensor Based on Optical Absorption, *IEEE Transactions on Biomedical Engineering*, 39, 531–537.

Zhang, Z., Grattan, K.T.V. and Palmer, A.W. (1993). Phase-Locked Detection of Fluorescence Lifetime and its Thermometric Applications, *Proceedings SPIE*, 1885, 228–239.

8. A LASER-BASED FIBER-OPTIC FLUOROMETER FOR *IN SITU* SEAWATER MEASUREMENTS

ROBERT F. CHEN

Environmental, Coastal & Ocean Sciences (ECOS),
University of Massachusetts-Boston (UMass Boston),
100 Morrissey Boulevard, Boston, MA 02125–3393, USA

8.1 INTRODUCTION

Marine chemists study ocean processes on various temporal and spatial scales, from nanosecond photochemical reactions to climate shifts over thousands or millions of years, and from microscale examinations of the sediment-water interface to global studies of ocean circulation. Except for a few well-established sensors such as: conductivity, temperature, pressure, oxygen, pH, and redox potential, most marine chemists have relied on traditional bottle sampling and analysis either onboard ship or more commonly in the laboratory. Sample numbers, and therefore spatial and temporal coverage, have therefore been limited. Artifacts due to sampling handling (especially filtration) and storage have to be considered and sometimes obfuscate the true nature of oceanic processes. While many researchers have concentrated on miniaturizing and redesigning instruments and techniques (nutrients, CO_2, trace metals, filtration systems) for *in situ* measurements, others are using seawater's natural properties to learn about the chemistry of some of its constituents. Seawater optical properties have become useful for examining phytoplankton biomass (chlorophyll fluorescence), total suspended matter (beam attenuation or backscatter), and dissolved organic matter (absorbance).

Nearly half a century ago, von Kurt Kalle (1949) observed that seawater fluoresced blue when irradiated by an ultraviolet light. However, only in the last decade or so have a multitude of researchers used this property of seawater to study a host of organic geochemical processes in the ocean such as oceanic organic carbon cycling (Duursma, 1974; Hayase *et al.*, 1988; Coble *et al.*, 1990; Chen, 1992; Momzikoff *et al.*, 1992; Mopper and Schultz, 1993; Determann *et al.*, 1994; Sierra *et al.*, 1994; Karabashev, 1996; Coble, 1996), river water mixing (Zimmerman and Rommets, 1974; Willey and Atkinson, 1982; Laane, 1981), photochemical degradation (Kramer, 1979; Determann *et al.*, 1994, Chen and Bada, 1992), sediment-water exchange (Chen *et al.*, 1993; Skoog *et al.*, 1996), identification of various components of DOM (Coble, 1996), and the influence of ocean margins on oceanic cycling (Chen and Repeta, 1994). With an increasing knowledge of the fluorophores present in seawater combined with rapid advances in the technology of ocean optical instrumentation, fluorescence of seawater should play a major role in marine chemistry. In this chapter, a laser-induced fluorometry system is described with some examples of field data to demonstrate the usefulness of an *in situ* sensor system. The chapter ends looking forward to technological and chemical developments that should make fluorescence an even more popular mode of analysis.

With the advent of commercially available DOM fluorometers [for example: SeaTech Corp., Corvallis, USA; Chelsea Instruments Ltd., Surrey, UK; WETLabs, Philomath, USA], which can be attached to standard CTD systems or towed-vehicles (tow-yo or undulating vehicles), fluorescence coverage has increased dramatically. An increasing demand has been placed on this technique with higher spatial and temporal scales needed for studying biogeochemical cycling in dynamic areas such as coastal regions. The system described in this chapter has a probe tip of less than 1 mm in diameter and a response time of a few seconds so that both spatial and temporal coverage are high. As will be discussed, this system also has the advantage that spectral resolution and time-resolved fluorescence measurements are possible.

8.2 DISSOLVED ORGANIC MATTER FLUORESCENCE

Dissolved organic matter (DOM) in seawater is one of the largest reactive carbon reservoirs on earth, and as such, its cycling and reactivity are important components of the global carbon cycle. However, DOM is made up of an extremely complex mixture of components, each with a different reactivity so that a simple understanding of the behavior of total DOM is not possible. Some easily measured components such as amino acids, monosaccharides, and individual lipids have been identified, and their cycles understood for the most part, but the majority of oceanic DOM components cannot be well-characterized. Even the total amount of DOM measured as dissolved organic carbon (DOC) in seawater was not well-established until recently (Sharp, 1997) due to the difficulty in measuring the small amount of organic matter in the large quantities of salt in seawater. There has been a distinct need for identifying and quantifying consistent fractions of DOM that behave similarly in the ocean. Several studies have begun to examine these larger groups of DOM components with somewhat well-defined characteristics and cycles. For example, the polysaccharide fraction may be the most important component of DOM (Benner *et al.*, 1992; Amon and Benner, 1994) and may be produced as a biopolymer (Aluwihare *et al.*, 1997). The isotopic composition of various fractions of particulate organic matter has been used to show that different classes of organic matter cycle at different rates with different sources and fates (Wang *et al.*, 1996), and similar studies of DOM may be quite revealing. In this chapter, the chromophoric (or colored) fraction of DOM (cDOM) will be used to learn about several aspects of dissolved organic matter cycling.

Chromophoric or colored dissolved organic matter (cDOM) is an important (and easily measured) fraction of the total DOM. cDOM absorbs energetic photons to initiate most of the photochemical reactions in the ocean. This absorbed energy can result in the production of reactive intermediates such as hydrated electrons, hydrogen peroxide, superoxide radicals, and hydroxy radicals (Zafiriou *et al.*, 1984) that control metal speciation (Moffett and Zika, 1987), photodegradation of refractory macromolecules (Mopper *et al.*, 1991), and production of trace gases (Conrad and Seiler, 1980) in surface waters. cDOM also affects the quality and quantity of light reaching photosynthetic cells. cDOM appears to be higher in riverine and estuarine waters and therefore may represent a significant portion of the DOM that is exported to the open ocean. cDOM affects the overall ocean color as seen by satellite remote sensing efforts of CZCS, TOCS and SeaWiFS. Knowledge of distributions and characterization of cDOM is essential for retrieving accurate chlorophyll-*a* concentrations in the world's oceans from satellite measurements (Hoge *et al.*, 1993). Luckily, the very characteristic that makes cDOM important, it's interaction with light, makes cDOM easily detectable.

However, by simply measuring the absorbance of a seawater sample at a single wavelength much of the information about the cDOM is not recovered. Spectral absorption measurements are better as they more adequately represent the complex mixture of chromophores present with their differing (but overlapping) spectra. The absorption of light in seawater due to DOM can generally be described as an exponential decrease in absorption with increasing wavelength (Armstrong and Boalch, 1961). Instruments designed to specifically measure the spectral absorption of seawater *in situ* are now being developed and are commercially available, but are still somewhat limited by the fact that the absorption measurement is not all that sensitive or selective.

One of the first recognized and least characterized properties of oceanic DOC is that it emits a blue fluorescence when irradiated by a UV light (Kalle, 1949). This is an extremely sensitive measurement as photons are counted against a dark background rather than counting the difference in photons against a bright background. The fluorescence of seawater provides a bulk characterization of seawater organics on an extremely small seawater sample and makes possible the use of fiber optic probes to deliver light to small places. Fluorescence measurements of DOM are free of the problems inherent in extraction and derivatization procedures. Slight variations in observed fluorescence intensities can be used to study the sources and sinks of these molecules, and the rates governing these processes. While a few studies of various aspects of seawater fluorescence have been carried out, the overall understanding of the global cycling of fluorescent organic matter is rather primitive (Figure 8.1) (Kalle, 1949; Kramer, 1979; Karabashev and Agatova, 1984; Hayase *et al.*, 1987; Coble *et al.*, 1990; Chen and Bada, 1992).

Fluorescence of terrestrial humic substances has been used to study river mixing processes in coastal waters and estuaries (Dorsch and Bidleman, 1982; Hayase *et al.*, 1987), as a water mass tracer (Zimmerman and Rommets, 1974; Willey and Atkinson, 1982; Cabaniss and Shuman, 1987) and as an indicator of total organic carbon (TOC) (Smart *et al.*, 1976), DOC (Laane and Koole, 1982), and various dissolved organic components (Karabashev and Agatova, 1984).

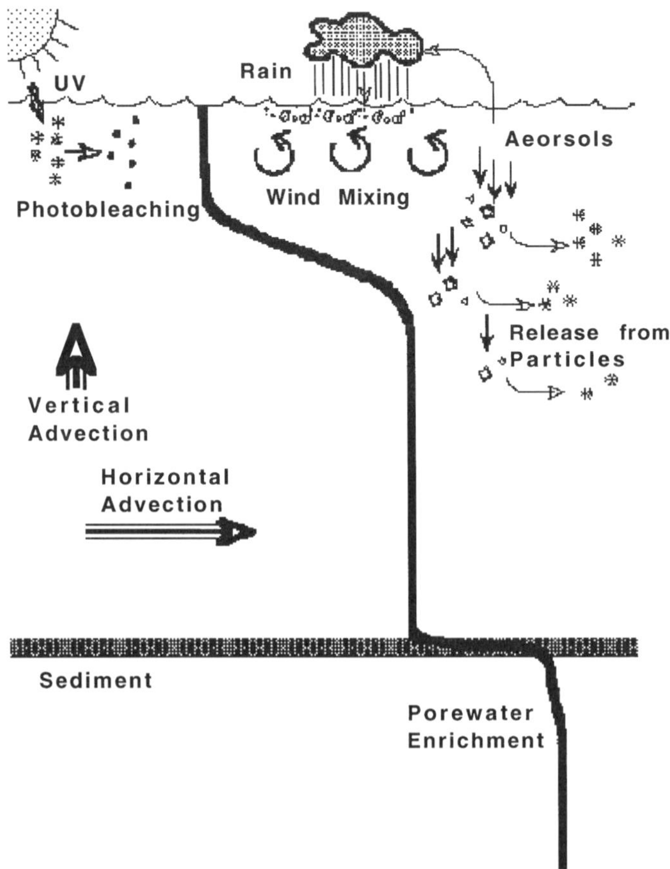

Figure 8.1 The oceanic cycle of the fluorescent dissolved organic matter (from Chen and Bada, 1992). Major controlling factors appear to be photobleaching in surface waters and production from particles in mid-depth waters. Release from particles could be a product of solubilization of insoluble macromolecules or aggregates or the production of fluorescent components by bacteria. Diffusion of highly fluorescent porewater FDOM is a small flux on a global scale, but could be significant in local areas such as anoxic basins and estuaries.

Marine humic substances are chromophores involved in photochemical processes in the surface ocean (Momzikoff, 1983; Zafiriou *et al.*, 1984; Zepp *et al.*, 1985; Zika, 1987; Amador *et al.*, 1990), but the molecular makeup of these complex molecules in seawater fluorescence is not well-known. Pteridines and flavins have been identified as fluorescent components in reefs, coastal waters, and the Black Sea (Dunlap and Susic, 1985; Coble, 1990). Photodegradation has been suggested as the destruction mechanism for fluorescence in surface waters (Kramer, 1979; Dunlap and Susic, 1986; Hayase *et al.*, 1987; Hayase *et al.*, 1988;

Chen and Bada, 1989; Chen and Bada, 1992). Bulk fluorescence measurements have proven quite useful, but Figure 8.2 shows that different excitation and emission wavelengths allow several discrete fluorescent components to be identified. Most of the work described in this chapter and the references within refer to an excitation wavelength between 300 and 350 nm and an emission from 400 to 450 nm. However, it is clear that by scanning over various wavelengths, the fluorescence technique can be used to study several independent DOM components (Coble, 1996).

The usefulness of the fluorescent measurement together with the ease of measurement makes fluorescence an ideal tool to study DOM, especially in

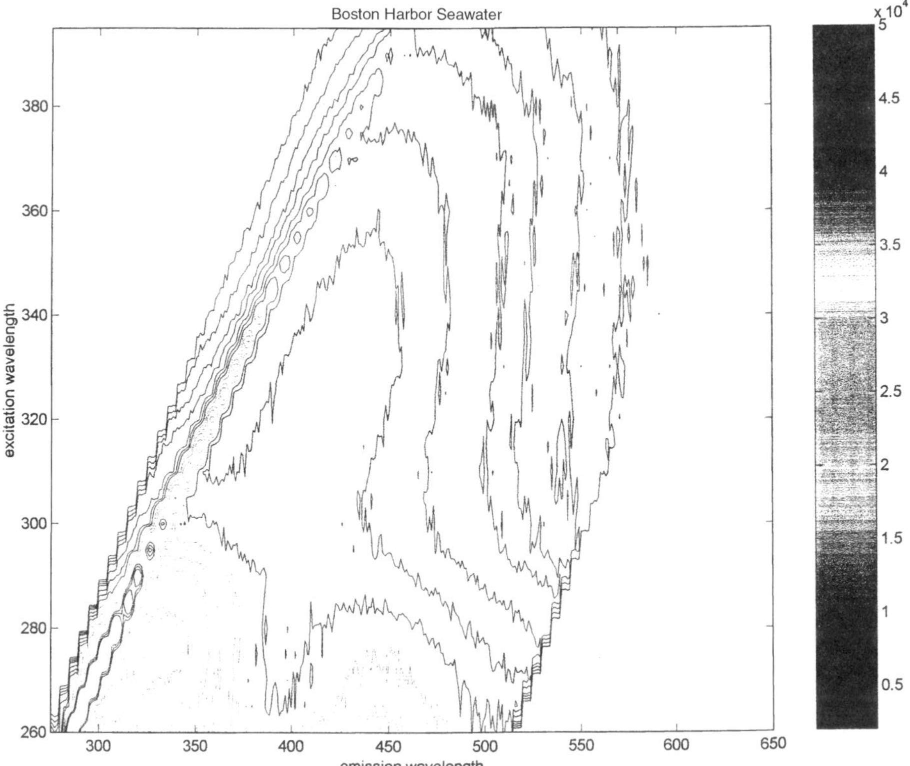

Figure 8.2 A fluorescence excitation-emission matrix spectrum of Boston Harbor Seawater. Emission scans are taken for every 5 nm of excitation with a Photon Technologies QM1 spectrofluorometer. Spectra are corrected for varying instrumental with both excitation and emission wavelength. The appearance of three peaks centered at (340, 440 nm), (275, 325 nm), and (260, 450 nm) indicates three separate dissolved fluorescent components. The (340, 440 nm) and (260, 450 nm) components are attributed to "humic substance" fluorescence while the (275, 325 nm) peak is due to protein (especially tryptophan) fluorescence.

coastal oceans where gradients are large, sources of DOM are very different, and transformations of DOM are quite dynamic. High temporal and spatial resolution are needed for these studies which can be supplied by the fluorescence technique.

8.3 FLUORESCENCE OF PETROLEUM HYDROCARBONS

In addition to the various natural DOM components that fluoresce, anthropognic inputs of petroleum include aromatic compounds such as polycyclic aromatic hydrocarbons (PAH) that are known to fluoresce at similar wavelengths to DOM. Because the impact of petroleum on the marine environment can be severe, and some PAH have been shown to be carcinogenic (Grimmer, 1983), several investigators have attempted to use the fluorescence properties of PAH to measure oil in seawater (Filippova *et al.*, 1993; Karyakin and Galkin, 1995). Most of these studies have concentrated on measuring the fluorescence of solvent extracts (Theobald, 1989; Ehrhardt and Petrick, 1989; Alarie *et al.*, 1993) because PAH are only present in water at low concentrations due to low solubilities. A comparison of fluorescence with GCMS data suggests that extracted fluorescence is due to more than just the PAH (Ehrhardt and Knap, 1989). Recent developments have used time-resolved fluorescence to detect PAH directly in seawater without extraction and have relied on differing lifetimes of PAH from background DOM (Niessner and Panne, 1991; Lieberman *et al.*, 1992; Romanovskaya and Lebedeva, 1993; Bublitz *et al.*, 1995; Schade and Bublitz, 1996; Chen *et al.*, 1997). This approach applied to Boston Harbor seawater is presented here with *in situ* applications, but a more detailed description of this technique is presented elsewhere (Rudnick and Chen, 1998).

8.4 HIGH RESOLUTION MEASUREMENT TECHNIQUES

The complex nature of coastal systems cannot be adequately studied using traditional discrete sampling methods. New, real-time, *in situ* measurement systems are necessary to address the sources, interactions, and fates of both natural and anthropogenic organic compounds at the appropriate temporal and spatial scales. While traditional environmental chemical analyses require discrete sampling and detailed, costly, time-consuming analyses in the laboratory, recent efforts have concentrated on real-time, *in situ* techniques to measure both organic and inorganic compounds (Snow, 1987; Lieberman *et al.*, 1992; DeGrandpre *et al.*, 1995). In the traditional method, sampling and storage may result in misleading artifacts. Often, initial analyses warrant further sampling, and multiple sampling expeditions become expensive and logistically difficult. Also, sampling strategies may underestimate, overestimate, or entirely miss small scale, but highly significant variations in concentrations of natural or anthropogenic compounds. Many of these problems are alleviated by the advent of fiber-optic based, real-time, *in situ* chemical sensors. Continuous monitoring and large scale mapping with small sensors allow areas to be rapidly, thoroughly screened and pinpoint areas that warrant further, more detailed studies. Measuring analytes in their natural environment minimizes possible artifacts and eliminates storage problems. Immediate analytical results allow a guided sampling strategy and immediate responses.

Fluorescence is a highly sensitive and versatile mode of detection. In the last two decades or so, a number of instrument systems have been developed based on the fluorescence technique. A few *in situ* submersible fluorometers have been constructed and used for florescence profiling (Karabashev and Solovyev, 1973; Klinkhammer, 1994), chlorophyll fluorescence distributions (Gieskes *et al.*, 1978), and oil dispersion studies (Genders, 1988), and *in situ* pumping systems have also been used (Kouassi, 1986; Coble *et al.*, 1990). Below, a shipboard instrument with a small, flexible fiber optic probe and capabilities of spectral resolution as well as time resolution is described.

8.5 LASER-INDUCED FLUOROMETRY (LIF) SYSTEM

Several laboratories have developed laser-induced fluorescence systems for seawater analysis (Niessner and Panne, 1991; Bublitz *et al.*, 1995; Schade and Bublitz, 1996), but to my knowledge, only three have been deployed at sea (Lieberman *et al.*, 1992; Karabashev, 1996; Chen *et al.*, 1997). The components of the LIF system are all standard off-the-shelf components and many of the building blocks of our system can easily be replaced to create a LIF system for different specific purposes.

Basically, our system is comprised of a UV laser, a sensitive detector, appropriate electronics, and a 30 m fiber optic cable (Figure 8.3). The nitrogen laser radiation (Photon Technologies, International, Monmouth, New Jersey; 337 nm, 600 ps pulse width) is focused into a fused silica core, fused silica clad fiber optic and is transmitted through a cable to the probe tip where fluorophores are excited. The optimal geometry for the probe has been found to consist of a single 400 μm core diameter fiber for excitation and 9 × 200 μm fibers in a concentric circle as emission fibers. This allows optimal collection of emission radiation which is then spectrally dispersed by an imaging spectrograph and detected on a two-dimensional intensified charge-coupled device detector (ICCD). A second fiber intercepting a small fraction of the laser radiation, allows rapid optical triggering of the detector, which can gate the intensifier ON and OFF on the nanosecond timescale. Time-resolved fluorescence measurements on the order of 1 to 2 nsec resolution are possible. Time-gating has the additional advantage of eliminating any interference due to ambient (continuous) light sources.

The probe end of the laser-induced fluorometry (LIF) system was attached to a Conductivity-Temperature-Depth sensor system (Seabird 911 plus CTD) cage and the package was towed through the water (Boston Harbor) at ~0.5 m depth. Standard measurements (0 ns delay, 50 ns gate width) were typically made every

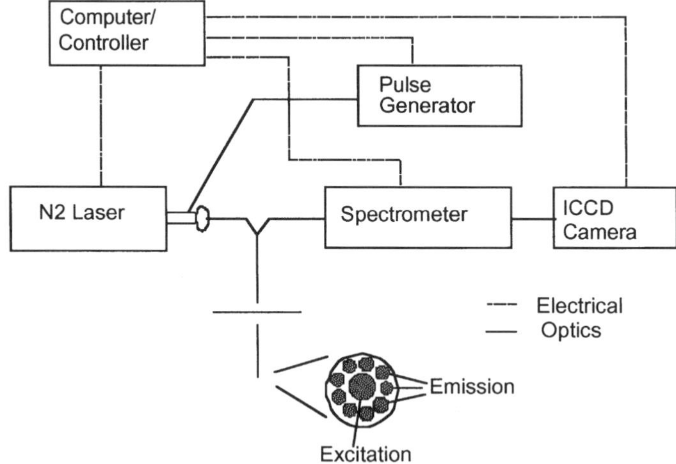

Figure 8.3 The laser-induced fluorescence system (LIF) developed at UMassBoston. A fiber optic cable 30 to 50 m long allows remote measurements. The entire system can be set up in about 2 hours.

10–20 seconds for ~30 m resolution at 4 knots. For longer cruises, the probe tip was placed in a flow-through debubbling cell onboard ship and fluorescence spectra were measured every 5 minutes for ~1.3 km resolution at 10 knots. For Mid-Atlantic Bight deployments, a set of time-resolved fluorescence spectra (0, 1, 2, 4, 8, 16, 32, 64 ns) were taken every hour. For Boston Harbor deployments, the full set of time-resolved spectra were taken at vertical profile stations and 32 ns, 64 ns, or 128 ns spectra were taken when warranted in surface waters.

The zero degree backscatter arrangement has been chosen as the most convenient geometry with which to work. A right angle between the excitation and emission fibers would reduce scattering due to water and particles, but the probe tip would then increase in size. In order to maximize sensitivity, the "9 around 1" geometry has been chosen that retains a small probe tip and increases signal over a two fiber probe by a factor of approximately 2-fold.

The major limitations to this system are the cost and the limited length of cable possible. Using the N_2 laser at 337 nm, 50 m is the maximum length of a fiber cable before attenuation in the fiber reduces the excitation radiation to background. Also, the system relies on rather high power and space demands on a ship or floating platform so that time-series are difficult and moorings or long-term studies are not possible. Only through miniaturization of the laser, electronics, and detection system can these surface applications be extended to deep sea, mooring, AUV, ROV, and long-term applications.

Calibration of an *in situ* system is critical and is carried out with a number of methods. The easiest method is normalization to the Raman scattering peak of water. In this standardization, water acts as an internal standard that defines the

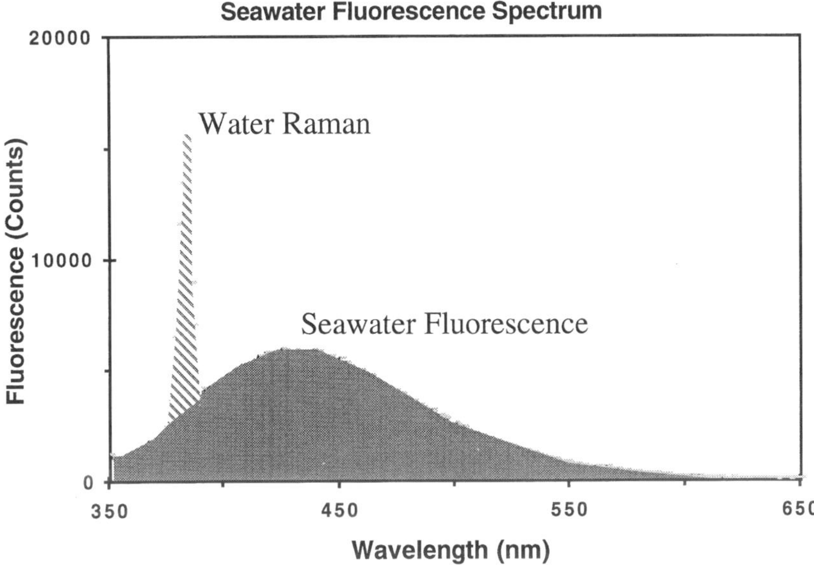

Figure 8.4 A typical seawater fluorescence spectrum. Integrated seawater fluorescence (shaded) can be normalized to the water Raman scattering band. This ratio can be calibrated against a fluorescent standard such as quinine sulfate so that measurements on different instruments can be compared.

volume of seawater analyzed. By taking the ratio of fluorescence (350–600 nm integrated) to the Raman peak (Figure 8.4), variations in laser intensity, absorption and scattering in water, and detector sensitivity are all accounted for. Compensation for spectral variations and calibration to known standards is also carried out before and after the LIF system is deployed so that data recorded can be replicated in the laboratory or by other investigators (Hoge *et al.*, 1993). Comparisons with discrete samples are useful to show this system's data in light of other work.

8.6 INTERPRETATION OF FLUORESCENCE MEASUREMENTS

Fluorescence of dissolved organic matter has been used as a tracer for DOC in coastal seawater (Willey and Atkinson, 1982; Zimmerman and Rommets, 1974) and for studying DOC components in open ocean waters (Chen and Bada, 1992; Momzikoff *et al.*, 1992) but few studies have compared the two measurements (Smart *et al.*, 1976; Laane and Koole, 1982; Vodacek *et al.*, 1995). Correlations appear to be fairly robust in surface coastal waters (Moran *et al.*, 1991; Vodacek *et al.*, 1995), but show some differences or even negative correlations in deeper waters (Chen, 1992). For some recent studies in Boston Harbor and the Mid-Atlantic Bight (Figure 8.5a), a consistent relationship between DOC and fluorescence can be observed from 0 to 35 PSU salinity. The relationship is similar to that on a transect off Delaware Bay across the shelf (Vodacek *et al.*, 1995) in that

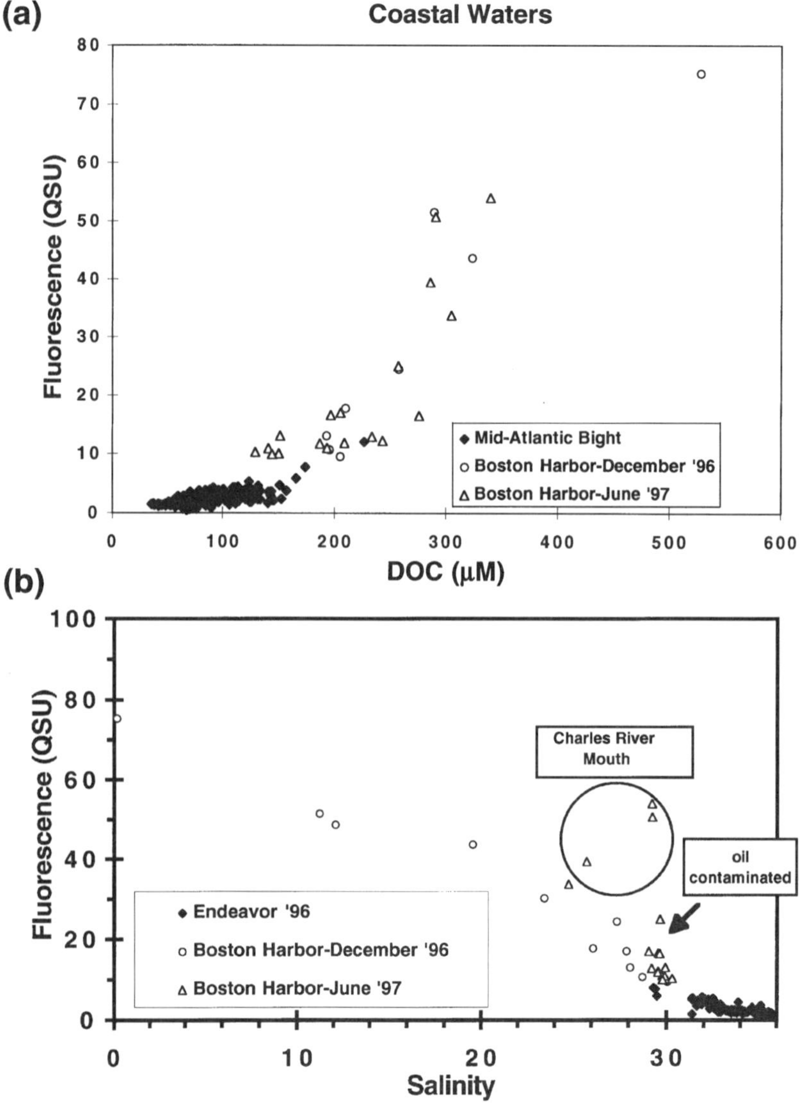

Figure 8.5 The relationship between fluorescence and DOC (a) and salinity (b). Measurements were made on discrete samples from 3 cruises. Additional sources of fluorescence in Boston Harbor associated with petroleum inputs are evident.

at lower DOC concentrations, the fluorescence is fairly linear ($r^2 = 0.49$ for $n = 255$). However, when the DOC increases above ~150 μM, there is a shift to higher fluorescence per unit DOC. While other freshwater endmembers need to be examined, it appears that the high DOC from rivers is more fluorescent than the equivalent DOC in marine systems consistent with past observations of higher fluorescence yields per unit carbon for terrestrial humics than marine humics. In addition, the fluorescence on the shelf cannot be explained simply by mixing of freshwaters with seawater across the shelf, but rather a significant fraction of the fluorescent DOM is apparently lost in estuaries. This observation is consistent with humic acid flocculation at low to moderate salinities (Sholkovitz *et al.*, 1978).

This relationship is consistent on two cruises across the Mid-Atlantic Bight (March, 1996-RV Endeavor; June, 1997-RV Seward Johnson) and two cruises in Boston Harbor (December, 1996; June, 1997). Because of high sensitivity, low cost, and the ability to make rapid *in situ* measurements, fluorescence determinations in seawater provide a high resolution method of studying DOC cycling in coastal waters. For monitoring and screening purposes, UV fluorescence is a rapid and easy way to obtain semi-quantitative information on DOC. With real-time knowledge, several discrete samples can be taken to ground-truth the field measurements.

8.7 RELATIONSHIP OF FLUORESCENCE TO SALINITY

The majority of the fluorescence variability in coastal waters can be explained by simple mixing between a high fluorescence freshwater endmember and a low fluorescence marine endmember (Figure 8.5b; Nieke, 1997). This relationship holds throughout the Mid-Atlantic Bight and Boston Harbor. Similar to fluorescence/DOC relationships, the fluorescence increase per unit salinity decrease is higher for the lower salinity endmembers. While there is surely a wide variation in zero salinity endmembers, this data suggests that fluorescent DOM is indeed flocculating at intermediate salinities. Nonetheless, the fairly tight ($r^2 = 0.56$, $n = 262$) correlation in the Mid-Atlantic Bight, and the very tight correlation for data in a given regime at a given time (Boston Harbor, June, 1997: $r^2 = 0.988$, $n = 2040$, data not shown) suggests that freshwater is indeed controlling fluorescent DOM (fDOM) distributions in coastal waters. On the other hand, several instances of seawater with fluorescence to salinity characteristics that do not fall on the mixing line between the major freshwater source and seawater can be quite revealing. It appears that production by phytoplankton and photobleaching affect the fluorescence-salinity relationships at some times and in some areas (Chen, 1998). In addition, three samples taken in the Charles River area (Figure 8.5b) show a higher than expected fluorescence, possibly due to fluorescence of petroleum hydrocarbon contamination or some other local source of fDOM. Finally, the fluorescence to salinity relationship changes somewhat with season (data not shown) and source so that the lower correlation found in the Mid-Atlantic Bight ($r^2 = 0.8$) with its numerous sources is not unexpected.

8.8 RESULTS FROM VARIOUS FIELD STUDIES

8.8.1 Boston Harbor/Massachusetts Bay

Boston Harbor is noted for the severe impacts brought on by the introduction of sewage effluent from 43 surrounding communities. Nearly 50% of the total freshwater input to the Harbor comes from primary treated effluent which is presently introduced near the entrance to the Harbor at Deer Island. Concerns over these impacts have led to the construction of the country's most expensive sewer plant/outfall, which by the end of the century will move the discharge 9 miles offshore into adjacent Massachusetts Bay, and provide secondary treatment (MWRA, 1993). While the cessation of sewage sludge dumping in 1995 has led to dramatic improvements to the condition of the Harbor, sewage effluent still dominates the inputs of dissolved organic matter and organic and trace metal contaminants in the area. The Charles River, the second major source for freshwater to the Harbor is also very heavily impacted. In recent surveys in Boston Harbor with the LIF system, we have found good inverse correlations between salinity and fluorescent DOM concentrations (see Figure 8.5). From a physical perspective, Boston Harbor can be characterized as varying between a well-mixed and partially-mixed estuary. A tidal range of approximately 3 meters, associated with the Gulf of Maine/Bay of Fundy system, provides a vigorous exchange between the Harbor and Massachusetts Bay. Without this exchange, environmental conditions within the Harbor might be much worse.

Figure 8.6(a) shows a deployment of our LIF system in Boston Harbor in June, 1996. An increase in fluorescence can be seen at Deer Island due to the sewage effluent, and fluorescence generally increases towards the Inner Harbor where highly fluorescent freshwater from the Charles River dominates other sources. Notice that the resolution of these measurements is approximately 20 to 30 m, thus allowing small features such as the sewage plume to be well-characterized. Data are presented as a simple integration of all wavelengths with a zero nanosecond delay. Full spectral resolution is also gathered but not presented.

8.8.2 Mid-Atlantic Bight

Seawater upwells around Georges Bank and flows south along the northeast coast of the United States (Beardsley and Boicourt, 1981). About 50% of the water is simply mixed horizontally across the shelf break as water travels south. The other 50% travels the length of the shelf and is forced offshore at Cape Hatteras. This water is subject to several biogeochemical influences during its several month transit time and can increase its carbon content through increases in dissolved inorganic carbon (DIC), dissolved organic carbon (DOC), or particulate organic carbon (POC). Phytoplankton increase the DOC and POC during photosynthesis at the expense of DIC. While DIC levels can equilibrate with the atmosphere and therefore are similar in source waters and water flowing off

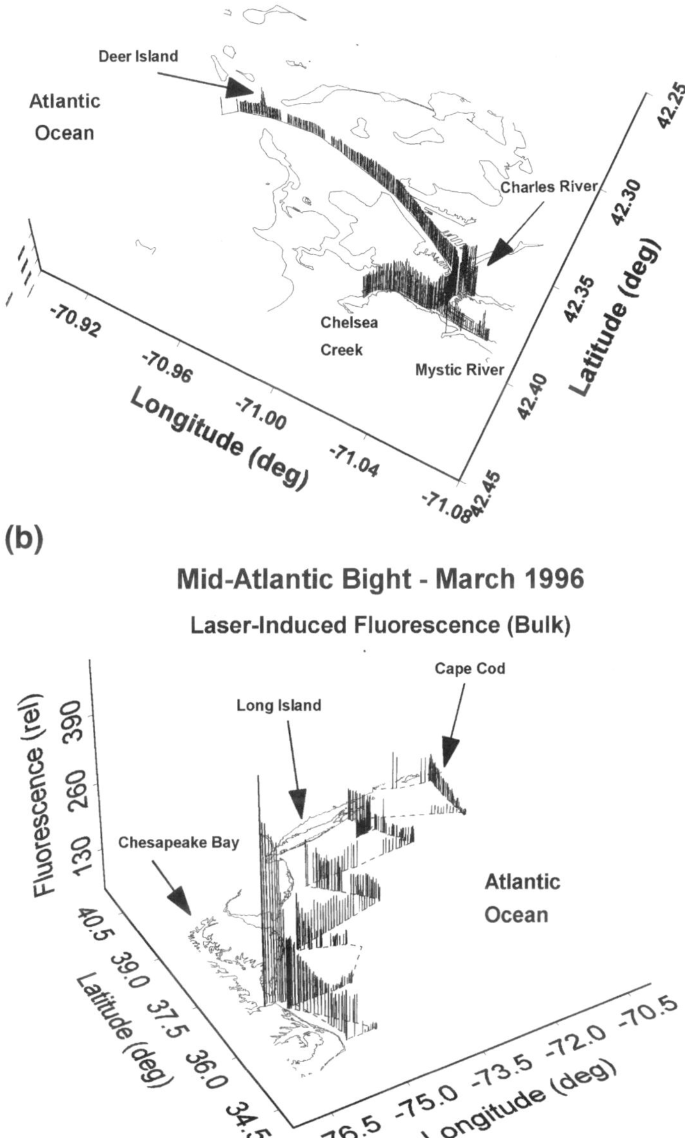

Figure 8.6 LIF measurements in Boston Harbor (a) and the Mid-Atlantic Bight (b). Measurements shown are integrated across all wavelengths with a 0 ns gate delay. The fiber optic probe was towed behind a small ship in Boston Harbor and placed in a seawater flow cell aboard the RV Endeavor in the MAB.

Cape Hatteras, DOC and POC are not subject to air-sea exchange and can accumulate. Other sources of DOC and POC include sediment resuspension and riverine inputs. It has been found that POC fluxes are minor in the carbon exchange across the shelf (Biscaye *et al.*, 1994). However, DOC has only recently been considered to be a major reactive component of the carbon cycling on ocean margins. Riverine inputs are suggested here to be a major source of "excess carbon" to mid-Atlantic Bight shelf waters. Quantification of source water DOC, DOC exchange across the shelf break, and fluxes of DOC offshore at Cape Hatteras are essential to understanding the role that ocean margins play in the global carbon cycle.

Figure 8.6(b) shows the LIF system deployed on a 10 day cruise in March, 1996. Horizontal resolution is about 1.3 km and shows that fluorescence is generally conservative across the shelf. Fluorescence increases from north to south and towards shore revealing that the major sources of fDOM on the shelf are riverine. High spatial resolution on the similar scales as physical measurements of temperature and salinity allow detailed modeling of DOM transport across the shelf. Once again, only bulk average, non-delayed fluorescence are summarized in this figure. Full spectral resolution and time-resolved spectra taken every hour reveal other trends not apparent in the bulk data (manuscript in preparation).

8.8.3 Time Series

Figure 8.7 shows data gathered during a 2-day deployment of a cDOM fluorometer (SeaTech) off the University of Massachusetts Boston (UMB) dock in August of 1996. This time series shows that fluorescence can rapidly change in the course of minutes and that the changes in a restricted area of an estuary are not necessarily well-correlated with tidal height. In the case of the UMB dock, water flow is restricted by exposed mud flats and only during the rising tide is highly fluorescent freshwater being pulled from the nearby Neponset River. As the tide continues to rise, high salinity water from Boston Harbor can cross the exposed mud flats and rapidly lower the local fluorescence. Only with high temporal resolution instrumentation can these details be flushed out.

8.8.4 Pyrene Detection

Because pyrene has a long fluorescent lifetime ($\sim$128 ns) while naturally occurring fluorophores have lifetimes of 0.1 to 10 ns, time-resolved fluorescence can be used to detect pyrene within the much higher background DOM fluorescence signal. Figure 8.8(a) shows a spiked seawater sample where the 0 ns delay data shows the dominant DOM fluorescence signal (LIF system excitation wavelength is 337 nm). If the detector is delayed 32 ns, the 4 peak pyrene spectrum is clearly visible and the DOM and Raman scattering signals are no longer detected.

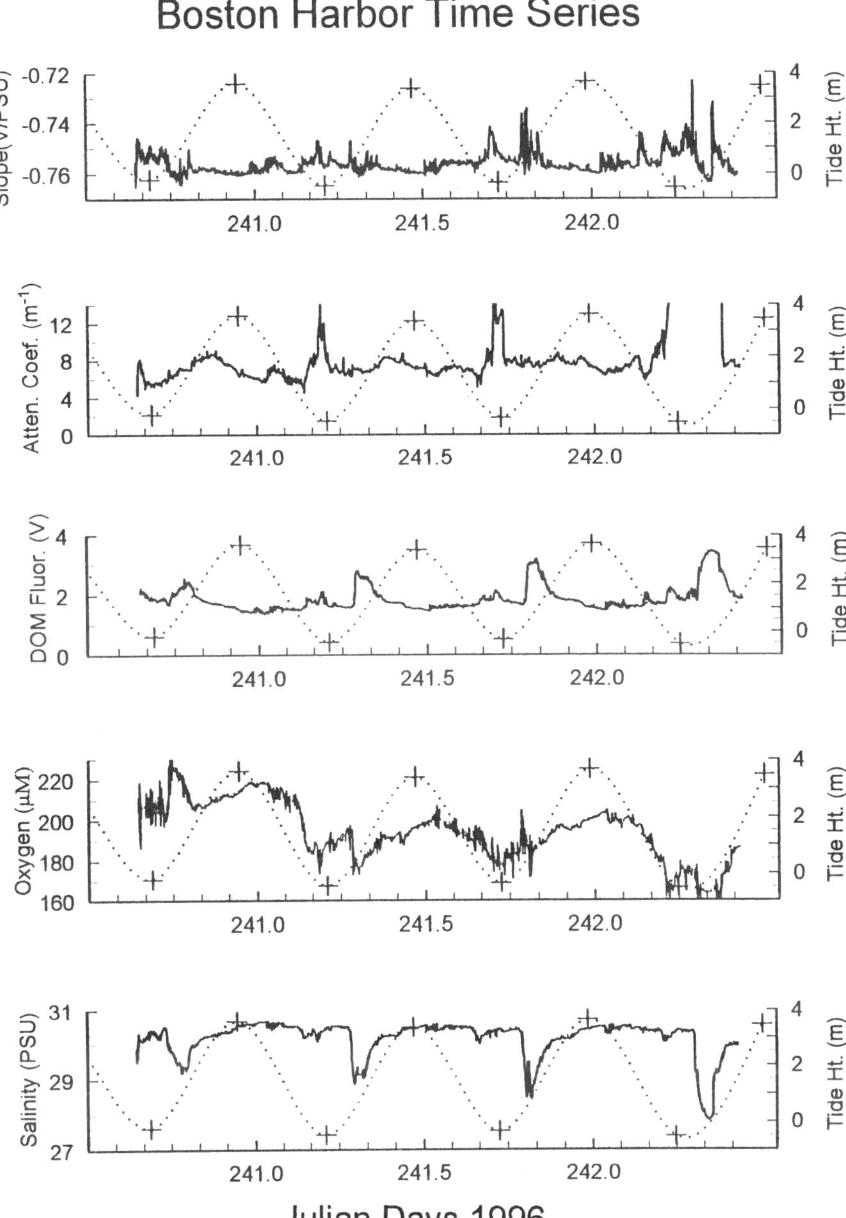

Figure 8.7 Data recovered from a time series deployment in Savin Hill Cove, Boston Harbor. DOM fluorescence shown here are from a SeaTech DOM fluorometer (LIF data not shown). Local tides are plotted as dotted lines. Restricted flow in Savin Hill Cove alters the expected sinusoidal tidal influence. DOM fluorescence normalized to salinity is shown as "slope" which shows that other factors besides freshwater affect seawater fluorescence.

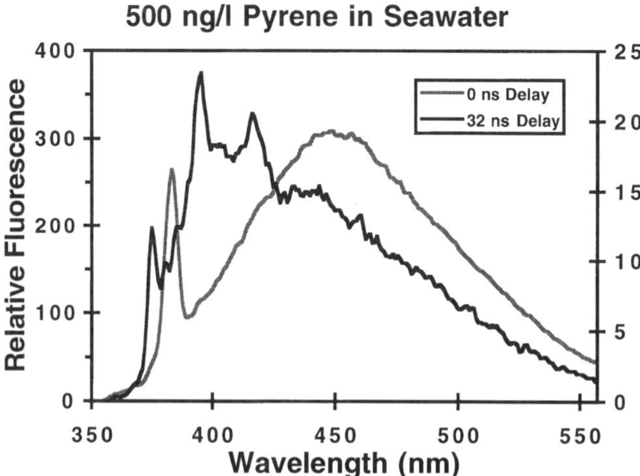

Figure 8.8 An example of time-resolved fluorescence spectroscopy. The shaded line shows the normal steady state fluorescence of seawater with a water Raman peak. No pyrene is detectable. However with a 32 ns delay (dark line, y-axis scale on the right), the natural fluorescence of seawater (due to "humics") is selected against and the longer lived pyrene fluorescence with its characteristic peaks can be seen. Notice the change in scale between the two curves.

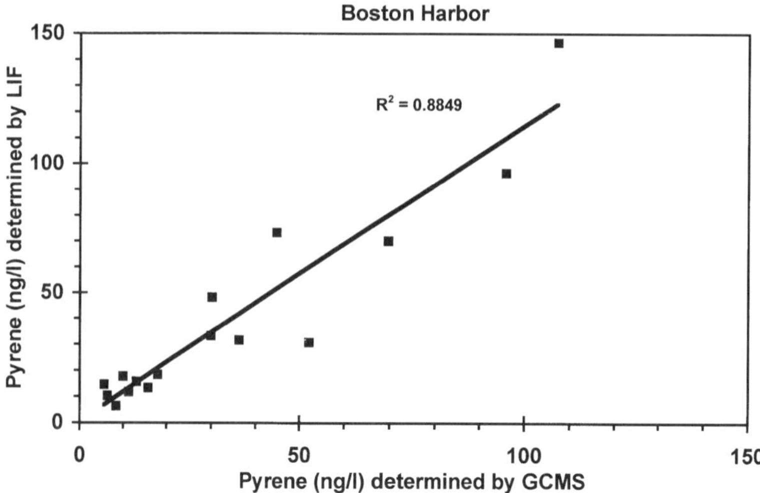

Figure 8.9 Agreement between gas chromatography/mass spectrometry and LIF determinations of pyrene in seawater.

Figure 8.8 shows the pyrene concentrations determined from the 128 ns delayed signal of 24 Boston Harbor seawater samples compared with pyrene concentrations determined by extraction and analysis by gas chromatography/mass spectrometry (GCMS). The advantages of the LIF system are that data are gathered in real-time (20 seconds) *in situ* so that no sampling errors are possible. Also, while GCMS analysis includes dissolved and colloidal PAH (whatever passes through a filter), the LIF analysis only detects the truly dissolved PAH because the fluorescence of PAH bound to colloids is quenched. Thus LIF gives a more accurate determination of the "free" or bioavailable PAH rather than some operationally defined fraction of the PAH obtained due to a filtration step.

8.9 DISCUSSION

The LIF system presented in this chapter is still in its infancy. It appears to represent a great advance in our ability to understand the cycling of the fluorescent fraction of DOM in seawater. With time resolution of a few seconds, spatial resolution of a few meters (or cm in certain applications), spectral resolution of less than 1 nm, and fluorescence lifetime resolution of about 1 ns, the LIF system shows its greatest strengths in large scale (continental shelf) mapping of surface fDOM and time series in rapidly changing areas such as tidally influenced estuaries. It also has the ability to measure, *in situ*, polycyclic aromatic hydrocarbons, especially pyrene due to its high absorption at 337 nm, its relatively high solubility (for a PAH), and its long lifetime (128 ns).

On the other hand, it is not the ideal chemical sensor system for seawater for a number of reasons:

(1) It is limited to surface water applications. UV attenuation, even in the highest quality fused silica core, fused silica clad fiber limits the practical working length to < 50 m.
(2) It is expensive. The system costs about $80,000 in components and takes about 1 year to assemble and become familiar with its operation.
(3) It is fairly large and consumes a significant amount of power (not amenable to battery operation). The total package weighs about 50 kg plus the computer and takes about 3 m of bench space. The current requirements include a refrigeration unit for the ICCD, a computer, and a nitrogen tank per day for the laser. Many of these problems could be alleviated with the addition of an air cooled camera, a nitrogen generator or sealed nitrogen laser, a notebook computer, and repackaging of all the components, but the system would still be a ship-based system.
(4) The system is currently limited to one excitation wavelength (337 nm) which happens to be optimum for DOM and pyrene, but does not allow detection of chlorophyll or proteins, nor does it allow excitation-emission matrix scans (Coble, 1996).
(5) Biofouling has not yet been investigated. However, over 10 days, there was no indication of any interference, and perhaps because of the high peak power over the small fiber area, biofilms may be discouraged from forming.

(6) Data acquisition and reduction is not automatic. As we are just developing the system, much more information is given than the average marine chemist may want. Further studies should allow us to design different LIF systems specifically for different chemical species and algorithms will be developed to reduce data to DOC or pyrene equivalents.
(7) Fluorescence systems measure an inherent optical property (fluorescence) of some subset of the total dissolved organic pool of molecules with a continuum of varying wavelengths and quantum efficiencies. Therefore LIF gives some apparent carbon content of average optical property that is difficult to relate to DOM's unknown molecular structures. Continual groundtruthing is needed to give confidence in the meaning of a particular fluorescence measurement.

8.10 FUTURE APPLICATION AND MODIFICATIONS

By developing the ability to send light down a fiber into seawater, the LIF system shows much promise for studying seawater chemistry. Many of the limitations mentioned above can be overcome in the near future. Most of the components have low power equivalents and can be miniaturized, and thus the entire system can be made submersible making the system autonomous rather than tethered to a ship. In this way, the LIF system would be compatible with AUVs, ROVs, moorings, towed vehicles, and benthic landers. Many of the component costs for the miniaturized versions are significantly less expensive than their larger versions making a commercial product in the $10,000 to $20,000 a possibility. Dye lasers, multiple lasers, Raman shifted lasers, and laser diodes are possibilities to alter the excitation wavelength appropriately. The 2 major signals that the LIF system can differentiate in its current configuration are DOM and pyrene. Rather than report entire spectra, integrated signals using a simple photomultiplier tube and filters is sufficient to estimate DOM or photochemical bleaching. Pyrene can be targeted with long gate delays without taking full time-resolved spectra. So overall, most if not all of the limitations of the LIF system can be overcome.

But more importantly, with a delivery system for light into seawater, fiber optics chemical sensors will allow a whole host of other analytes to be measured. There are several methods for detecting non-fluorescent molecules. One approach is to measure the change of fluorescence when a compound associated with the fiber interacts with the compound of interest. For example, organic nitro-compounds such as 2,4,6-trinitrotoluene (TNT), 2,4-dinitrotoluene (DNT), and hexahydro-1,3,5-trinitro-1,3,5-triazine (RDX) have been measured fluorometrically based on the quenching of a fluorophore incorporated in a membrane (Jian and Seitz, 1990). Immobilization of antibodies onto the fiber optic creates biosensors that can measure specific toxins (Hobbs, 1992). Possibly, a glucose biosensor already developed (Moreno-Bondi *et al.*, 1990) can be modified to supply a measurement of the carbohydrate fraction of the DOC, a fraction that is rather reactive and may make up 50% of the total DOC (Benner *et al.*, 1992).

Another approach is to release a reagent that specifically binds the analyte and changes its fluorescence characteristics. In the case of dissolved metals, non-fluorescent indicator molecules such as para-Tosyl-aminoquinoline may form fluorescent complexes with metals such as zinc and cadmium resulting in very high sensitivities. A pressurized-membrane reagent delivery system has already been developed (Inman *et al.*, 1988), so with the proper indicator, many non-fluorescent analytes can be made to fluoresce. Amino acids have been measured in seawater using a similar fluorescence system (Wing *et al.*, 1990). Several nitrogenous pesticides have been detected by photolyzing the analyte and then reacting it with ortho-phthaldialdehyde-2-mercaptoethanol (OPA-MERC) (Patel *et al.*, 1990). The resulting isoindoles are highly fluorescent. Dissolved nucleic acids have also been measured in seawater with a fluorescent dye (Sakano and Kamatani, 1992), and so is a candidate for *in situ* measurement.

Finally, the development of fiber optic chemical sensors is advancing at a very high rate. Medical applications, the use of sol gels, hydrogels, immunosensors, and fiber bundles may allow similar LIF systems to monitor a whole host of components of seawater. Along with submersible flow-injection systems and electrochemical sensors, fiber optic fluorescence systems will help marine chemists analyze more seawater components faster and more sensitively.

References

Alarie, J., Vo-Dinh, T., Miller, G., Ericson, M., Maddox, S., Watts, W., Eastwood, D., Lidberg, R. and Dominguez, M. (1993). *Rev. Sci. Instrum.*, 64, 2541–2546.

Aluwihare, L.I., Repeta, D.J. and Chen, R.F. (1997). *Nature*, 387, 166–169.

Amador, J.A., Milne, P.J., Moore, C.A. and Zika, R.G. (1990). *Mar. Chem.*, 29, 1–17.

Amon, R. and Benner, R. (1994). *Nature*, 369, 549–552.

Armstrong, F.A.J. and Boalch, G.T. (1961). *J. Mar. Bio. Ass. U.K.*, 41, 591–597.

Beardsley, R.C. and Boicourt, W.C. (1981). *In Evolution of physical oceanography*, (Ed. Wunsch, B.A.W.a.C.) MIT Press, Cambridge, MA, pp. 198–233.

Benner, R., Pakulski, J.D., McCarthy, M., Hedges, J. and Hatcher, P. (1992). *Science*, 255, 1561–1564.

Biscaye, P.E., Flagg, C.H. and Falkowski, P.G. (1994). *Deep-Sea Res. Part II*, 41, 231–252.

Bublitz, J., Dichenhausen, M., Gratz, M., Todt, S. and Schade, W. (1995). *Appl. Optics*, 34, 3223–3233.

Cabaniss, S. and Shuman, M. (1987). *Mar. Chem.*, 21, 37–50.

Chen, R.F. (1992). University of California, San Diego.

Chen, R.F. (1998). *Org. Geochem.*, in press.

Chen, R.F. and Bada, J.L. (1989). *Geophys. Res. Lett.*, 16, 687–690.

Chen, R.F. and Bada, J.L. (1992). *Mar. Chem.*, 37, 191–221.

Chen, R.F., Bada, J.L. and Suzuki, Y. (1993). *Geochim. Cosmochim. Acta*, 57, 2149–2153.

Chen, R.F., Chadwick, D.B. and Lieberman, S.H. (1997). *Org. Geochem.*, 26, 67–77.

Chen, R.F. and Repeta, D.L. (1994). *EOS Transactions*, 75, 311.

Coble, P., Green, S., Blough, N. and Gagosian, R. (1990). *Nature*, 348, 432–435.

Coble, P.G. (1990). *MIT/WHOI*, WHOI-90-05.

Coble, P.G. (1996). *Mar. Chem.*, 51, 325–346.

Conrad, R. and Seiler, W. (1980). *EEMS Microbiol. Lett.*, 9, 61–64.
DeGrandpre, M.D., Hammar, T.R., Smith, S.P. and Sayles, F.L. (1995). *Limnol. Oceanogr.*, 40, 969–975.
Determann, S., Reuter, R., Wagner, P. and Willkomm, R. (1994). *Deep-Sea Res.*, 41, 659–675.
Dorsch, J.E. and Bidleman, T.F. (1982). *Est. Coast. Shelf Sci.*, 15, 701–707.
Dunlap, W.C. and Susic, M. (1985). *Mar. Chem.*, 17, 185–198.
Dunlap, W.C. and Susic, M. (1986). *Mar. Chem.*, 19, 99–107.
Duursma, E.K. (1974). *In Optical Aspects of Oceanography*, (Ed. Nielsen, N.G.J.a.E.S.) Academic Press, New York, pp. 237–256.
Ehrhardt, M. and Knap, A. (1989). *Mar. Chem.*, 26, 179–188.
Ehrhardt, M. and Petrick, G. (1989). *Mar. Poll. Bul.*, 20, 560–565.
Filippova, E.M., Chubarov, V.V. and Fadeev, V.V. (1993). *Can. J. Appl. Spectroscop.*, 33.
Genders, S. (1988). *Oil Chem. Poll.*, 4, 113–126.
Gieskes, W., Kraay, G. and Tijssen, S. (1978). *Neth. J. Sea Res.*, 12, 195–204.
Grimmer, G. (1983). Environmental carcinogensL polycyclic aromatic hydrocarbons, CRC Press, Boca, Raton.
Hayase, K., Tsubota, H., Sunada, I., Goda, S. and Yamazaki, H. (1988). *Mar. Chem.*, 25, 373–381.
Hayase, K., Yamamoto, M., Nakazawa, I. and Tsubota, H. (1987). *Mar. Chem.*, 20, 265–276.
Hobbs, J. (1992). *Laser Focus World*, 83–86.
Hoge, F.E., Vodacek, A. and Blough, N.V. (1993). *Limnol. Oceanogr.*, 38, 1394–1402.
Jian, C. and Seitz, W.R. (1990). *Anal. Chim. Acta*, 237, 265–271.
Kalle, K. (1949). *Deut. Hydrogr. Z.*, 2, 117–124.
Karabashev, G.S. (1996). *Oceanol.*, 36, 149–156.
Karabashev, G.S. and Agatova, A.I. (1984). *Oceanol.*, 24, 680–682.
Karabashev, G.S. and Solovyev, A.N. (1973). *Oceanol.*, 13, 361–366.
Karyakin, A.V. and Galkin, A.V. (1995). *J. Anal. Chem.*, 50, 1078–1080.
Klinkhammer, G. (1994). *Mar. Chem.*, 47, 13–20.
Kouassi, A.M. (1986). *Univ. of Miami.*
Kramer, C.J.M. (1979). *Neth. J. Sea Res.*, 13, 325–329.
Laane, R. (1981). *Neth. J. Sea Res.*, 15, 88–99.
Laane, R.W.P.M. and Koole, L. (1982). *Neth. J. Sea Res.*, 15, 217–227.
Lieberman, S.H., Inman, S.M. and Theriault, G.A. (1992). *In Oceans '91*, pp. 507–514.
Moffett, J. and Zika, R. (1987). *In Photochemistry of Environmental Aquatic Systems*, pp. 116–130.
Momzikoff, A. (1983). *Mar. Chem.*, 12, 1–14.
Momzikoff, A., Dallot, S. and Pizay, M.-D. (1992). *Deep-Sea Res.*, 39, 1481–1498.
Mopper, K. and Schultz, C. (1993). *Mar. Chem.*, 41, 229–238.
Mopper, K., Zhou, X., Kieber, R.J., Kieber, D.J., Sikorski, R.J. and Jones, R.D. (1991). *Nature*, 353, 60–62.
Moran, M.A., Wicks, R.J. and Hodson, R.E. (1991). *Mar. Ecol. Prog. Ser.*, 76, 175–184.
Moreno-Bondi, M., Wolfbeis, O., Leiner, M. and Schaffar, B. (1990). *Anal. Chem.*, 62, 2377–2380.
MWRA (1993). The state of Boston Harbor, MWRA, Boston, MA.
Niessner, R. and Panne, U. (1991). *Anal. Chim. Acta*, 255, 231–243.
Patel, B., Moye, H. and Weinberger, R. (1990). *J. Agric. Food Chem.*, 38, 126–134.
Romanovskaya, G.I. and Lebedeva, N.A. (1993). *J. Anal. Chem.*, 48, 1400–1405.
Rudnick, S.M. and Chen, R.F. (1998). *Talanta*, 47, 907–919.

Sakano, S. and Kamatani, A. (1992). *Mar. Chem.*, 37, 239–255.
Schade, W. and Bublitz, J. (1996). *Environ. Sci. Technol.*, 30, 1451–1458.
Sharp, J.H. (1997). *Mar. Chem.*, 56, 265–277.
Sholkovitz, E.R., Boyle, E.A. and Price, N.B. (1978). *Earth Planet. Sci. Lett.*, 40, 130–136.
Sierra, M.D.S., Donard, O., Lamotte, M., Belin, C. and Ewald, M. (1994). *Mar. Chem.*, 47, 127–144.
Skoog, A., Hall, P.O.J., Hulth, S., Paxeus, N., Loeff, M.R.v.d. and Westerlund, S. (1996). *Geochim. Cosmochim. Acta*, 60, 3421–3431.
Smart, P.L., Finlayson, B.L., Rylands, W.D. and Ball, C.M. (1976). *Water Res.*, 10, 805–811.
Snow, J. (1987). *Sea Technol.*, 10–13.
Theobald, N. (1989). *Mar. Poll. Bul.*, 20, 134–140.
Vodacek, A., Hoge, F., Swift, R., Yungel, J., Peltzer, E. and Blough, N. (1995). *Limnol. Oceanogr.*, 40, 411–415.
Wang, X.C., Druffel, E.R.M. and Lee, C. (1996). *Geophys. Res. Lett.*, 23, 3583–3586.
Willey, J. and Atkinson, L. (1982). *Est. Coast. Shelf Sci.*, 14, 49–59.
Wing, M.R., Stromvall, E.J. and Lieberman, S.H. (1990). *Mar. Chem.*, 29, 325–338.
Zafiriou, O.C., Jousset-Dubien, J., Zepp, R.G. and Zika, R.G. (1984). *Environ. Sci. Technol.*, 18, 358A-371A.
Zepp, R., Schiotzhauer, P. and Sink, R. (1985). *Environ. Sci. Technol.*, 19, 74–81.
Zika, R.G. (1987). *Rev. Geophys.*, 25, 1390–1394.
Zimmerman, J.T.F. and Rommets, J.W. (1974). *Neth. J. Sea Res.*, 8, 117–125.

9. ELECTROCHEMICAL SENSORS FOR *IN SITU* SMALL-SCALE, FAST TEMPORAL MEASUREMENTS OF ORGANIC MOLECULES IN SEAWATER

PAUL A. MOORE[a] and GREG A. GERHARDT[b]

[a]*Laboratory for Sensory Ecology, Department of Biological Sciences, Bowling Green State University, Bowling Green, OH 43403 and*
[b]*Departments of Psychiatry and Pharmacology, Rocky Mountain Centre for Sensor Technology, University of Colorado Health Science Centre, Denver, CO 80262*

9.1 THE IMPORTANCE OF ORGANIC MOLECULES FOR AQUATIC ORGANISMS

Chemical cues are important in the natural history of virtually every aquatic organism. For example, for growth and reproduction, aquatic primary producers rely on chemical cues e.g., nitrogen and phosphorus. Many aquatic larvae use chemical signals to find suitable settlement sites. For macroscopic animals, chemical signals are used in predator-prey interactions, foraging ecology, and mate selection. Great inroads have been made by many scientists in identifying what chemicals are involved in these different interactions, but a complete ecological picture of these interactions is missing mainly due to the lack of knowledge about the fine-scale spatial and temporal distribution of chemicals in aquatic habitats.

For macroscopic animals there is considerable morphological, electrophysiological, and behavioral evidence for chemosensory location of prey and predators by marine organisms on scales of millimetres to metres (e.g. Carr and Derby, 1986; Legier-Visser *et al.*, 1986; Price *et al.*, 1988), and the importance of near-field chemosensory orientation is widely accepted. While the sensory basis of such behavior in laboratory settings is fairly well understood for some benthic marine organisms like lobsters (e.g. Atema, 1985; Derby and Atema, 1982; Moore *et al.*, 1991), very little is known about chemosensory behavior in natural environments. Analogies from coastal species in the laboratory are not always applicable to the size and time scales of benthic or pelagic environments. Physical processes dispersing chemical signals are different for near-shore and deep benthic environments as compared with laboratory settings and these natural processes have rarely been measured and quantified. The water column and benthic environments offer several potential cues for behavioral orientation, ranging from large-scale gradients of pressure, temperature, and light through meso-scale changes in food and chemical concentrations to small-scale and transient visual, chemical, hydrodynamic, and electrical signals generated by animals. Each of these sensory signals has properties and characteristics, which convey different types of information over different temporal and spatial scales (Atema *et al.*, 1988). Phenomena, such as migration and swarming that involves large numbers of organisms over large distances, rely on large-scale sensory cues, while inter-individual events like predation, escape, and mating are mediated by short-range and short-lived biogenic signals (Maier and Müller, 1986; Atema and Engstrom, 1971; Gleeson *et al.*, 1987; Miller, 1979).

Aquatic chemical signals can convey specific information over a long range, but have been shown to be highly patchy over short time and space scales (Atema, 1985; Moore and Atema, 1991; Atema *et al.*, 1991; Moore *et al.*, 1992). The importance of chemical signals to deep-sea animals has been amply demonstrated by numerous studies using baited traps to catch or photograph sparse and elusive scavengers (e.g. Hessler *et al.*, 1972; Busdosh *et al.*, 1982; Hargrave, 1985; Ingram and Hessler, 1983; Sainte-Marie, 1986; Wilson and Smith, 1984). It is clear from these studies that chemical information can travel considerable distances and guide predators or scavengers quickly and accurately to a food source. On a smaller scale, Hamner and Hamner (1977) have demonstrated precise scent tracking by neritic sergestid shrimp. It is possible that the midwater and benthic environments are laced with plumes and trails of chemical signals left, for varying lengths of time, by passing individuals or groups. In addition to food odors, algal metabolites (e.g. Maier and Müller, 1986), pheromones (e.g. Atema and Engstrom, 1971; Gleeson *et al.*, 1987) or sperm attractants (e.g. Miller, 1979) might play crucial roles in the biology of widely dispersed organisms, regulating feeding, aggregation, and spawning. In all of these instances, it has been shown quite convincingly that fine-scale chemical signals play an important role in the behavior of marine animals, but virtually nothing is known of the spatial and temporal characteristics of chemical signals in the open ocean environment. All of these studies show the importance of chemical signals, which are small organic molecules, to the ecology, physiology, and behavior of organisms.

To understand how organisms use chemical signals in their natural environment we must be able to "see" their odor world. With the development of new technology to investigate the fine-scale distribution of chemical signals in aquatic environments, it has been possible to further our understanding of the role of chemical signals in ecosystem dynamics. In order to do this, it is important to develop sensors and recording systems that sample at the same spatial and temporal scales used by these organisms.

9.2 APPROPRIATE SPATIAL AND TEMPORAL SCALES FOR MEASUREMENT

The physical processes involved in the distribution of chemicals in the aquatic habitat range from molecular diffusion to large-scale turbulence. Within this range, biologically relevant spatial and temporal scales must be estimated from behavioral and electrophysiological studies. Most arthropods orient to chemical signals using sensory input from antennae or antennules (Reeder and Ache, 1980; Devine and Atema, 1982). These appendages usually have small hairs or sensilla that are permeable to odors and contain the dendrites of the primary receptor cells (Ghiradella *et al.*, 1968; Laverack, 1988). In addition, male algal gametes can track signals from female gametes over distances of 300–400 microns (Maier and Muller, 1986). To match the spatial sampling area associated with these organisms, we can choose electrochemical electrodes with diameters of 8–150 μm. In contrast, the temporal sampling scales for aquatic organisms is not well understood and must be estimated from a limited number of physiological studies. Neurons, both peripheral and CNS, in two moth species can give distinct responses to odor pulses presented as fast as 10 Hz (Kaissling *et al.*, 1987; Christensen and Hildebrand, 1988). Similar studies in the lobster *Homarus americanus* have found some peripheral chemoreceptor cells that can follow 4 Hz pulses (Gomez *et al.*, 1994) and others that begin to adapt in 500 ms (Voigt and Atema, 1990). Algal gametes have been shown to make behavioral decisions on the order of 100 ms (Maier and Muller, 1986). From these studies, we can estimate a temporal sampling scale between 100 and 500 ms. In addition, frequency spectra of aquatic odor signals measured at sampling rates of 10, 25, and 200 Hz (Moore and Atema, 1988; 1991) have shown that most of the odor signal fluctuations lie below 10 Hz. Thus, a sampling rate of 25–50 Hz in an aquatic medium will give a biologically realistic resolution of odor pulses, although a more thorough understanding of the temporal dynamics associated with the turbulent dispersal of chemical signals requires a much higher sampling rate.

9.3 BASIC THEORY OF ELECTROCHEMICAL MEASUREMENT OF ORGANICS IN SEAWATER

The techniques for electrochemical recordings in seawater are based on the principles of electrochemical measurements at solid electrodes (Adams, 1969). In general terms, the working electrode (described below) is polarized in an aqueous solution using a potentiostat. In solution, molecules that physically come into contact with the exposed surface of the working electrode can donate or receive electrons from the electrode's surface, provided that the polarization is sufficient to cause the molecule to oxidize or reduce. If the molecule donates electrons, oxidation of the molecule has occurred and if the molecule receives electrons then reduction has happened. The resulting current flow due to the exchange of electrons at the working electrode surface is directly proportional to the number of molecules that contact the surface of the working electrode. A calibration of the electrode prior to *in situ* measurements is used to convert the current flow into concentration units.

This process is illustrated in Figure 9.1 for the detection of dopamine in seawater. Thus, *in situ* electrochemical measurements entail causing oxidation or reduction reactions at the surfaces of electrodes and measuring the amount of

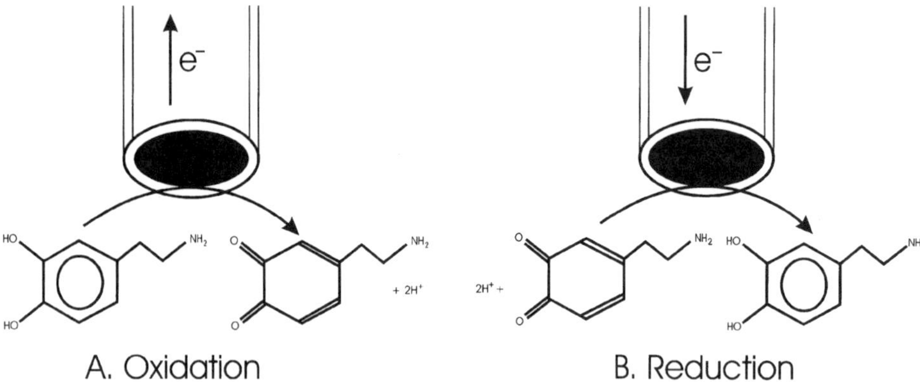

Figure 9.1 Oxidation and reduction reactions that occur at the surface during electrochemical measurements of dopamine in seawater. During the oxidation phase (A) dopamine is converted to a quinone and two electrons per molecule of dopamine are received by the electrode. During the reduction phase (B), some or all of the quinone can be converted back to dopamine and 2 electrons per molecule of the quinone are received from the electrode surface.

current flow that results from these chemical reactions. By quantifying the current flow, it is possible to quantify the number of molecules that contact the electrode's surface.

There are numerous recording techniques for carrying out electrochemical recordings *in situ*. These methods differ in the way in which the potential applied to the working electrode is varied. One method is called chronoamperometry and involves varying the polarization or applied potentials to the electrode's surface using a square-wave or voltage step (see below). The resulting chemical reactions at the electrode surface are measured as a function of time. Thus, chronoamperometry is a time (chrono) versus current (amperometry) measurement.

9.4 CHRONOAMPEROMETRY

Chronoamperometry is the electrochemical method that uses a square-wave voltage to detect chemicals at the electrode surface (Figure 9.2). This can be explained best by describing the technique that is used to measure dopamine in seawater using the IVEC-10 electrochemical system at a sample rate of 10 Hz. A positive voltage is applied to a working electrode versus a Ag/AgCl reference electrode. The reference electrode is using a small gauge silver wire that has been plated with chloride. This is a common method for the construction of reference electrodes. For detection of dopamine, the oxidation voltage is set to +0.55 V for

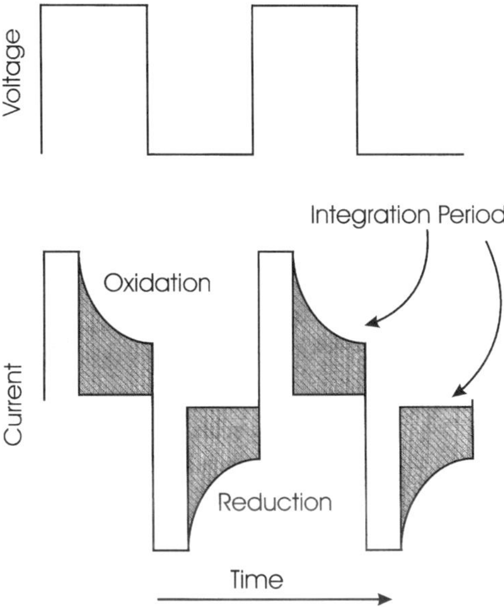

Figure 9.2 Schematic showing the voltage steps applied to the working electrode (upper trace) and the resulting current measured at the electrode. Shaded area is the part of the current response that is integrated for the concentration measurement.

50 ms. During this phase of the measurement, analog-to-digital conversions of the current from the working electrode occur at 4,000 Hz. These 4,000 Hz samples are averaged for the entire 50 ms and stored as a single measurement of charge (integrated current) for this time epoch. The voltage is then stepped down to 0.0 V for the reduction phase of the measurement. Again, analog-to-digital conversions occur at 4,000 Hz and are averaged for the 50 ms epoch. The average is stored as the single measurement of charge for the reduction phase of measurement. The 50 ms oxidation step along with the 50 ms reduction step is repeated for the entire measurement period. This process results in a single time point for the oxidation measurement and a separate time point for the reduction measurement each measured at a rate of 10 Hz. The time length of the oxidation and reduction step sets the overall sampling rate of chemical signals. Currently, the IVEC-10 system can sample using this technique at an upper rate of 100 Hz.

Chronoamperometry has several advantages over other recording methods. The short time periods of the applied potential allow for rapid sampling of environmental signals (See examples of environmental signals below). In addition, the technique of stepping voltages decreases the possibility of chemicals adhering to the working electrode and compromising the sensitivity of the electrode. Using the currently available equipment, it is possible to measure nanomolar, micromolar, and millimolar levels of certain molecules in seawater.

9.5 AMPEROMETRY

Another method for electrochemical recordings involves applying a constant polarizing voltage to the electrode surface and is called amperometry (current monitoring). This approach is used when very high temporal resolution recordings are needed. This approach is very similar to the one outlined above except that a constant voltage is applied to the electrode instead of a square-wave. Only oxidation reactions occur when using the constant voltage technique at +0.55 V using most carbon electrodes. Analog-to-digital conversions occur at 4,000 Hz and are averaged for the time epoch appropriate for the desired sampling rate. The average is stored as the single measurement of charge for the oxidation measurements. Using the constant potential recordings, temporal resolutions of 10,000 Hz are easily achieved. We have used this type of recording method to measure aquatic tracer signals in the deep ocean (Atema *et al.*, 1991) and the fine structure of aquatic tracer signals in a flume (Moore *et al.*, 1992). An example of these high-resolution signals recorded from 2300 feet from the ocean surface is seen in Figure 9.3.

There are, however, several problems with this methodology. First, the constant application of the polarizing potential can cause the recording electrodes to

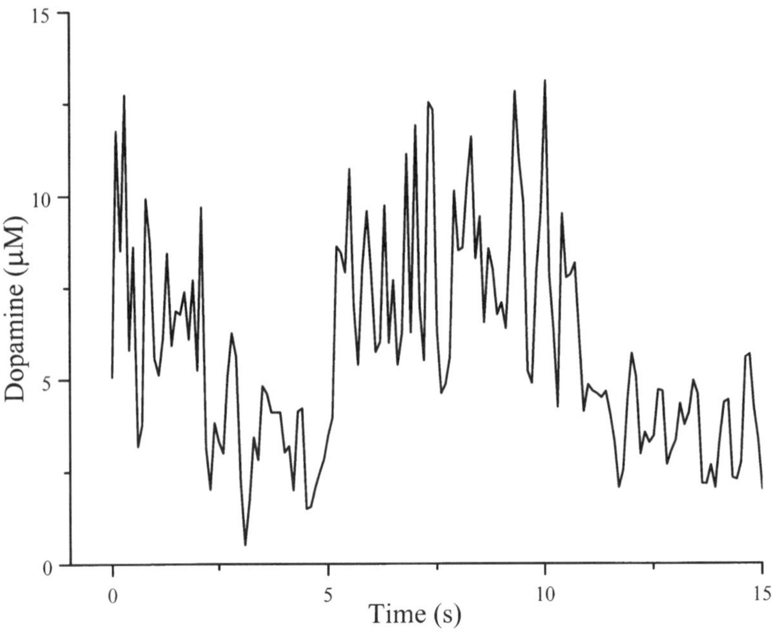

Figure 9.3 High-speed (200 Hz) recordings from a deep-water dive using the Johnson Sea-Link showing high frequency fluctuations of chemicals as measured using amperometry.

adsorb reaction products from the electrochemical reactions and this will result in loss of sensitivity for the recordings. Second, the rapid measures preclude signal-averaging methods from being used and therefore the method is less sensitive. For example, a temporal resolution of 10,000 Hz means a sampling bin of 0.1 ms. If we have analog-to-digital conversions at a rate of 100,000 Hz, this means that only 10 oxidation measurements are averaged for the final concentration estimate.

9.6 SELECTIVITY OF THE ELECTRODE AND FINGERPRINTING OF THE SIGNAL MOLECULE

There are several techniques that can be used to either further identify signal chemicals or to increase the specificity of the electrodes for target molecules. Each of these techniques has advantages and disadvantages, and the selection of whether the technique should be used is dependent upon the experimental situation.

First, the specificity of the electrodes for dopamine or other compounds depends on the electrical characteristics of their oxidation and reduction reactions. Molecules are oxidized and reduced at different voltages depending upon the surrounding conditions (pH, etc.) and type of sensor. These voltages can be determined in laboratory trials, so that by selecting the optimal voltages, one can include certain compounds and exclude others. For example, by selecting stepping voltages between 0.0 V and +0.55 V for dopamine, it is possible to create a voltage window that will exclude all of those compounds that are oxidized at higher voltages. This technique does not compromise the temporal or spatial recording characteristics of the electrode and is the preferred first step to increase electrode selectivity.

Second, the relative strength of the oxidation and reduction currents within any one sampling epoch can be used as a characteristic signature of the chemical species. By measuring this ratio (reduction to oxidation current ratio) during the sampling of *in situ* chemical signals, it is possible to further identify the chemical (or chemicals) that are contributing to the currents measured by the electrode. Each organic chemical has a characteristic oxidation to reduction ratios recorded using the IVEC-10 system. For example, using the graphite epoxy electrodes (described below), dopamine has a characteristic ratio of 0.5–0.7 and serotonin has a ratio of 0–0.1 at pH of 7.4. By determining these ratios in the laboratory beforehand, it is possible to further identify the chemical composition of signals during *in situ* measurements.

Finally, it is possible to coat the surface of the electrode with ion-selective membranes. Nafion-coated electrodes have been used in the past to increase selectivity for dopamine over other potentially inferring molecules within the central nervous system (Gerhardt *et al.*, 1984), but has not been needed for oceanographic applications. Other membranes may be used to increase selectivity for other compounds, but have not been extensively tested or employed (Friedemann *et al.*, 1996). Surface membranes have a drawback of decreasing the temporal response of the electrode by adding another diffusional step to the detection process.

9.7 CURRENT INSTRUMENTATION

The present recording system that is being used to measure the spatial and temporal distribution of aquatic chemical signals is called the IVEC-10. The In Vivo Electrochemistry Computer system (IVEC-10) was developed by Dr. Greg A. Gerhardt at the University of Colorado Health Sciences Center and is currently being manufactured and sold by Medical Systems Corp. Greenvale, NY (800–654–5406). This system was originally developed and used extensively by Dr. Gerhardt to measure nanomolar to micromolar quantities of neurotransmitters within the CNS of mammalian systems (Adams, 1969; Adams and Marsden, 1982; Gerhardt and Adams, 1982; Gerhardt *et al.*, 1984; 1987; Gerhardt and Palmer, 1987; Gratton *et al.*, 1988). This system has been adapted for field use in conjunction with Dr. Gerhardt to measure concentrations of tracer compounds in seawater with very high temporal resolution (Moore *et al.*, 1989; Moore and Atema, 1991; Atema *et al.*, 1991; Moore *et al.*, 1992). The electrodes were originally developed for the detection of monoamines (e.g., dopamine, norepinephrine and serotonin). Use of dopamine has the advantage that it does not occur naturally in seawater at levels detectable by our system and other chemicals do not interfere with its detection by the sensors. It has the drawback, however, of being a tracer and not a naturally occurring signal chemical. An additional drawback of the current system is that the methodologies to measure endogenous aquatic signal compounds (such as tyrosine, tryptophan, glutamate, glutathione, nitrates, or phosphates) have not been developed.

9.8 SENSOR CONSTRUCTION

The current electrodes are small glass capillary tubes either filled with a graphite epoxy (GEC) or with a single carbon fiber (Fiber) as small as 8 μm (Figure 9.4). The glass capillary for the electrodes is shaped on a common glass electrode puller using sodalime glass. For the carbon fiber electrodes, a single fiber is sealed in the glass capillary tube with an epoxy and baked to harden the epoxy. Both the GEC and Fiber electrodes are back-filled with graphite epoxy until the upper part of the capillary is filled with conducting epoxy. A copper wire is inserted into the epoxy for electrical connection and this assembly is baked again. The electrodes are connected to the IVEC-10 circuitry via a small connector for laboratory work. The sensors can currently be purchased through the Rocky Mountain Center for Sensor Technology and directly from Quanteon, Denver CO. (303–315–8650).

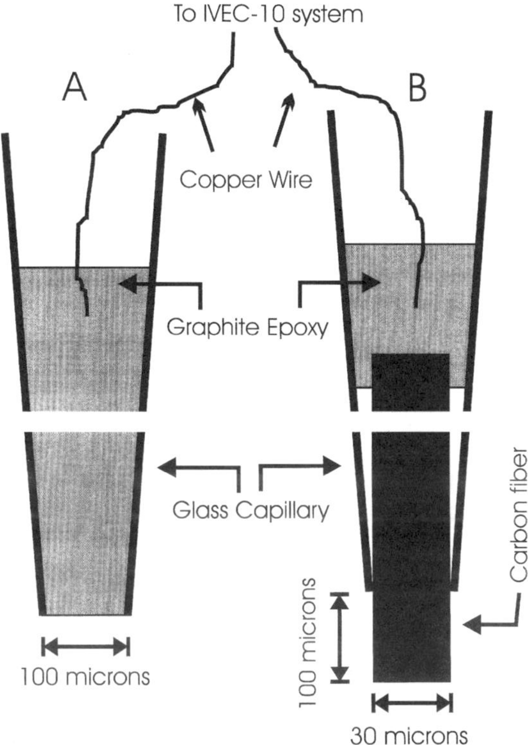

Figure 9.4 Drawing showing the two basic types of electrodes used with the IVEC-10 electrochemical system. The length of the electrodes is not drawn to the same scale as the width and the middle section of the electrode (white strip) has been removed for clarity. The graphite-epoxy electrode (A) consists of a glass capillary that is back-filled with graphite epoxy. A copper wire, which is inserted into the epoxy before hardening, is used to connect the electrode to the main IVEC-10 system. The detecting surface of the electrode is a flat disk. The carbon-fiber electrode (B) has a carbon fiber that is used as the surface of the electrode. The carbon fiber can be extended out from the end of the capillary, so that the detecting surface of the electrode is a cylinder.

For fieldwork, we have connected the electrodes to submersible cables and then sealed all of the connections using silicon and epoxy. These are connected to a submersible headstage, which amplifies the signal. This signal is fed through a submersible cable that connects it to the IVEC-10 system.

9.9 EXAMPLE OF USES

The present detection system has been used in a number of laboratories for both laboratory and field measurement of chemical signals in seawater. The main emphasis of all of these examples is measuring organic signals in flowing seawater at small spatial scales and high temporal resolutions.

9.9.1 Measurement of Chemical Signals in Aquatic Field Settings

We have performed a number of studies using the electrochemical recording techniques and the IVEC-10 system to measure concentrations of potential signal molecules (Moore *et al.*, 1989; Moore and Atema, 1991; Atema *et al.*, 1991; Moore *et al.*, 1992; Moore *et al.*, 1994). These studies have provided insight into the fine-scale structure of chemical signals dispersed by turbulent flow and allowed us to make predictions on the role of chemical signals in guiding behavior and foraging. Results from previous studies (Moore and Atema, 1991; Moore *et al.*, 1992; Moore *et al.*, 1994) have demonstrated three major consequences of different flow velocities and flow regimes on the distribution of chemical signals. First, slower flow velocities allow the initial energy of an odor plume to carry the main axis of the odor plume farther into the water column before the carrier flow begins to transport the chemical downcurrent. The main axis of an odor plume contains the most odor pulses and exhibits the largest fluctuations in concentration. Second, there is a general relationship between plume microstructure and the magnitude and degree of penetrants of turbulence into the boundary layer. Increased turbulence serves both to disperse odors and to distribute odor molecules into smaller and more discrete patches. Conversely, less turbulent flows result in greater average odor concentrations and more uniform distribution of odor further downstream from the source. Third, increased turbulence generally results in more predictable relationships between microscale plume structure and distance to the odor source. At higher flow velocities, plumes remain patchier as they are advected downstream as compared to slower velocities. These smaller and more discrete patches will have higher concentration gradients between patch/non-patch boundaries. In general, odor pulses will have higher slopes at higher flow velocities. At slower speeds, odor patches take a longer time to move downcurrent. This increase in transport time is correlated with a smoothing of the concentration gradient at patch boundaries, resulting in decreased slope values. Thus, the distribution of pulse slopes, and to a lesser extent pulse heights, shows a stronger correlation at higher flow speeds with downstream distance from the odor source within all boundary layer regions. At slower flow velocities, there is rarely any relationship between distance and plume microstructure.

9.9.2 Measurements in the Open Ocean Using Submersibles

The first field measurements of odor plume structure were made at a site 900 m deep near St. Croix, U.S. Virgin Islands with a field version of the IVEC-10 called "Subnose I". The study mapped the concentration fluctuations within an odor plume at a sampling rate of 10 Hz (Atema *et al.*, 1991). Subnose-I was mounted on the Johnson Sea-Link and used to map the concentration of a dopamine-labeled plume at distances up to 50 m from its source. It recorded very high micro-scale variability in concentration due to the distribution of small eddies within the fluid flow. The high degree of patchiness seen in laboratory tests was confirmed by the *in-situ* electrode measurements. Many patches near the source had steep patch gradients and onset slopes, which may be important spatial information for orienting animals.

An updated version, "Subnose-II", was used for a second study east of Cape May, NJ that included concentration mapping at 50 Hz, simultaneous recordings from dual sensors at 200 Hz, and searching for naturally occurring chemical signals (Moore *et al.*, in prep.). Recordings made at 50 Hz showed additional high frequency concentration fluctuations not seen in laboratory tests or the previous field study. Simultaneous recordings from two electrodes spaced 4 cm apart revealed microscale differences within the odor structure that may be an important source of information for aquatic animals. In addition, several recordings were made in which naturally occurring (non-dopamine) chemical signals were detected. Exact identification of the molecular species was not possible, but the redox ratio combined with the use of high oxidation potentials support the hypothesis that the signals were composed of low levels of tyrosine.

9.9.3 Measurements Around Chemosensory Appendages

Other studies have mounted the electrodes on organisms or within a particular appendage (Moore *et al.*, 1991; Schneider *et al.*, in press). A preliminary study demonstrated that the boundary layers around the chemosensory appendages of the lobster, *Homarus americanus*, significantly alter the spatial distribution of chemical signals arriving at the sensory hairs and that the distinct morphology of different sensory organs has a differential effect on the filtering of incoming odor signals. The electrode placed within the boundary layer of the lateral antennule detected lower concentrations as compared to recordings in either the medial antennule or the walking leg. This was true for all flow velocities, although it was more apparent at the lower flow rates. This demonstrates that the distinct morphology of each receptor structure results in a distinct boundary layer (Moore *et al.*, 1991). Previous studies on other relevant flow situations (Vogel, 1983; Cheer and Koehl, 1987) have also shown that the fluid flow around and through sensory appendages is significantly altered by the morphology of the appendage. Under stationary conditions, odor access to all parts of the sensory sensillum is limited in the lateral antennule. Under active sampling conditions, the fluid dynamic conditions are changed and chemical signals have equal odor access to all regions of the aesthetasc hairs. The boundary layer surrounding a sensory organ acts as a smoothing filter for odor signals; temporal aspects of the signal are changed by the boundary layer. Since the thickness and structure of the boundary layer will depend upon the morphology of the chemoreceptor appendage and the relative fluid velocities, identical odor pulses in the environment will be different in the microscale environment of morphologically different chemosensory appendages. Each of the lobster's appendages, lateral and medial antennule and walking legs, has a distinct morphology and therefore each appendage will have a unique boundary layer.

9.10 FUTURE DEVELOPMENT

The current IVEC-10 system is constantly under development for implementation in oceanographic work. Future development of additional techniques and hardware are expected to make the current system easier to use and to broaden its application. The new developments fall into three categories. First, given the constant advances in microcircuitry, the system will probably be redesigned to be smaller and more portable, on-board battery sources of power will be added, and smaller and faster A-to-D cards will be used. This will increase the number of different field uses for the current system. Second, electrodes are currently being developed that are multisite electrodes (Figure 9.5), which will allow for

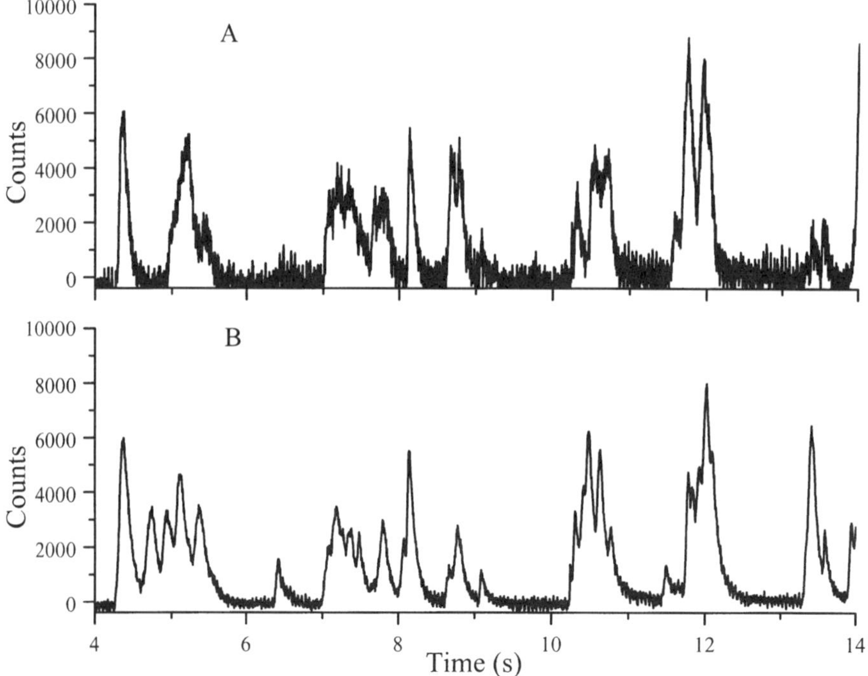

Figure 9.5 Examples of 10 s chemical records measured at two sensors on a multisite semiconductor-based probe. Sensors used in these recordings are spaced 600 microns apart. This probe had five sensors in a linear array. Each sensor was 200 microns apart. (A) Sensor #1 and (B) Sensor #4.

multiple measurements at electrodes 100–200 microns apart. Finally, additional methods are being developed that will allow for a larger range of organic molecules to be detected with high sensitivity and selectivity against background levels.

References

Adams, R.N. (1969). Electrochemistry at Solid Electrodes, Marcel Dekker.

Adams, R.N. and Marsden, C.A. (1982). Electrochemical detection methods for monamine measurements *in vitro* and *in vivo*. In Inversen, S.D. and Snyder, S.H. (eds), *Handbook of Psycopharmocalogy*, Vol. 15. New York: Plenum Press.

Atema, J. (1985). Chemoreception in the sea: adaptation of chemoreceptors and behavior to aquatic stimulus conditions, *Soc. Exp. Biol. Symp.*, 39, 387–423.

Atema, J. and Engstrom, D.G. (1971). Sex pheromone in the lobster, *Homarus americanus. Nat.*, 232, 261–263.

Atema, J., Popper, A.N., Fay, R.R. and Tavolga, W.N., eds. (1988). Sensory Biology of Aquatic Animals, *Springer-Verlag*.

Busdosh, M., Robillard, G.A., Tarbox, K. and Beehler, C.L. (1982). Chemoreception in an arctic amphipod crustacean: a field study, *J. exp. Mar. Biol. Ecol.*, 62, 261–269.

Carr, W.E.S. and Derby, C.D. (1986). Behavioral chemoattractants for the shrimp, *Palaemonetes pugio*: identification of active components in food extracts and evidence of synergistic mixture interactions, *Chem. Senses*, 11, 49–64.

Cheer, A.Y.L. and Koehl, M.A.R. (1987). Paddles and rakes: fluid flow through bristled appendages of small organisms, *J. Theor. Biol.*, 129, 17–39.

Christensen, T.A. and Hildebrand, J.G. (1988). Frequency coding by central olfactory neurons in the Sphinx moth *Manduca sexta, Chem. Senses*, 13, 123–130. Denny, M.W. 1993. Air and Water: the biology and physics of life's media. Princeton University Press, New Jersey.

Derby, C.D. and Atema, J. (1982). The function of chemo- and mechanoreceptors in lobster (*Homarus americanus*) feeding behavior, *J. Exp. Biol.*, 98, 317–327.

Devine, D.V. and Atema, J. (1982). Function of chemoreceptor organs in spatial orientation of the lobster, *Homarus americanus*: differences and overlap, *Biol. Bull.*, 163, 144–153.

Friedemann, M.N., Robinson, S.W. and Gerhardt, G.A. (1996). o-Phenylenediamine-modified carbon fiber electrodes for the detection of Nitric Oxide, *Analytical Chem.*, 68, 2621–2628.

Gerhardt, G.A. and Adams, R.N. (1982). Battery-powered apparatus for chronoamperometric measurements, *Anal. Chem.*, 54, 1888–1889.

Gerhardt, G.A. and Palmer, M.R. (1987). Characterization of the techniques of pressure ejection and micro-iontophoresis using *in vivo* electrochemistry, *J. Neuroscience Methods.*, 22, 147–159.

Gerhardt, G.A., Oke, A.F., Nagy, G., Moghaddam, B. and Adams, R.N. (1984). Nafion-coated electrodes with high sensitivity for CNS electrochemistry, *Brain Res.*, 290, 390–395.

Ghiradella, H., Case, J.F. and Cronshaw, J. (1968). Structure of aesthetascs in selected marine and terrestrial decapods: chemoreceptor morphology and environment, *Am. Zool.*, 8, 603–621.

Gleeson, R.A., Adams, M.A. and Smith, A.B. (1987). Hormonal modulation of pheromone-mediated behavior in a crustacean, *Biol. Bull.*, 172, 1–9.

Gomez, G., Voight, R. and Atema, J. (1994). Frequency filter properties of lobster chemoreceptor cells determined with high resolution stimulus measurement, *J. Comp. Physiol.*, 174, 803–811.

Gratton, A., Hoffer, B.J. and Gerhardt, G.A. (1988). Effects of electrical stimulation of brain reward sites on release of dopamine in rat: an *in vivo* electrochemical study, *Brain Res. Bull.*, 21, 319–324.

Hamner, P. and Hamner, W.M. (1977). Chemosensory tracking of scent trails by the planktonic shrimp, *Acetes sibogae australis*. *Sci.*, 195, 886–888.

Hargrave, B.T. (1985). Feeding rates of abyssal scavenging amphipods (*Eurythenes gryllus*) determined *in situ* by time-lapse photography, *Deep-Sea Res.*, 32, 443–450.

Hessler, R.R., Isaacs, J.D. and Mills, E.L. (1972). Giant amphipods from the abyssal Pacific Ocean, *Sci.*, 175, 636–637.

Ingram, C.L. and Hessler, R.R. (1983). Distribution and behavior of scavenging amphipods from the central North Pacific, *Deep-Sea Res.*, 30, 683–706.

Kaissling, K.E., Zack-Straussfeld, C. and Rumbo, E. (1987). Adaptation processes in insect olfactory receptors: mechanisms and behavioral significance, p. 104–112 *In* Roper, S. and Atema, J. (eds), Olfaction and Taste IX N.Y. *Acad. Sci.*, 510.

Laverack, M.S. (1988). The diversity of chemoreceptors, p. 287–312 *In* Atema, J., Popper, A.N., Fay, R.R. and Tavolga, W.N. (eds), Sensory Biology of Aquatic Animals, Springer-Verlag.

Legier-Visser, M., Mitchell, J.G., Okubo, A. and Fuhrman, J.A. (1986). Mechanoreception in calanoid copepods: a mechanism for prey detection, *Mar. Biol.*, 90, 529–536.

Maier, I. and Müller, D.G. (1986). Sexual pheromones in algae, *Biol. Bull.*, 170, 145–175.

Miller, R.L. (1979). Sperm chemotaxis in the Hydromedusae II. Some chemical properties of the sperm attractants, *Mar. Biol.*, 53, 115–124.

Moore, P.A. and Atema, J. (1988). A model of a temporal filter in chemoreception to extract directional information from a turbulent odor plume, *Biol. Bull.*, 174, 355–363.

Moore, P.A. and Atema, J. (1991). Spatial information in the three-dimensional fine structure of an aquatic odor plume, *Biol. Bull.*, 181, 408–418

Moore, P.A., Atema, J. and Gerhardt, G.A. (1991). Fluid dynamics and microscale odor movement in the chemosensory appendages of the lobster, *Homarus americanus*. *Chem Senses*, 16, 663–674.

Moore, P.A., Gerhardt, G.A. and Atema, J. (1989). High resolution spatio-temporal analysis of aquatic chemical signals using microelectrochemical electrodes, *Chem. Senses*, 14, 829–840.

Moore, P.A., Weissburg, M.J., Parrish, J.M., Zimmer-Faust, R.K. and Gerhardt, G.A. (1994). Spatial distribution of odors in simulated benthic boundary layer flows, *J. Chem. Ecol.*, 20, 255–279.

Moore, P.A., Zimmer-Faust, R.K., BeMent, S.L., Wiessburg, M.J., Parrish, J.M. and Gerhardt, G.A. (1992). Measurement of microscale patchiness in a turbulent aquatic odor plume using a semiconductor-based microprobe, *Biol. Bull.*, 183, 138–142

Price, H.J. Paffenhofer, G.-A., Boyd, C.M., Cowles, T.J., Donaghay, P.L., Hamner, W.M., Lampert, W. and Quetin, L.B. (1988). Future studies of zooplankton behavior: Questions and technological developments, *Bull. Mar. Sci.*, 43, 853–872.

Reeder, P.B. and Ache, B.W. (1980). Chemotaxis in the Florida spiny lobster, *Panulirus argus*. *Anim. Behav.*, 28, 831–839.

Sainte-Marie, B. (1986). Effect of bait size and sampling time on the attraction of the Iysianassid amphipods *Anonyx sarsi* Steele and Brunel and *Orchomenella pinguis* (Boeck), *J. Exp. Mar. Biol. Ecol.*, 99, 63–77.

Schneider, R.W.S., Price, B.A. and Moore, P.A. (1997). Antennae Morphology as a Physical Filter of Olfaction: Temporal Tuning of the Antennae of the Honeybee, *Apis mellifera*. *J. Insect Physiology*. (In press).Vogel, S. 1983. How much air passes through a silkmoths's antenna, *J. Insect Physiol.*, 29, 597–602.

Voigt, R. and Atema, J. (1990). Adaptation in chemoreceptor cells III. Effects of cumulative adaptation, *J. Comp. Physiol. A*. 166, 865–874.

Wilson, R.R. and Smith Jr. K.L. (1984). Effect of near-bottom currents on detection of bait by the abyssal grenadier fishes Coryphaenoides spp. recorded *in situ* with a video camera in a free vehicle, *Mar. Biol.*, 84, 83–91.

10. ELECTROCHEMICAL MONITOR FOR NEAR REAL-TIME DETERMINATION OF DISSOLVED TRACE METALS IN MARINE WATERS

ERIC P. ACHTERBERG, CHARLOTTE BRAUNGARDT
and DAVID J. WHITWORTH

Department of Environmental Sciences, University of Plymouth, Plymouth PL4 8AA, UK

10.1 INTRODUCTION

Measurements of chemical constituents in marine waters present many challenges as a result of the distinctive composition of seawater, the large temporal and spatial variabilities in marine systems and the problematic accessibility of study areas. With an increasing global environmental awareness, there is now a greater demand for instrumentation that can be used for chemical marine monitoring. The most used chemical marine sensor systems today are based upon electrochemical principles and are for the determination of oxygen and pH. However, electrochemical methods have also been developed for the determination of trace metals in seawater. Monitoring of trace metals in marine systems is important for overseeing the health of our seas. For example, monitoring programmes for trace metals and other chemical constituents in the North Atlantic and Arctic Oceans (including the North Sea) have been internationally agreed in the Oslo-Paris conventions (Ospar). Ospar represents 15 European countries which have coastlines on or rivers discharging into the north-east Atlantic.

The concentrations of dissolved metals in unpolluted oceanic waters are generally very low, typically at nanomolar (10^{-9} mol l^{-1}) levels or less. Copper, Zn and Ni, for example, occur at levels of between 0.5 and 5×10^{-9} mol l^{-1} in waters of the Atlantic Ocean (Bruland and Franks, 1983; Buckley and van den Berg, 1986; Jickells and Burton, 1988; Kremling and Pohl, 1989), whilst the concentrations of these metals are approximately ten times higher in coastal waters of the Irish sea and North Sea (Tappin *et al.*, 1995; Achterberg and van den Berg, 1996). Levels of dissolved Cu, Zn and Ni in some mine polluted estuarine systems in southern Spain (Huelva) and south-west England (Restronguet Creek) can however reach levels of 10^{-6} to 10^{-5} mol l^{-1} (Leblanc *et al.*, 1995; Bryan and Langston, 1992; Rijstenbil *et al.*, 1991). Different approaches are taken to monitor metals in the marine environment. The US Mussel Watch programme utilises mussels which are collected in coastal areas (Lauenstein *et al.*, 1990; Larsen, 1992; Stephenson and Leonard, 1994). The programme has been successful in highlighting pollution hot-spots, by determining metals (and other pollutants) in the mussel tissue. As mussels are filter feeders, the pollutant levels in mussels largely reflect the levels in suspended particulate matter and not in the dissolved phase (by definition the dissolved phase passes through a 0.4–0.45 μm membrane filter and the particulate phase is retained). The pollutant levels in mussels provide a picture about perturbation in their environment over a longer period, and can therefore be used to check the health of that environment. A picture of historical pollution levels in marine systems can also be obtained from determinations of pollutants in sediments. Metal levels at different depth in the sediments can be related to historical time periods, but care must be taken that processes like bioturbation and metal mobility through redox processes have not disturbed the vertical metal profiles. A great deal of progress has been made in recent years in using behavioural and functional responses of organisms to pollutants (including metals and xenobiotic organic compounds) in natural waters (Baldwin and Kramer, 1994). The use of such bio-indicators provide a near-instantaneous feed-back on toxic pollutant levels in the water. However, more work will need to be done with respect to the selectivity and sensitivity of these methods.

The most reliable methods for monitoring dissolved trace metal concentrations in the marine environment still involve chemical determinations. These analyses provide information on the dissolved form of metals which, for example, may affect the level of growth of phytoplankton and bacteria in the sea. Dissolved metal monitoring can also identify metal containing waste discharges into marine waters. Traditionally, sea water samples are collected in a discrete manner. From a survey vessel discrete surface water samples can be obtained using a pump with a bottom-weighted hose, and in deeper waters with the use of samplers (e.g. Go-Flo or Niskin bottles) which are attached to a hydrowire and lowered in the sea with the use of a winch. The samples are usually filtered on-board ship and subsequently analysed in a land-based laboratory. Commonly used laboratory techniques for dissolved trace metal analysis in sea water include Graphite Furnace Atomic Absorption Spectrometry (GFAAS) and Inductively Coupled Plasma Mass Spectrometry (ICP-MS) (after solvent or solid-phase extraction for trace metal pre-concentration and matrix removal), chronopotentiometry, colorimetry, chemiluminescence and stripping voltammetry. This approach of laboratory based analyses of discrete samples is time-consuming and therefore expensive. Only a limited number of samples can be collected using discrete sampling techniques and as a result important changes in water quality may be missed. Trace metal levels in estuarine, coastal and oceanic waters are often low. Inadequate sampling and sample treatment techniques have for a long time prevented the uncontaminated collection and analysis of sea water samples. Sample contamination may arise from components of the sampling gear and from sample handling operations. Concentrations of elements like Ni, Cu, Zn and Fe which were determined in sea water samples collected prior to the mid-seventies and sometimes early eighties were therefore in many cases affected by sample contamination. Nowadays, the use of PTFE coated samplers, free of internal metal components, and with as few external metal components as possible, greatly improves contamination-free sample collection. Improved understanding of post-sampling contamination has resulted in the introduction of ultra-trace working practices (Morley *et al.*, 1988). Sample bottles (preferably High Density Polyethylene; HDPE), filters and filtration equipment are acid cleaned prior to use, and all sample handling is performed in a clean environment (class-100 laminar flow hood in a clean room) (Howard and Statham, 1993). These precautions against sample contamination are essential in order to obtain high quality trace metal data, but greatly reduce the number of samples that can be processed by a research worker.

10.2 ELECTROCHEMICAL TECHNIQUES: POTENTIOMETRY

pH is the most commonly measured chemical parameter in natural waters and knowledge of pH is necessary for the understanding of speciation of trace elements in these waters. The determination of pH is most often performed using potentiometry, whereby the potential over an electrode pair is measured without current flow. A reference electrode (Ag/AgCl or calomel) and ion-selective electrode (glass electrode) are used for this purpose. Potentiometry is also used for the determination of trace metals in natural waters using metal-selective solid state electrodes. Such electrodes are available for Cu, Ag, Pb and Cd and determine the activity of the metals in solution. The metal-selective electrodes incorporate membranes fabricated from insoluble crystalline materials and these contain the metal ion for which the electrode is selective. The application of the electrodes is hampered by poor sensitivity and accuracy: metal-selective electrodes usually exhibit a non-Nernstian behaviour at analyte concentrations below 10^{-7} mol l^{-1}. Another major drawback for application of metal-selective electrodes to oceanography is that serious interferences are often produced by the major ions in seawater.

10.3 ELECTROCHEMICAL TECHNIQUES: STRIPPING VOLTAMMETRY

The most suitable electrochemical methods for the determination of low levels of trace metals in sea water make use of stripping techniques: anodic and adsorptive cathodic stripping voltammetry (ASV and ACSV, respectively). Important advances have been made during the past 20 years in the application of stripping voltammetry to marine trace metal measurements. The strength of stripping voltammetry is in its extremely low detection limits (10^{-10}–10^{-11} mol l^{-1}). Analytical developments have resulted in our ability to determine a wide range of trace metals in seawater (over 20 metals including: Co, Cu, V, U, Fe, etc.; van den Berg, 1989; 1991), and the instrumentation has been computerised and made portable.

The basic voltammetric equipment for marine applications consists of a voltammetric analyser, a three-electrode cell (working electrode, reference electrode and counter electrode) and a computer for automated measurements and data acquisition. Modern voltammetric analysers are simple, low-cost and able to perform a range of scan forms. The reference electrode is often an Ag/AgCl electrode and the counter electrode may be a platinum wire or a carbon rod. The most popular working electrode for environmental trace metal analysis is a hanging mercury drop working electrode (HMDE), but glassy carbon and carbon fibre electrodes on which a mercury film is plated are also used in stripping voltammetry. The advantage of an HMDE is that with the formation of each new drop, a new electrode surface is produced, which is important for unattended trace metal monitoring activities. The drops generated by modern mercury drop electrodes are very small (e.g. VA Stand 663 from Metrohm (Switzerland) produces drops with an area of 0.52 mm^2), and safe storage and recycling of the used mercury will ensure minimal environmental and health risks.

Anodic stripping voltammetry has been applied successfully for trace measurements of Cu, Cd, Pb and Zn in seawater. Other elements can be determined using this technique, but their seawater concentrations are too low for ASV, or the analysis is hampered by interferences. A deposition, or pre-concentration, step is carried out under conditions of forced convection (e.g. solution stirring or flow). The deposition potential should be ca. 0.3–0.4 V more negative than the reduction potential of the metal. During the deposition step of an ASV analysis, metal ions are collected in the mercury drop by reduction (to metallic state) and amalgamation with the mercury (see equation 10.1).

$$M^{n+} + ne^- + Hg \leftrightarrow M(Hg) \tag{10.1}$$

Only a small fraction of the metal is actually being deposited during the deposition step. The sensitivity of ASV is improved by using a mercury film rather than a mercury drop electrode because the smaller volume of the mercury film results in a greater concentration factor of the metals collected into the mercury. The deposition is followed by a voltammetric scan towards more positive potentials during which the metal in the mercury is oxidised and the current produced is determined. The resultant current potential stripping voltammogram provides quantitative information: the height of the peak is proportional to the metal concentration; and also qualitative information: the potential of the peak is an indication for the metal analysed. Different scan forms are being applied during the measurement of trace metals in seawater to improve the sensitivity of the methods. The most basic scan form is linear sweep, but pulse-voltammetric waveforms (e.g. differential pulse and square wave) are more useful as they effectively correct for background current contributions. The limit of detection for ASV analysis of Cu and Cd in seawater is typically 10^{-11} and 10^{-10} mol l^{-1}, respectively. Depending on the encountered trace metal levels in a marine system, ASV allows for simultaneous determination of more than one metal.

Adsorptive cathodic stripping voltammetry makes use of a specific ligand (L), which is added to the water sample and forms an adsorptive complex with the trace metal(s) under investigation (equation 10.2).

$$M^{n+} + zL \rightarrow ML_z^{n+} \tag{10.2}$$

$$ML_z^{n+} \rightarrow ML_z^{n+} \text{ (adsorbed)} \tag{10.3}$$

$$ML_z^{n+} \text{ (adsorbed)} + e^- \rightarrow M^{(n-1)+} + zL \tag{10.4}$$

A pH buffer is used to control the pH of the sample as the formation of the metal-ACSV ligand is pH dependent. Generally, ACSV is carried out using a hanging mercury drop electrode. A minute fraction of the metal-ligand complex is adsorbed on the surface of the mercury drop (equation 10.3) and a potential scan is carried out. The adsorption step is carried out at a carefully controlled potential as it determines the adsorption efficiency. In most cases, an adsorption potential is chosen which is slightly more positive (ca. 0.1 V or more) than the reduction potential of the metal-ligand complex. The scan direction is towards more negative potentials and the resulting current is determined (see Figure 10.1).

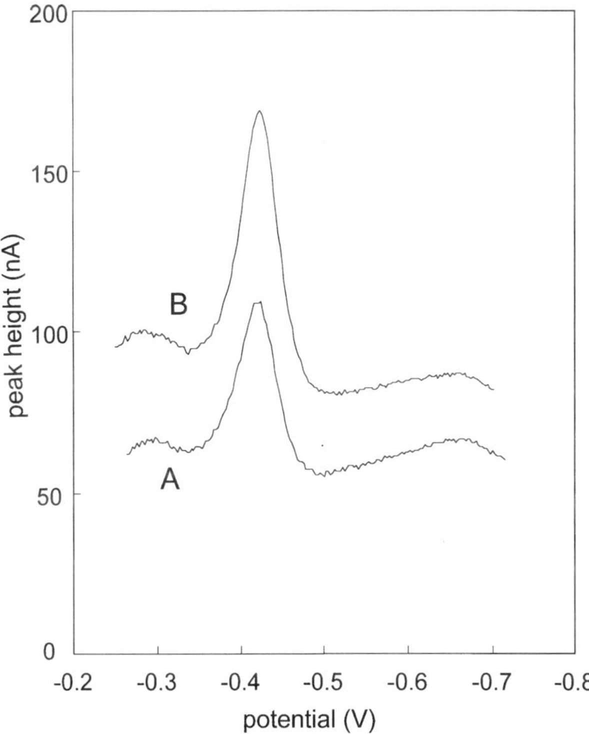

Figure 10.1 Voltammetric scan of dissolved Cu in seawater. Oxine (0.15 mM) was used as the ACSV ligand, Hepes (10 mM) was used as pH buffer (pH 7.7). Voltammetric conditions were: 20 s deposition at -1 V, 8 s equilibration at -0.25 V and 200 Hz square wave scan towards more negative potentials. Scan A is for sample, and scan B is for sample plus standard addition (5×10^{-9} mol l^{-1} final concentration).

The reduction current is the result of the reduction of a reducible group on the ligand or of the metal itself in the adsorbed complex (equation 10.4). The scan forms applied during ACSV include linear sweep, but fast pulse-voltammetric waveforms (e.g. differential pulse and square wave) are also used if the reduction of the metal-ligand complex is electrochemically reversible. The limit of detection of ACSV for metals is typically on the order of 10^{-9}–10^{-11} mol l^{-1}. Even lower metal concentrations (down to 10^{-12} mol l^{-1} for Pt, Ti, Co) can be determined by using a catalyst in order to enhance the reduction current. Multi-elemental ACSV methods have been developed recently (Colombo and van den Berg, 1997), whereby with the use of mixed ACSV ligands up to 6 trace metals (Cu, Pb, Cd, Ni, Co and Zn) can be determined simultaneously in coastal and estuarine waters. Voltammetric trace metal analysis commonly makes use of the standard addition method for the quantification of metal concentrations in water samples.

Voltammetric techniques have the advantage that they allow determination of trace metals directly in sea water, without a separate pre-concentration step. Alkali metals in seawater do not interfere with trace metal determinations, but actually increase the sensitivity. In addition, the high sensitivity and selectivity allow the determination of trace metal speciation (Achterberg and van den Berg, 1994a). Speciation analysis involves the determination of different physico-chemical forms of trace metals. In the case of ACSV, ligand competition is used for speciation measurements whereby the added ACSV ligand competes for trace metals with naturally occurring ligands (Figure 10.2). The competition conditions

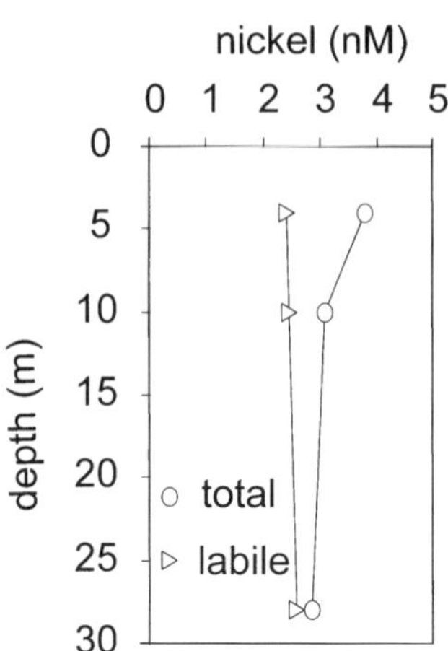

Figure 10.2 Depth profile of dissolved Ni speciation in the Gulf of Cadiz (November 1996), determined using ACSV. Labile and total Ni were determined using DMG (dimethylglyoxime, ACSV ligand) concentrations of 2×10^{-5} and 2×10^{-4} mol l^{-1}, respectively. Labile Ni was determined at sea, whereas total Ni (after UV-digestion of the samples) was analysed in the laboratory. Largest difference between total and labile Ni was observed in the surface waters and can probably be attributed to presence of enhanced Ni complexing organic matter, produced by primary producers.

can be carefully manipulated by choosing a suitable ACSV ligand (with known conditional stability constants for the metal under investigation) and an appropriate ACSV ligand concentration. For example, ACSV ligands used for speciation measurements of Cu in seawater include Tropolone, Salicylaldoxime and Oxine, with Tropolone being the weakest Cu complexing ACSV ligand and Oxine the strongest. Metal speciation is becoming more important because of the recognition that data on total dissolved metal concentrations does not yield sufficient information about the toxicity, bioavailability and geochemical behaviour of trace metals in natural waters. For many metals (including Cu, Zn, Cd and Ni) the free aqueous form is reported to be the most bioavailable and toxic (Tessier and Turner, 1995), but only a few analytical techniques (including stripping voltammetry and chemiluminescence) are sensitive enough to determine labile/free aqueous metal fractions in sea water. It is important for metal speciation measurements to be performed as soon as possible upon sampling, as chemical equilibria are readily disturbed during sample storage. The application of *in-situ* (including ship-board) techniques is therefore required.

10.4 SHIP-BOARD VOLTAMMETRIC TECHNIQUES

In recent years there has been a move towards trace metal analysis on-board ship. This approach not only reduces the risk of sample contamination, but often also results in a higher sample through-put and hence an increased amount of environmental data. Analytical instrumentation for ship-board use will preferably have a limited weight and size, in other words be portable. This precludes the use of GFAAS and ICP-MS techniques, because the instrumentation is bulky. In addition, instrumentation for GFAAS and ICP-MS is very sensitive to the constant vibrations caused by the ship's engines. Electrochemical techniques often make use of portable instrumentation. The purchase and running costs of such instrumentation are much lower than for GFAAS and ICP-MS, and make the electrochemical techniques very suitable for field monitoring of dissolved trace metals. Most methods for dissolved trace metal analysis using voltammetry can be operated in an automated batch or flow-analysis mode. In this case sample and reagent transport and metal standard additions are performed using pumps. The application of computers in such systems allows fully automated sample and reagent transport, standard additions, metal analysis, data acquisition and treatment (Achterberg and van den Berg, 1994a). The automated batch or flow-analysis approach not only reduces the risk of sample contamination and increases the sample through-put, it also enhances the quality of the data by the fully computerised operation. As pointed out before, mercury drop electrodes are commonly used and have the advantage over other types of electrodes that a fresh electrode surface is formed each time a new drop is made. It is our experience that this type of electrode behaves very well at sea. Even during force 10 storms (on the Beaufort scale), or on ships with strong engine vibrations, the mercury drops do not dislodge during the collection or scanning steps. With the

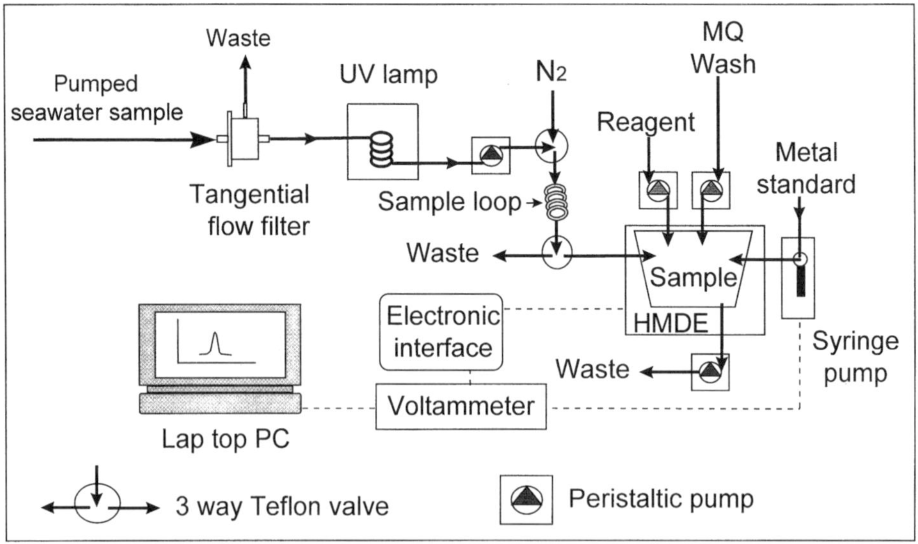

Figure 10.3 Diagram of automated voltammetric system with dotted lines representing electrical connections and solid lines representing sample and reagent flows. HMDE: hanging drop mercury electrode.

use of fast scan forms (e.g. square wave at 300 Hz) any movement of the vessel goes unnoticed during the voltammetric scan and high quality scans are obtained.

An example of an automated voltammetric metal monitor which is used for both land-based laboratory and near real-time *in-situ* analysis of trace metals is shown in Figure 10.3. The monitor comprises of an μAutolab voltammetric analyser (EcoChemie, The Netherlands) and a Metrohm hanging mercury drop electrode (VA Stand 663, Switzerland). The sample and reagent transport is performed using peristaltic pumps and metal standard additions are made to the voltammetric cell using a syringe pump (Cavro). Three-way inert Teflon® valves (Cole-Parmer) are used to fill and empty a sample loop (ca. 10 ml). The voltammeter and peripheral instruments are controlled using a portable PC. All tubing used in the voltammetric system is made of Teflon®, with the exception of pump tubing, which is Santoprene®. On-line filtration of the seawater is performed using a tangential flow filtration system (Figure 10.4). A membrane filter (0.45 μm pore size, 47 mm diameter) is placed in the filter holder and sample is obtained at a rate of a few ml min^{-1}. A filter can be used for a prolonged period of time, even in turbid estuaries, because the filtration system is self-cleansing. Particulate material is constantly removed from the filter by a pronounced cross-filter sea water flow (up to ca. 2–3 l min^{-1}). An on-line UV-digestion unit (Figure 10.4) is used for the break-down of surfactants and natural metal-complexing organic ligands (Achterberg and van den Berg, 1994b). The surfactants need to

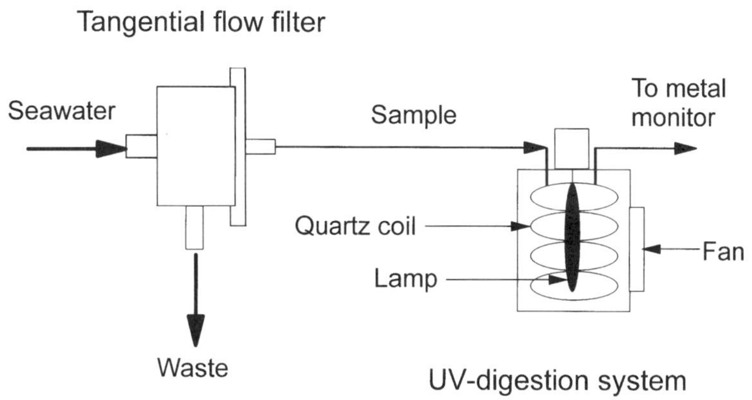

Figure 10.4 Diagram of on-line tangential flow filtration system and on-line UV-digestion system.

be removed because they may interfere with the voltammetric analysis by fouling the working electrode surface. Metal-complexing ligands occur naturally in sea water and are thought to be released by phytoplankton (algal exudates), bacteria (e.g. siderophores), but also include breakdown products of marine organisms (e.g. porphyrins) and humic substances from land-run off. The complexation of metals by the natural ligands reduces the electrochemically labile metal concentration and hence the voltammetric signal, and their destruction releases the metals and results in the determination of total dissolved metal concentrations. The UV-digestion unit contains a medium pressure mercury vapour lamp (100 to 400 W lamps are used) and a quartz glass coil (inner diameter 1 mm, length 3–4 m) and is cooled using a fan (Figure 10.4). The optimal temperature for UV-digestion of organic compounds is ca. 70–80 °C.

The voltammetric system presented in Figure 10.3 operates in an automated batch-mode, with the analysis of 10 ml aliquots at a rate of one complete measurement every ca. 10–20 min. Each sample is fully calibrated, resulting in high quality data required for biogeochemical and pollution studies. The use of a lower calibration frequency would increase the sample through-put, but may pose problems in coastal and estuarine waters where important variations in the sample matrix would result in pronounced changes in the sensitivity of the voltammetric analysis. Dedicated soft-ware was produced for the voltammetric system for data acquisition, treatment and storage. The soft-ware is self-decisive and intelligent: it is able to reject failed scans, to perform additional scans when required and to make additional metal standard addition in case the initial peak increase was insufficient. The software capabilities have resulted in stand-alone monitoring system, allowing unattended 24 h operation. Table 10.1 shows the sequence of operation for an automated trace metal analysis using the system outlined above.

Table 10.1 Operational sequence for automated voltammetric metal measurements. Valve 1 and 2 are positioned before and after the sample loop, respectively (see Figure 10.3).

Stage	*Valve 1 Position*	*Valve 2 Position*	*Reagent Pump*	*Wash Pump*	*Sample Pump*	*Voltammeter*	*Comments*
Sample acquisition	Open to sample	Open to waste	Off	Cell rinsed 3 times	520 s at 1.5 ml min^{-1}	Idle	Wash cycle undertaken
Sample transport	Open to nitrogen	Open to cell	20 s at 0.75 ml min^{-1}	Off	Off	Idle	Addition of sample and reagents
Sample purge	Open to sample	Open to waste	Off	Off	520 s at 1.5 ml min^{-1}	Purgin for 180 s	Acquisition of new sample
Measurement	Open to sample	Open to waste	Off	Off	On	Measuring	10 s purge between measurements
Quantification	Open to sample	Open to waste	Off	Off	On	Measuring	10 s purge between measurements
Wash cycle	Open to sample	Open to waste	Off	Cell rinsed 3 times	Off	Idle	New cycle

Total cycle time ~10–15 min.

10.5 UNDERWAY PUMPING

A very recent development in marine trace metal studies is the application of analytical monitoring equipment on-board ship for near real-time measurements of surface waters using voltammetry (Achterberg and van den Berg, 1996). This new methodology uses underway pumping as a means of sample collection and thereby obviates the need for the vessel to halt for the collection of discrete samples. Near real-time dissolved trace metal determinations may be performed using on-line voltammetry. Sample contamination is prevented by eliminating contact of the sea water with metal components by using inert materials (e.g. Teflon®, Polyvinyl Chloride, Polyethylene). An effective underway pumping system can be designed using a peristaltic or Teflon®-bellows pump and a long (20–60 m) and strong Polyethylene or Polyvinyl Chloride hose. The hose is hung overboard and attached to a "fish" (torpedo-like structure, KIPPER-1) which is towed from a strong cable attached to a winch (Figure 10.5). The design and weight (ca. 40 kg) ensures that KIPPER-1 stays at a constant depth (ca. 3–4 m) even at speeds over 10 knots. KIPPER-1 is made of solid carbon steel with an inlet at the front and a hole through the middle for the sampling hose. The fish is coated with a non-metallic epoxy-based paint. All the tubing used is rapidly

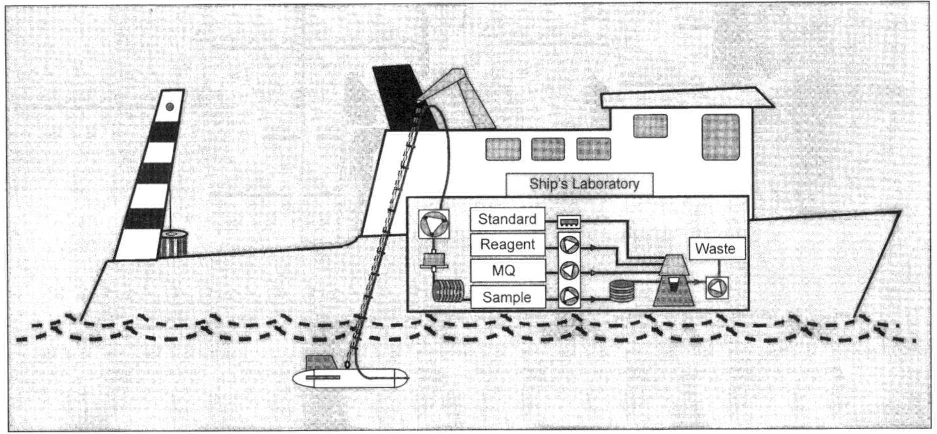

Figure 10.5 Schematic drawing showing the underway pumping system with KIPPER-1, the path of the sample through on-line filtration, on-line UV-digestion and to the metal monitor in the ship's laboratory.

equilibrated with the sea water as it is automatically and continuously rinsed during sample collection. This underway method of sampling is therefore largely self-cleansing. The fish should be positioned away, below and forward of the ship, so that water is collected that has not been in contact with the ship's hull. Ship's hulls are known to release trace metals, especially after treatment with metal based paints to reduce attachment of barnacles and plankton.

Advantages of this method of underway sampling include a minimization of changes in chemical speciation of trace metals in sea water due to rapid analysis upon sampling, and a reduced risk of sample contamination. This monitoring approach results in enhanced sampling frequencies and is therefore an important tool for biogeochemical and pollution studies in marine systems which require high-resolution measurements. The data can, for example, be used in numerical computer models for modelling metal distributions and behaviour in marine systems. The near real-time analysis also provides the opportunity for an interactive sampling campaign, because the results of the measurements are directly available and can be evaluated on-board ship whilst the vessel is steaming.

10.6 OPEN OCEAN APPLICATION OF THE AUTOMATED VOLTAMMETRIC METAL MONITOR; NI IN THE ATLANTIC OCEAN

The low concentrations of trace metals in open ocean waters require very sensitive monitoring instrumentation. For dissolved Ni, the lowest levels (nanomolar or subnanomolar) occur in the surface waters, as a result of biological removal. The low inputs of trace metals to the open ocean and the effectiveness of the removal mechanisms result in rather uniform surface ocean concentrations for Ni (and also for e.g. Cu, Cd and Zn). The spatial variation in concentrations of Ni and many other trace metals is therefore not as large in surface waters of the open ocean, compared with coastal waters. Figure 10.6 shows results of near real-time determinations of Ni in the Atlantic Ocean. The automated voltammetric instrumentation was used on-board *RRS Discovery* during an OMEX (Ocean Margin Exchange) cruise in the Atlantic shelf and Channel region in August/ September 1995. OMEX investigates fluxes of trace metals, nutrients an organic compounds over the Atlantic shelf waters. The automated metal monitor used during the OMEX cruise is similar to the system displayed in Figure 10.3. Surface sea water was pumped up with the use of a Teflon®-bellows pump from the fish which was positioned at a depth of 3–4 m. The seawater was not subjected to on-line filtration and UV-digestion and therefore "electrochemically-labile" Ni was determined. However, because the added chelating ligand (DMG) that was used forms a very strong complex with Ni and the organic matter and suspended particle concentrations were low in these waters, the observed labile concentrations were most likely close to the total dissolved Ni concentrations.

Each data point in Figure 10.6 is individually calibrated, resulting in high quality data. A simultaneous determination of Ni and Cu was performed on the Atlantic Ocean (Cu data not presented here). Only a fraction of the measurements performed during the 3 week cruise are presented here. The results show low Ni concentrations (between 1.2 and 2.5×10^{-9} mol l^{-1}) in the Atlantic Ocean, with oscillating Ni levels due to the vessel repeatedly crossing the shelf break towards open Atlantic waters (decrease in Ni) or towards shelf waters (increase in Ni).

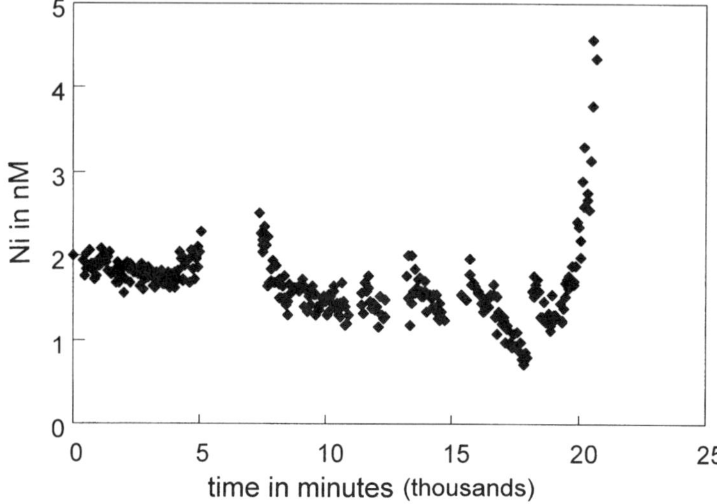

Figure 10.6 Labile Ni concentrations versus time in surface waters of the Atlantic shelf region between southern Ireland and western France (OMEX cruise, September 1995).

A further increase in Ni concentration was observed (to ca. 4.5×10^{-9} mol l^{-1}) as the vessel steamed into the English Channel towards Southampton. This increase in the concentration of Ni can be attributed to mixing of clean Atlantic waters with waters from the North Sea, and riverine and atmospheric inputs into coastal waters in the vicinity of the British mainland. The gap in the graph (between ca. 5000 and 7000 min) is caused by a break in monitoring activities due to a hurricane.

10.7 COASTAL WATER APPLICATION OF THE AUTOMATED VOLTAMMETRIC METAL MONITOR; CU IN THE GULF OF CADIZ

High population density and industrialization in coastal areas have generated demand on marine resources, and have resulted in pollutant inputs into coastal waters. Sources of metals to coastal environments include effluents from industry, mining and ore processing activities, as well as the dumping of sewage sludge and industrial wastes at sea. Metal concentrations in coastal waters are variable as a result of the changing strength of point and diffuse sources, and seasonal variations in metal removal mechanisms. Tidal movement and currents influence local as well as long-range metal distributions in coastal seas.

The strong variability in coastal waters requires a high spatial and temporal resolution in the design of sampling strategies. The movement of currents and tides has to be considered in order to avoid sampling a moving parcel of water repeatedly at adjacent sampling stations. Seasonal variations may affect not only biological and geochemical processes but also the direction and strength of prevailing and wind-induced currents. Data sets acquired during separate cruises can only be compared with each other if tidal movement and salinity distributions are taken into account.

Traditionally, a large number of discrete samples are taken and processed during coastal surveys, but such studies are expensive and time-consuming and can be viewed as "snap-shot" exercises. Ship-time and other economic considerations frequently limit the number of discrete samples that can be taken. The application to coastal monitoring of automated, near real-time ship-board monitors for metal analysis in surface waters addresses some of the challenges encountered in coastal sampling: (a) high spatial resolution can be achieved at slow steaming speeds, (b) calibrated measurements can span a wide concentration range, (c) precious steaming time between sampling stations is utilised, and (d) the scientist on duty has the opportunity to perform other tasks while overseeing the correct functioning of the metal monitor. The use of the automated voltammetric instrumentation is illustrated by its application in coastal waters of the Gulf of Cadiz, south-west Spain. The research is part of the TOROS project, which is an European Union funded ELOISE project and investigates metal fluxes in the Gulf of Cadiz. The shelf waters of the Gulf of Cadiz are enriched in dissolved metals, especially Cu, Zn, Pb and Cd. Studies conducted in the small, but strongly polluted Tinto and Odiel rivers have indicated that these rivers are important sources of metals to the coastal waters (Van Geen *et al.*, 1991; 1997; Leblanc *et al.*, 1995). The Tinto and Odiel rivers are characterised by low and seasonal variable water discharge (combined annual average 15 $m^3 s^{-1}$), and fresh water metal concentrations of up to 11 $\times 10^{-3}$ mol l^{-1} Fe, 6.1 $\times 10^{-4}$ mol l^{-1} Zn, 4.6 $\times 10^{-4}$ mol l^{-1} Cu and low pH values (pH 2.5–3) (data from November '96).

Figure 10.7 illustrates the advantages of high-resolution monitoring for this coastal system. Discrete samples (denoted by stars) were taken in the Gulf of Cadiz between the coast and the 500 m isobath on a grid of approximately 10–15 km. Discrete sampling was performed using a modified Niskin samplers on a CTD rosette. The continuous underway sampling approach resulted in a much better coverage of the coastal area compared with discrete sampling. The underway pumping system operated almost continuously during steaming and station time for 10 days. The ship's speed was 8 knots, and the resolution of the automated on-line metal analyses (after on-line filtration and UV-digestion) of surface samples was between 3.3 and 4.5 km. The distribution of dissolved Cu in Figure 10.8 shows enhanced metal levels in the coastal region between the mouths of the Huelva (15 $\times 10^{-9}$ mol l^{-1} Cu) and the Guadalquivir (20 $\times 10^{-9}$ mol l^{-1} Cu) estuaries. The data used in this plot were obtained during the first 4 days of the cruise in June 1997, and have not been corrected for tidal movement. A decrease in Cu concentrations with increasing distance from the coast was observed. This can be explained by the mixing of metal-polluted estuarine with cleaner Atlantic waters.

In Figure 10.9, four separate surveys on successive days in June 1997 are shown, and the size of the station markers is proportional to the dissolved Cu

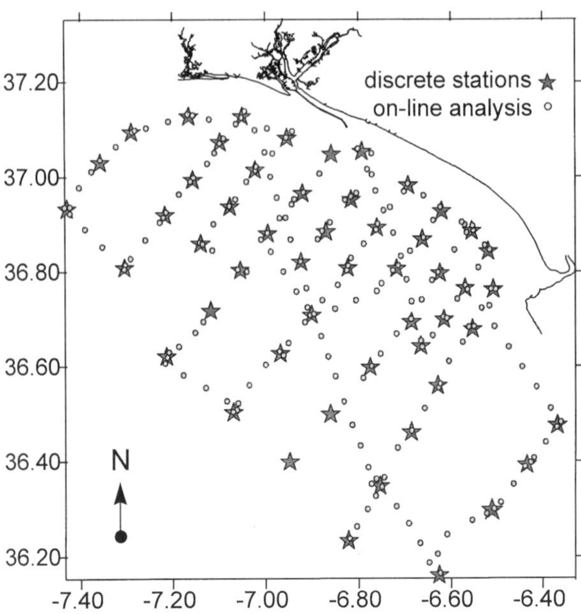

Figure 10.7 Discrete and on-line sampling during a four day survey in the Gulf of Cadiz, Spain (June 1997).

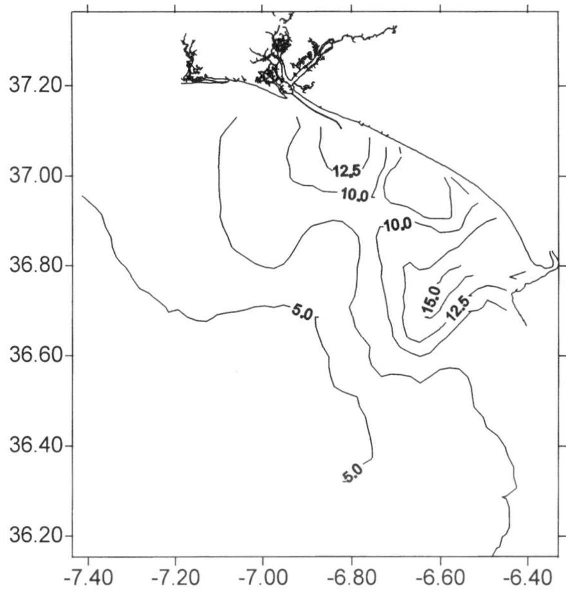

Figure 10.8 Contour plot of dissolved Cu in surface waters of the Gulf of Cadiz; automated on-line analysis during four day survey (June 1997).

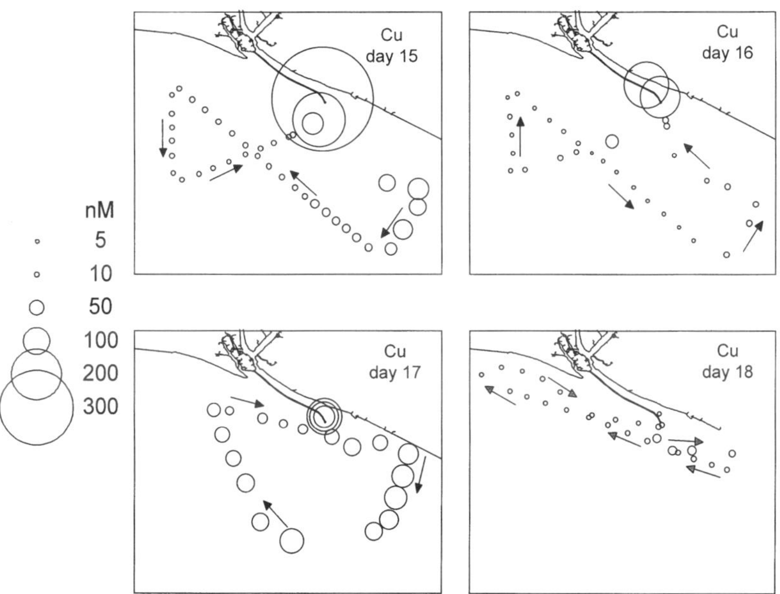

Figure 10.9 Dissolved Cu in coastal waters at the mouth of the Huelva estuary at different tidal stages (June 1997).

concentration measured. During the 4 days the ship's motion was not interrupted for discrete sampling exercises and sampling and analysis were carried out continuously at a speed of 4 knots. This resulted in high resolution data with a distance between sampling points of ca. 1.5–2.5 km, clearly showing the tidally dependent development of the Huelva estuarine plume. On day 15, elevated dissolved Cu concentrations ($60–76 \times 10^{-9}$ mol l^{-1}) were measured two to three hours after low water (LW) in an area to the south-east of the Huelva estuary. One day later (day 16), around the time of high water (HW), Cu concentrations between 13 and 14×10^{-9} mol l^{-1} were observed in this area. A steep increase in dissolved Cu concentrations to levels above 200×10^{-9} mol l^{-1} was observed during both days upon returning to the estuary about one hour ahead of LW. On day 17, the research vessel remained anchored at the mouth of the estuary for a 2.5 hour period, and left this position at the time of LW. Dissolved Cu levels increased during this period from 79 to 139×10^{-9} mol l^{-1}. On the subsequent semicircle around the estuary's mouth (radius approximately 10–12 km), elevated Cu levels were observed to the south-east ($68–85 \times 10^{-9}$ mol l^{-1} Cu) and south (90×10^{-9} mol l^{-1} Cu), with a decreasing trend to the west of the estuary. The three short surveys show the variability in the development of the metal plume in the Gulf of Cadiz and therefore, the data illustrates the value of high-resolution ship-board metal monitoring in tidally influenced waters.

10.8 APPLICATION OF AN AUTOMATED METAL MONITOR TO AN ESTUARINE ENVIRONMENT; NI IN THE TAMAR

Estuaries are highly reactive zones, where fluvial discharges mix with sea water and dissolved elements interact with organic material and particles in the water column. Thus constituents in river water undergo chemical and physical transformations during tidal mixing, and as a consequence only a limited proportion of trace metals carried by river water reach the open ocean. Enhanced concentrations of dissolved trace metals occur in many estuaries and are attributed to inputs from natural and anthropogenic sources. The concentration of dissolved trace metals is subject to large temporal and spatial variations as a result of variability in the extent of run-off, biological activity, tidal movement and anthropogenic discharges. An understanding of the fate of trace metals in estuaries is important for an evaluation of their impact on estuarine organisms, and fluxes into the oceans.

The study of trace metals in estuaries is complicated by the strong physico-chemical gradients (salinity, major ions, pH and turbidity) and high variability in trace metal concentration noticed in these systems. Large changes in water chemistry present challenges which impair the performance of most analytical techniques. A typical style of surveying in estuaries involves sampling along an axial transect of an estuary. This is undertaken using a vessel which travels along the centre of an estuary from one water end-member to another e.g. from coastal marine water past the tidal limit to fresh river water (or vice versa). Discrete samples are collected by manually submerging sample bottles for obtaining surface water, by using a peristaltic pump with a bottom weighted hose or Go-Flo or Niskin samplers for deeper samples. Salinity is generally accepted as the main index of mixing of sea water with river waters. A higher sampling frequency is usually adopted in the upper estuary where large salinity gradients occur and where variations in trace metal concentrations are more pronounced. Measurements of estuarine master variables (salinity, pH, dissolved oxygen and temperature) are usually carried out at the time of sampling using portable meters and probes, this procedure allows an interactive sampling campaign. Analysis of the samples is then undertaken upon return to a laboratory, the time between sampling and analysis could be in the order of days which may compromise the integrity of the water samples.

A very suitable approach for trace metal studies in estuarine environments involves the application of automated metal monitors, yielding high temporal resolution measurements of trace metals. Figure 10.10 shows the instrumental set-up utilised for automated analysis of total dissolved Ni by ACSV during a tidal cycle study carried out on the Tamar estuary, south-west England. The voltammetric metal monitor was transported in and operated from a regular medium sized town van, and powered by a 5 kW, 220–240 V portable generator. Surface water samples were collected using a float deployed in the estuarine channel. The float was attached to an anchor by a two metre nylon rope and a PVC hose was attached to the float and submerged to a depth of ca. 50 cm. Water was continuously pumped using a peristaltic pump (flow ca. 0.5–2 l min^{-1}) and on-line tangential filtration and UV-digestion was applied as described previously. A tidal cycle study describes a sampling style where samples are collected from a single geographical location on the estuary, and the changes in physicochemical conditions observed over the tidal cycle (ca. 13 hours) are particular to water mixing driven by tidal processes. This type of study may yield valuable information on biogeochemical processes occurring in the estuary, especially in estuaries with large salinity gradients. The main advantages of this approach include the ease of transportation and deployment of instrumentation and the relatively low costs of the studies compared with ship-board sampling exercises.

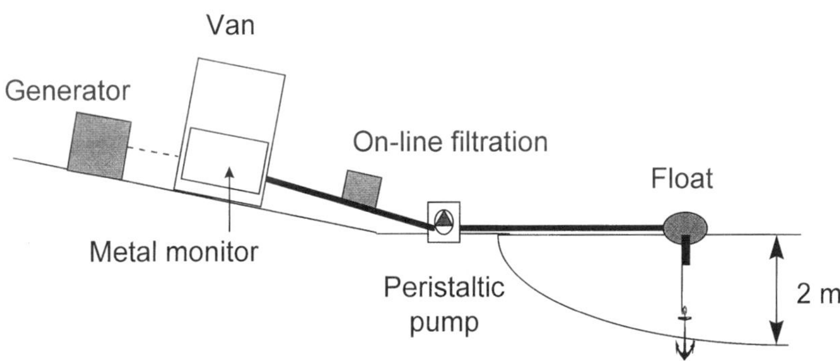

Figure 10.10 Monitoring set-up employed during tidal cycle study in the Tamar estuary.

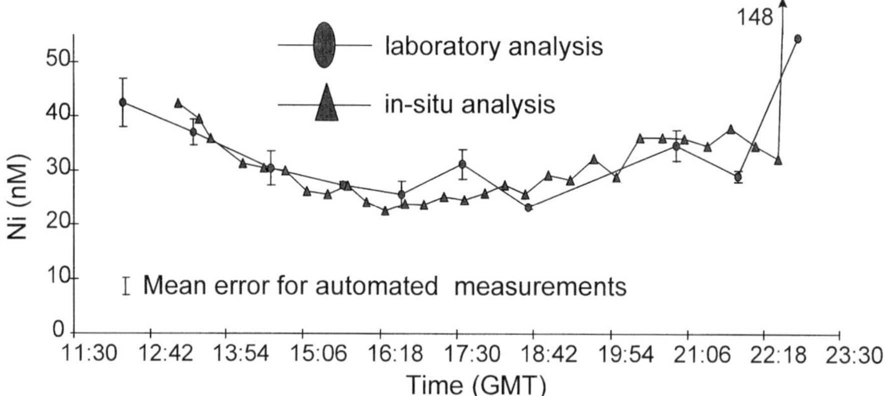

Figure 10.11 Results of tidal cycle study in the Tamar (July 1997), showing dissolved Ni concentrations from *in-situ* and laboratory analysis.

Figure 10.11 shows the results of automated total dissolved Ni measurements obtained from the *in-situ* application of the metal monitor on a bank of the Tamar Estuary. Figure 10.11 also shows values for total dissolved Ni obtained in discrete water samples collected from the estuary during the same study and analysed in the University laboratory. Quality of trace metal analysis in the laboratory was verified using certified reference materials. The *in-situ* and laboratory obtained data are within analytical uncertainty of each other, which demonstrates the high quality of the *in-situ* measurements. An important advantage of the *in-situ* monitoring approach includes the larger number of data points obtained during the automated study: 32 automated measurements compared with 10 discrete measurements. This allows a more thorough geochemical interpretation of the data. The trace metal measurements are presented in Figure 10.12, with complementary salinity and suspended particulate material (SPM) data. Salinity was obtained from conductivity measurements and SPM concentrations were obtained from weight of material collected on pre-weighed 0.45 μm membrane filters. Enhanced total dissolved Ni and SPM concentrations were observed at low salinities (up to 148 nM of Ni at a salinity of ~1) and low Ni and SPM concentrations were found at high salinities (23 nM of Ni at a salinity of ~24). Dilution of river water with enhanced Ni concentrations, with Ni depleted seawater therefore seems to be an important process determining the Ni behaviour in this estuarine system.

10.9 CONCLUSIONS AND FUTURE OF AUTOMATED MONITORING

The extremely low detection limits, coupled with its multi-element and speciation capabilities, high accuracy, modest cost and suitability to ship-board and flow analysis, have made stripping voltammetry an important technique for marine

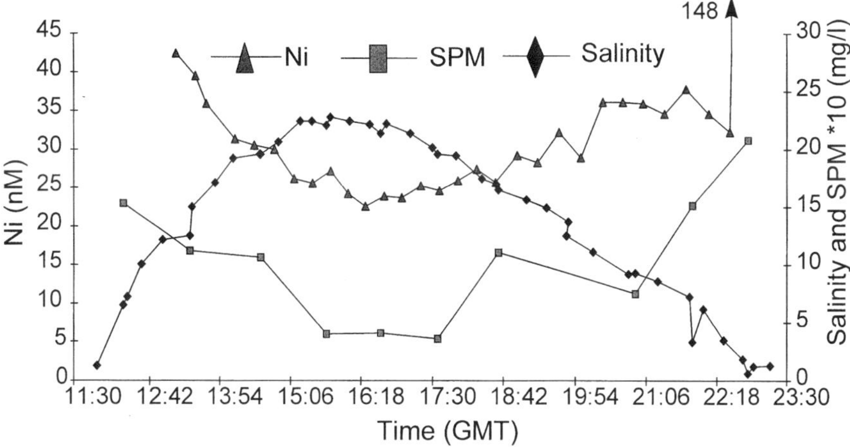

Figure 10.12 Results of tidal cycle study in the Tamar (July 1997), showing *in-situ* measured dissolved Ni concentrations with salinity and SPM (suspended particulate matter) data.

trace metal monitoring. The voltammetric trace metal monitor as described in this chapter, has been applied to different marine systems, ranging from unpolluted open ocean waters to metal polluted estuarine systems. The high resolution, high quality data obtained using the monitor are valuable for pollution control purposes, but also for biogeochemical modelling exercises. The automated voltammetric system, with on-line filtration and UV-digestion has recently been improved by incorporation of an on-line CTpH meter, for automated conductivity, temperature and pH measurements, complimenting the trace metal data. This low-cost approach of simultaneous automated data collection results in high quality data-sets which allow better interpretation of the chemical speciation in marine systems. The fully automated systems used on the Atlantic shelf, Tamar and Gulf of Cadiz perform ca. 3–6 fully calibrated metal determinations per hour. This number can be increased by using a flow cell fitted to an hanging mercury drop electrode. Up to 50–60 trace metal determinations per hour can then be performed, with calibration by switching to reagent with added metal standard at regular intervals (typically after 300 metal determinations) (Colombo and van den Berg, 1997). This approach is however less suitable for estuarine and near-coastal waters, where strong variations in salinity and dissolved organic matter concentrations will result in large changes in the sensitivity of the metal determination.

The trend in automated trace metal determinations is towards *in-situ* deployment of analytical instrumentation, with underwater application of sensors. This approach not only further reduces the risk of sample contamination, but also potentially leads to an improved way of trace metal speciation measurements as the *in-situ* analysis results in a minimal disturbance of the chemical equilibria.

The *in-situ* application of a hanging mercury drop electrode in a Swiss lake has been reported recently (de Vitre *et al.*, 1991; Tercier and Buffle, 1993). These workers used differential pulse ASV for the automated determination of Cu, Pb and Cd in Swiss lakes. A novel *in-situ* voltammetric profiling system reported by Belmont *et al.* (1996) makes use of a agarose membrane covered mercury plated Iridium-based micro-electrode for the determination of "ASV labile" Cu, Cd, Pb and Zn in fresh and marine waters. This system employs a coating on the electrode surface, and the diffusion of trace metals through this coating forms the time limiting step of the analysis, resulting in a measurement frequency of 2 h^{-1}. Wang *et al.* (1995) have reported the remote analysis of labile Cu in an estuarine system (San Diego Bay) using stripping potentiometry with a gold fibre electrode. This method has a reported limit of detection of ca. 5 nM, and is therefore suitable for estuarine and coastal waters with enhanced metal levels, but not for unpolluted ocean waters.

Acknowledgements

We would like to thank Dr. P.J. Statham (University of Southampton) for enabling EPA to participate in the OMEX cruise (D216). We thank the European Union for funding the TOROS project (Environment & Climate Programme ENV4-CT96-0217) and the University of Plymouth for funding DJW's studentship.

References

Achterberg, E.P. and van den Berg, C.M.G. (1994a). Automated voltammetric system for shipboard determination of metal speciation in sea water, *Anal. Chim. Acta*, 284, 463–471.

Achterberg, E.P. and van den Berg, C.M.G. (1994b). In-line ultra-violet digestion of natural water samples for trace metal determination using an automated voltammetric system, *Anal. Chim. Acta*, 291, 213–232.

Achterberg, E.P. and van den Berg, C.M.G. (1996). Automated monitoring of Ni, Cu and Zn in the Irish Sea, *Mar. Poll. Bull.*, 32, 471–479.

Baldwin, I.G. and Kramer, K.J.M. (1994). Biological Early Warning Systems. In *Biomonitoring of coastal waters and estuaries*, edited by Kramer, K.J.M. pp. 1–23. CRC Press.

Belmont, C., Tercier, M.L., Buffle, J., Fiaccabrino, G.C., Koudelka-Hep, M. (1996). Mercury-plated Iridium based microelectrode arrays for trace metals detection by voltammetry – optimum conditions and reliability, *Anal. Chim. Acta*, 329, 203–214.

Bruland, K. and Franks, R.P. (1983). Mn, Ni, Cu, Zn and Cd in the Western North Atlantic. In *Trace Metals in Seawater*, edited by Wong, C.S., Boyle, E.A., Bruland, K.W., Burton, J.D. and Goldberg, E.D. pp. 395–414. Plenum Press.

Bryan, G.W. and Langston, W.J. (1992). Bioavailability, accumulation and effects of heavy metals in sediments with special reference to United Kingdom estuaries: a review, *Envir. Poll.*, 76, 89–131.

Buckley, P.J.M. and van den Berg, C.M.G. (1986). Copper complexation profiles in the Atlantic Ocean, *Mar. Chem.*, 19, 281–296.

Colombo, C. and van den Berg, C.M.G. (1997). Simultaneous determination of several trace metals in seawater using cathodic stripping voltammetry with mixed ligands, *Anal. Chim. Acta*, 337, 29–40.

de Vitre, R.R., Tercier, M.-L. and Buffle, J. (1991). *In situ* voltammetric measurements in natural waters: the advantages of microelectrodes, *Anal. Proceedings*, 28, 74–75.

Howard, A.G. and Statham, P.J. (1993). *Inorganic trace analysis: philosophy and practice*, Wiley, p. 182.

Jickells, T.D. and Burton, J.D. (1988). Cobalt, copper, manganese and nickel in the Sargasso Sea, *Mar. Chem.*, 23, 131–144.

Kremling, K. and Pohl, C. (1989). Studies on the spatial and seasonal variability of dissolved cadmium, copper and nickel in North-East Atlantic surface waters, *Mar. Chem.*, 27, 43–60.

Larsen, P.F. (1992). Marine environmental quality in the Gulf of Maine, *Rev. Aquat. Sci.*, 6, 67.

Lauenstein, G.G., Robertson, A. and O'Connor, T.P. (1990). Comparison of trace metal data in mussels and oysters from a Mussel Watch Programme of the 1970s with those from a 1980s programme, *Mar. Poll. Bull.*, 21, 440–447.

Leblanc, M., Benothman, D., Elbaz-Poulichet, F. and Luck, J.M. (1995). Rio Tinto (Spain), an acidic river from the oldest and most important mining areas of Western Europe: Preliminary data on metal fluxes. In *Mineral deposits: From their origins to their environmental impacts, Proceedings of the third biennial SGA meeting, Prague*, edited by Pasava, J., Kribek, B. and Zak, K. pp. 669–670, A.A. Balkema.

Morley, N.H., Fay, C.W. and Statham, P.J. (1988). Design and use of a clean shipboard handling system for seawater samples, *Advances in Underwat. Technol.*, 16, 283–289.

Rijstenbil, J.W., Merks, A.G.A., Peene, J., Poortvliet, T.C.W. and Wijnholds, J.A. (1991). Phytoplankton composition and spatial distribution of copper and zinc in the Fal estuary (Cornwall, UK), *Hydrobio. Bull.*, 25, 37–44.

Stephenson, M.D. and Leonard, G.H. (1994). Evidence for the decline of silver and lead and the increase of copper from 1977 to 1990 in the coastal marine waters of California, *Mar. Pollut. Bull.*, 28, 148–153.

Tappin, A.D., Millward, G.E., Statham, P.J., Burton, J.D. and Morris, A.W. (1995). Trace metals in the Central and Southern North Sea, *Estuar. Coastal Shelf Sci.*, 41, 275–323.

Tercier, M.L. and Buffle, J. (1993). *In situ* voltammetric measurements in natural waters: future prospects and challenges, *Electroanalysis*, 5, 187–200.

Tessier, A. and Turner, D.R. (1995). *IUPC Series on Analytical and Physical Chemistry of Environmental Systems, Metal speciation and bioavailability in aquatic systems*, Vol. 3, Wiley, p. 679.

Van den Berg, C.M.G. (1989). Adsorptive cathodic stripping voltammetry of trace elements in sea water, *Analyst*, 114, 1527–1530.

Van den Berg, C.M.G. (1991). Potentials and potentialities of cathodic stripping voltammetry of trace elements in natural waters, *Anal. Chim. Acta*, 250, 165–276.

Van Geen, A., Boyle, E.A. and Moore, W.S. (1991). Trace metal enrichments in waters of the Gulf of Cadiz, Spain, *Geochim. Cosmochim. Acta*, 55, 2173–2191.

Van Geen, A., Adkins, J.F., Boyle, E.A., Nelson, C.H. and Palanques, A. (1997). A 120 yr record of widespread contamination from mining of the Iberian pyrite belt, *Geology*, 25, 291–294.

Wang, J., Foster, N., Armalis, S., Larson, D., Zirino, A. and Olsen, K. (1995). Remote stripping electrode for *in situ* monitoring of labile copper in the marine environment, *Anal. Chim. Acta*, 310, 223–231.

11. *IN SITU* CHEMICAL SENSOR MEASUREMENTS AT THE SEDIMENT-WATER INTERFACE

CLARE E. REIMERS[a] and RONNIE N. GLUD[b]

[a]*Institute of Marine and Coastal Sciences, Rutgers University, New Brunswick, New Jersey 08901–8521 USA and*
[b]*Marine Biological Laboratory, University of Copenhagen, Strandpromenaden 5, 3000 Helsingor*

11.1 INTRODUCTION

The sediment-water interface is a zone of the aquatic environment where the application of chemical sensors has led to significant scientific advances. Building on traditional approaches, a growing number of oceanographers are using chemical sensors *in situ* to characterize chemical, biological and physical processes at the benthic interface.

There are two basic requirements for chemical sensors in benthic research. The first requirement is for robust and rapidly responding microsensors that can map chemical distributions in and above sediments with a spatial resolution comparable to the scale of diffusive boundary layers and the chemical zones created by microorganisms. The second need is for chemical sensors with sufficient stability and specificity to track solute changes over time periods of days to weeks. Both these sensor types must also operate autonomously and from instrumentation packages small enough to be carried to the sea floor by "landers" and other unmanned or manned oceanographic vehicles (Tengberg *et al.*, 1995; Reimers *et al.*, 1999).

The object of this chapter is to review the specific sensors and instrumentation that have successfully met the requirements of benthic *in situ* experiments. The reader should not expect a comprehensive review of all benthic sampling methods, but rather insight into *in situ* sensor applications that have evolved since Revsbech and Jørgensen (1986) reviewed the use of microelectrodes in microbial ecology. We include descriptions of the operating principles of many standard and developing electrochemical sensors, sensor support equipment that is unique to deep-water benthic studies, and examples of noteworthy technical or experimental observations and problems. Additionally, we present some new optical sensor approaches, which are not yet applicable to oceanographic vehicles but which have a proven potential for resolving solute dynamics at benthic interfaces. We hope this chapter will unite many of the practical and insightful aspects of *in situ* chemical sensor measurements that are seldom emphasized in the literature. Kohl and Revsbech (1999) have prepared another review of microsensors that focuses primarily on laboratory developments.

11.2 *IN SITU* APPROACHES, SENSORS AND INSTRUMENTATION

11.2.1 *In Situ* Calibration and Polarographic Oxygen Microelectrodes

The primary reason that a variety of chemical sensors have been used successfully for sea floor studies of biogeochemical activity is that during these investigations, sensors operate generally under constant temperature, salinity and pressure conditions. Without these variables, changes in the response of most chemical sensors can be interpreted as a change in concentration that is easily calibrated by independent analyses of water samples. As an illustration, we begin by considering the application of polarographic O_2-electrodes via two instruments that have become workhorses in benthic chemical research: the *in situ* microprofiler (Reimers, 1987) and the benthic chamber (Berelson *et al.*, 1987; Jahnke and Christiansen, 1989; Tengberg *et al.*, 1995) (Figure 11.1: See colour plate section). The *in situ* microprofiler is an instrument that lowers or raises microsensors in small (e.g. 0.1 mm) vertical steps relative to the sediment-water interface while reading sensor outputs at each step. A flux chamber is usually an instrumented box or cylinder that is placed over an area of sea floor so that, over time, water enclosed above the sediment will register concentration changes driven by the exchange of solutes across the sediment-water interface. The expected experimental outcomes resulting from the use of these respective instruments are: solute microprofiles across the sediment-water interface (from

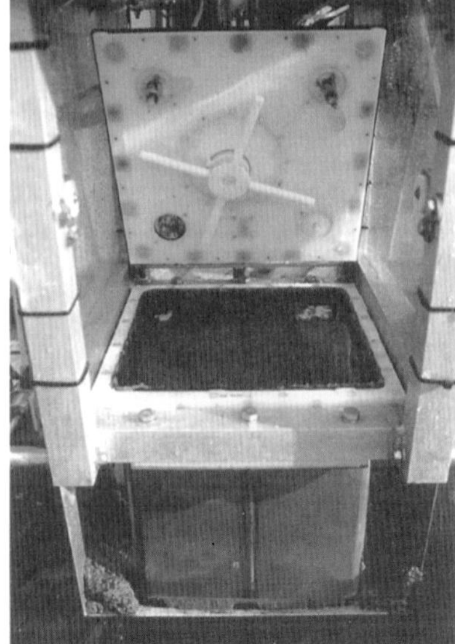

Figure 11.1 Photographs (A) microprofiling and (B) benthic chamber instruments. Sensors are mounted directly on the bottom end cap of the aluminum housing of the microprofiler and to the lid of the benthic chamber. *See* Color Plate 10.

which diffusive fluxes and chemical reaction rates are often derived), and total solute fluxes driven by any biogeochemical or transport process operating on the scale of the chamber (Glud *et al.*, 1994a; Reimers *et al.*, 1999). Figure 11.2a shows that microprofiler records obtained with polarographic O_2-microelectrodes have two straightforward calibration points: the constant signal in the bottom water and the constant low readings recorded in subsurface anoxic sediments. A comparison of the sensor readings in bottom water before and after the sensor enters the sediment can also be used to evaluate if the sensor has drifted or changed its sensitivity during the course of a profile. Similarly, when monitoring oxygen consumption within benthic chambers, O_2 electrode readings in bottom waters at the beginning of chamber deployments are usually combined with dissolved O_2 determinations from water samples retracted at discrete time intervals for sensor calibration (Tengberg *et al.*, 1995; Figure 11.2b).

The first deep-sea microprofiler measurements were achieved by Reimers *et al.* (1986) as an outgrowth of the manual profiling approaches developed by Revsbech (1983) for shallow benthic environments. The applied sensors were O_2 microelectrodes of a design adopted from the membrane-coated O_2 needle sensor of Baumgartl and Lubbers (1973; 1983). Revsbech and Ward (1983), and Revsbech and Jørgensen (1986) later developed a Clark-style O_2 microelectrode and Revsbech (1989) optimized this sensor by including a guard cathode (Figure 11.3). As with any polarographic O_2 electrode, this sensor's output is produced because of the reduction of dissolved O_2 ($4e^- + O_2 + H_2O \rightarrow 4OH^-$) at the metal surface of the sensor cathode that is held at a constant negative potential (usually ~ -0.8 V) relative to the reference electrode (or anode). The reference electrode is a AgCl-coated Ag wire which with the aid of an internal alkaline electrolyte completes the circuit (Figure 11.3). The guard cathode reduces interference from O_2 dissolved in the backing electrolyte solution, and thereby the zero-current is kept low and constant (Revsbech, 1989). With proper signal amplification and analog-to-digital conversion, it is possible to resolve changes in oxygen of $\sim 0.2\,\mu M\ kg^{-1}$ with this sensor, and it is now routinely used *in situ*. However, the greatest advantages of the Clark-style O_2 microelectrode result from of the small size of its tip. The micro-sized tip allows the sensor to be introduced into sediments with little physical disturbance or change in the surrounding oxygen tension due to the minimal O_2 consumption by the sensor. In addition, the small transport distance from the environment to the cathode insures a 90% response time (T_{90}) that is generally less than 2 seconds and may be as rapid as 0.2 seconds. Finally, the sensor has very little stirring sensitivity (usually $< 2\%$ in air-saturated sea water) because the supply of O_2 to the cathode is impeded mainly by diffusion over the transport distance from the sensor tip to the cathode. Oxygen macro-electrodes do not have these advantages and when used in benthic chambers or other applications are usually combined with pumps or stirring systems that maintain a constant flow velocity around the sensor (Sayles and Dickinson, 1991). The latter equipment adds another level of complexity to instrumented systems.

Recently a mathematical model has been developed that describes the signal current above background (I) from a Clark-style O_2 microelectrode as a function

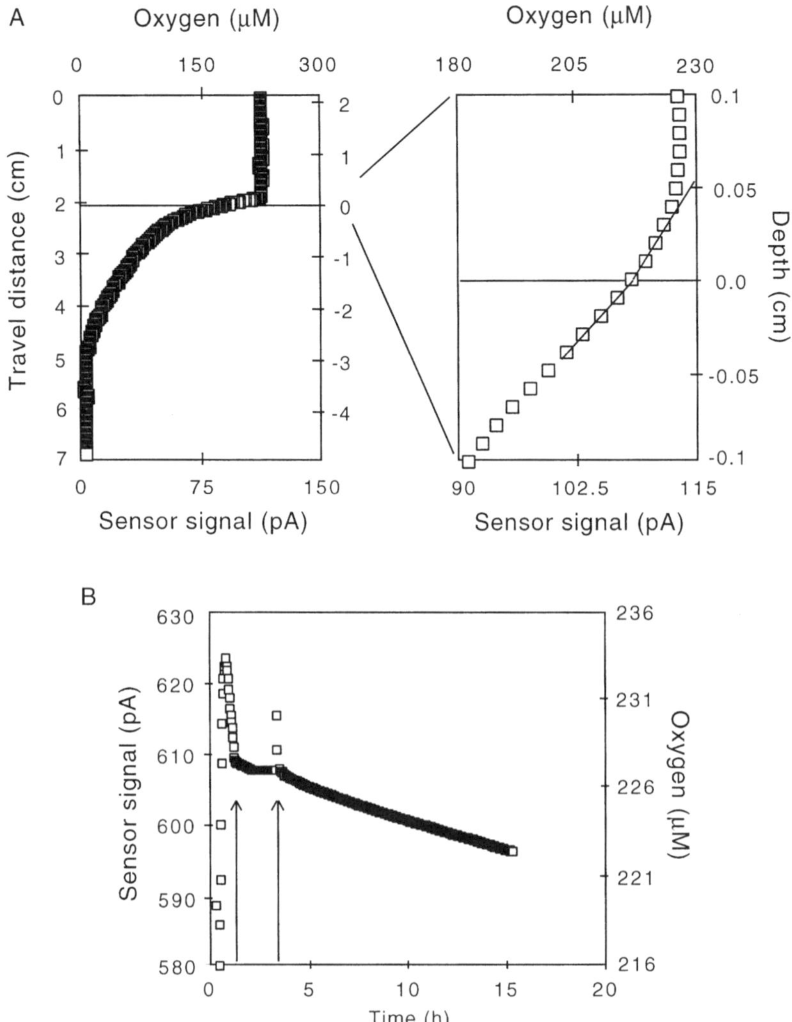

Figure 11.2 (A) An illustration of the calibration of an O_2 profile obtained at 3100 m depth in the SE Atlantic. The O_2 concentration scale is based on a linear calibration defined by microelectrode readings in the bottom water and in the anoxic sediments (after the sensor profiled over a vertical travel distance of 7 cm). The sediment-water interface is defined after identification of a linear gradient representing the diffusive sublayer immediately above the sediment (see enlargement). The point where the sensor crossed into the sediment is marked by a increase in the slope of the O_2 profile because of a reduction in the effective diffusion coefficient. (B) The readout of an O_2 electrode mounted in the lid of a benthic chamber deployed at the same location. The two arrows indicate the times when the lander landed and the chamber lid closed. The steady electrode signal between these points is assumed to correspond to the O_2 concentration of the bottom water (227 μM). Data from Glud *et al.* (1994a).

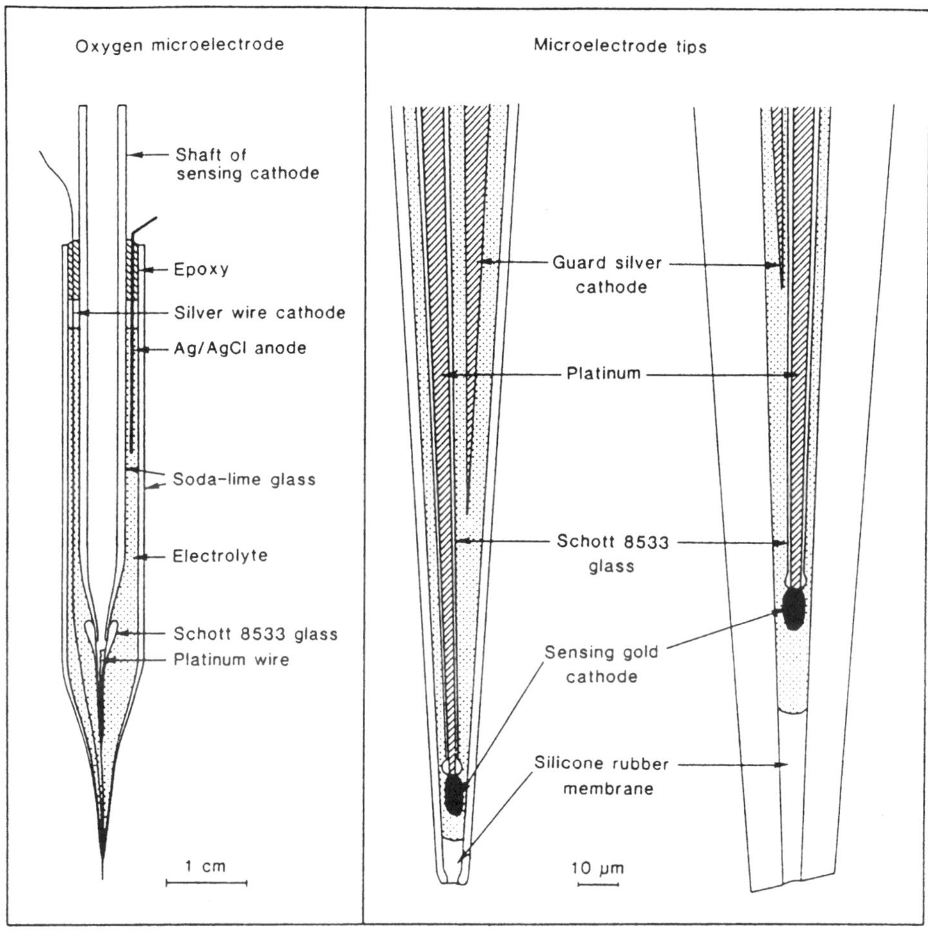

Figure 11.3 A Clark-style O_2 microelectrode with internal reference and guard cathode. Enhancement of the sensor tip is shown in the right panel (redrawn from Revsbech, 1989).

of the sensor dimensions, and a number of physical/chemical parameters that vary with the temperature and pressure of the environment (Gundersen *et al.*, 1998)

$$I = \phi \pi r_1 P \left(\frac{Z_m}{\psi r_2} + \frac{Z_e}{D_e S_e r_0} \right)^{-1} \tag{11.1}$$

Symbols:

ϕ = Generated current per mole O_2 reduced (3.86×10^5 A mol^{-1})
r_0 = Inner radius of the electrode capillary (in front of cathode)
r_1 = Inner radius of the electrode capillary (behind membrane)
r_2 = Radius of the sensor opening

P = Partial pressure of O_2 at the sensor surface
Z_m = Length of silicone membrane
Z_e = Distance from membrane to cathode
D_e = O_2 diffusion coefficient in electrolyte
S_e = O_2 solubility in electrolyte
Ψ = O_2 permeability of silicone membrane

This model helps emphasize that if classified according to the type of analytical signal measured, an O_2 electrode is an amperometric sensor. The only poorly defined variable in the equation above is the membrane permeability, and hence for work at conditions of constant temperature and pressure, this variable is derived empirically after determining the sensor's response at a single non-zero calibration point. Applying the sensor under conditions of variable pressure (e.g. for water column profiling) is confounded by changes in the membrane permeability due to compression effects.

The Clark-style microelectrode with internal reference and guard cathode has been proven to be stable for several days (drift $< 1-5\%$ of sensor signal in air-saturated sea water; Revsbech, 1989). In addition, a more rugged design of this sensor has been developed specifically for chamber measurements that very often are run for 1–7 days (Glud *et al.*, 1995). The sensor tip has an outside diameter of approximately 1 mm, but the membrane opening is less than 5 μm. Therefore the advantages of a microsensor are combined with a rugged structure. The long-term stability and life time of this sensor have been further improved by applying a small porous gold clump as a guard cathode instead of a silver wire (Glud *et al.*, 1995).

11.2.2 Potentiometric Microelectrode Measurements

The only potentiometric sensors that have been applied very widely for benthic studies are small versions of the glass membrane pH-, the Severinghaus pCO_2-, and the Ag/Ag_2S sulfide-electrode. As is explained by Turner and Whitfield (1981), potentiometric methods differ from other electrochemical electrode techniques by being equilibrium measurements of potentials established when there is no net current flow between working and reference electrodes. Electrochemical descriptions of common forms of the electrode cells employed for pH, pCO_2 and sulfide measurements in natural waters are given below:

$$\text{Ag–AgCl}|0.7\text{ M HCl} + \text{AgCl}(sat.)|\ \begin{matrix}\text{pH Glass} \\ \text{Membrane}\end{matrix}\left|\begin{matrix}\text{Lake, Sea or} \\ \text{Pore water}\end{matrix}\right|\begin{matrix}\text{Liquid junction} \\ \text{(ground glass)}\end{matrix}|3\text{ M KCl}|\text{AgCl–Ag} \quad (11.2)$$

$$\text{Pt–Ag–Ag}_2\text{S}|\text{Lake, Sea or Porewater}|\ \begin{matrix}\text{Liquid junction} \\ \text{(ground glass)}\end{matrix}|3\text{M KCl}|\begin{matrix}\text{Ag–AgCl} \\ \text{(positioned above the sulfide zone)}\end{matrix} \quad (11.3)$$

$$\text{Ag–AgCl}|0.7\,\text{M HCl} + \text{AgCl}(sat.)|$$

$$\begin{array}{r|c|l} \text{pH Glass} & 1\text{–}3\,\text{mM NaHCO}_3 + 0.7\,\text{M NaCl} & \\ \text{Membrane} & \text{separated from the Lake, Sea or Pore water} & \text{AgCl–Ag} \\ & \text{by a gas permeable silicone membrane} & \end{array} \qquad (11.4)$$

Although in principle these cells will produce potentials that are a combination of partial potentials (e.g. for pH, the inner reference electrode potential + inner glass surface potential + the outer glass surface potential + liquid junction potential + the reference electrode potential + $\cdots$), the usual assumptions in benthic work are that the ion-selective outer surface potential is the only potential that changes, and this response follows changes in the activity of the selected analyte according to an appropriate form of the Nernst expression. Thus, electrode signals that drift over time can often be traced to problems with the reference electrode. The reported properties of several designs of potentiometric sensors that have been used successfully in benthic studies are compiled in Table 11.1.

Calibration procedures for potentiometric sensors must include a reference to ambient conditions for the reason that most of the partial potentials mentioned above vary as a function of temperature, pressure and to a lesser extent salinity (Grasshoff, 1983). Briefly, the pH cell may be calibrated by assuming the electrode cell potential (E_z) recorded *in situ* at any depth below the sediment-water interface is described as:

$$E_z = E_{BW} + k_{pH}\{pH_z - pH_{BW}\}, \qquad (11.5)$$

where the subscript "BW" refers to bottom water values, and the constant k_{pH} is the temperature-dependent, but pressure invariant, electrode slope or Nernst coefficient (ideally –59.2 mV per pH unit @ 25 °C; Disteche, 1959; Archer *et al.*, 1989a; Cai and Reimers, 1993). In cases where pH microelectrodes are applied in fresh water, k_{pH} may be determined with NBS standards at known temperatures prior to field measurements. However, for sea water measurements, it is our practice to determine k_{pH} with buffers that have the ionic strength of sea water and to report pH according to the total hydrogen or sea water scales (Dickson, 1992). pH_{BW} may be determined by calculation from total alkalinity and total carbon dioxide concentration measurements or from spectrophotometric determinations of pH made at 1 atmosphere and 25 °C that are recalculated for *in situ* conditions (Clayton and Byrne, 1993).

Similarly, pCO_2 microelectrode measurements are referenced to bottom water values assuming

$$E_z = E_{BW} + k_{pCO_2}\{\log(pCO_{2z}) - \log(pCO_{2BW})\} \qquad (11.6)$$

Values of k_{pCO_2} are determined prior to field measurements by recording the electrode response to a range of gas mixtures of known $\%CO_2$ (e.g. 0.03–5%) bubbled into sea water or fresh water. A point to emphasize is that pCO_2 sensors are also sensitive to H_2S (de Beer *et al.*, 1997). Their lifetime, response time and

Table 11.1 Reported properties of potentiometric electrodes that have been developed for benthic studies.

Electrode (source)	*Tip Geometry and Design*	*Sensitivity*	*Response time*	*Comments on in situ use*
pH microelectrode (Revsbech *et al.*, 1983; Archer *et al.*, 1989a)	Cone-shaped, protruding tip with a diameter ranging from 50–750 μm	$>$ 97% Nernst response	T_{90} (1 pH unit change) 10–60 sec.	Applied widely and to full-ocean depths
pH mini-electrode (de Jong *et al.*, 1988)	Bulb-shaped tip with a diameter of 500–1000 μm	95–98% Nernst response	T_{90} 5–15 s	Application was to intertidal sediments
pH mini-electrode (Wilson *et al.*, 1989)	Bulb-shaped tip with a diameter of 6 mm surrounded by a steel cage	No information	No information	Used to profile sediments to $>$ 30 cm depth
pH microelectrode (Cai and Reimers, 1993; Cai *et al.*, 1995; Reimers *et al.*, 1996; Komada *et al.*, submitted)	Bulb-shaped tip with a diameter of 100–150 μm	$>$ 98% Nernst response	T_{98} (1 pH unit change) $<$ 2 min.	Has been applied *in situ* in coastal and deep-sea sediments
pCO_2 microelectrode (Cai and Reimers, 1993; Cai *et al.*, 1995; Komada *et al.*, submitted)	Glass pH microelectrode ~50 μm behind a silicone membrane at the tip of a micropipet with an outer diameter of ~300 μm	$>$ 93% Nernst response	T_{98} 5–25 min dependent on the magnitude and the direction of change in $\log(pCO_2)$	Has been applied *in situ* in coastal sediments and at a water depth of 4100 m

pCO_2 microelectrode (Zhao and Cai, 1997)	Liquid membrane pH microelectrode positioned ~5–30 μm behind a silicone membrane at the tip of a micropipet with an outer diameter of 50–300 m	Nernstian slope from ~500 μatm to 50 matm with a detection limit of ~200 μatm	T_{98} 2–4 min	*In situ* application will be attempted during the summer of 1998 by Reimers
pCO_2 microelectrode (de Beer *et al.*, 1997)	Liquid Ion Exchange membrane pH sensor positioned 3–10 μm from a 5 μm thick membrane in a pipet with a tip diameter of ~10 μm	Detection limit < 10 μM if carbonic anhydrase is used in filling solution, standard curves non-linear	T_{90} 10 s to a few min. Addition of carbonic anhydrase decreases response time.	Has been applied *in situ* at abyssal depths (Wenzhö f̄er pers. comm.)
Ag/Ag_2S sulfide-microelectrode (Revsbech and Jørgensen, 1986; Visscher *et al.*, 1991; Gundersen *et al.*, 1992)	Slightly recessed Ag/Ag_2S Pt wire surrounded by Pb glass, diameter ~200 μm	Standard curves often show greater than Nernstian slopes but are log-linear from ~10 to 3000 μM S_{total}	T_{98} 2–5 min.	Has been applied *in situ* in coastal and Guaymas Basin sediments

stability can be improved if a liquid pH membrane and aluminosilicate glass (rather than Pyrex or lead glass) are used in the construction (Zhao and Cai, 1997) and carbonic anhydrase is added to the internal electrolyte (de Beer *et al.*, 1997). In addition, Komada *et al.* (1998) found that a very durable gas permeable membrane can be made with a plug of silicone elastomer (Dow-Corning MDX4–4210). In operation, the pCO_2 microelectrode tends to be less sensitive to noise than their component pH microelectrodes because the outer casing and electrolyte act to shield the complete electrode circuit.

Since the potential of the Ag/Ag_2S sulfide-electrode is usually expressed as a direct function of the negative logarithm of the activity of sulfide ion (Berner, 1963), it is generally not feasible to reference the *in situ* readings of these sensors to bottom water activities. This is because most bottom waters are oxic and therefore sulfide ion activities will be indeterminate. Instead, a reference horizon in sulfide-rich anoxic sediments can be used to calibrate *in situ* profiles such that:

$$E_z = E_{REF} + ks\{pS_z - pS_{REF}\} \tag{11.7}$$

Values of pS_{REF} can be defined if pore water samples are retrieved (e.g. by fine-scale sectioning of cores) and analyzed for pH and concentrations of total-H_2S. In these calculations, values for the first and second dissociation constants for H_2S are needed as a function of temperature, pressure and salinity (see e.g. Millero, 1986; Smith and Martell, 1976). Values of k_s may be derived from laboratory calibrations run as a function of temperature in environmental waters having various H_2S concentrations and pH values. Computed values of pS_z may then be used with parallel *in situ* measurements of pH_z to calculate total H_2S at any depth. This myriad of calculations and large uncertainties associated with assignment of the second dissociation constant for H_2S make the application of Ag/Ag_2S sulfide-electrodes not very accurate and time consuming (Revsbech *et al.*, 1983). For this reason a new amperometric H_2S microsensor (Kühl *et al.*, 1998), and the voltammetric microelectrode of Brendel and Luther (1995) (discussed in section 2.3), are rapidly replacing the Ag/Ag_2S sulfide- electrode in benthic studies. The amperometric microelectrode has the same physical design and dimensions as the Clark-style O_2 microelectrode, but in this sensor the measuring Pt electrode (positioned just behind the membrane) is positively polarized and the internal electrolyte is an alkaline solution of 0.05 M $K_3Fe(CN)_6$ (ferricyanide). In operation, H_2S diffuses across the silicone membrane, deprotonates to HS^- and subsequently gets oxidized to S^0 by reaction with the ferricyanide in solution. The reduced $K_4Fe(CN)_6$ (ferrocyanide) formed as a product of this reaction is then reoxidized to ferricyanide at the polarized platinum anode (typically +85 to + 150 mV). The current generated from this process is directly proportional to the H_2S concentration outside the sensor. At the counter electrode situated in the bulk electrolyte within the shaft of the electrode, ferricyanide is reduced back to ferrocyanide. A guard anode, which is polarized like the measuring anode, prevents ferrocyanide produced at the counter electrode from accumulating at the sensor tip which ensures a constant background concentration of ferricyanide at the measuring anode.

Measurements of pH, pCO_2 and sulfide in pore waters or in benthic chambers are important as indicators of rates of organic matter decay and mineral (e.g. $CaCO_3$ and FeS) dissolution/precipitation reactions and transport processes (see e.g. Boudreau, 1991; Wallmann *et al.*, 1997). As examples, the chemical zonation of organic-rich sediments from Sturgeon Bay, Lake Michigan is shown in comparison to the zonation through a thick *Beggiatoa* mat and underlying sediment in a hydrothermal region of Guaymas Basin, Gulf of California (Figures 11.4 and 11.5). These measurements were all made *in situ* by assorted potentiometric microelectrodes, a Clark-type O_2 electrode, and thermistor probes (Reimers unpublished data, and Gundersen *et al.*, 1992). The first few millimeters of the lake pore water profiles reflect the rapid production of dissolved carbon dioxide by oxic organic matter decay and methane oxidation tempered by molecular diffusion. Deeper into the sediment, anaerobic organic matter decomposition by methanogenesis (the term used for the disproportionation of organic matter into CO_2 and CH_4) is the major chemical process responsible for the pH and pCO_2 distributions. In the Guaymas basin sediments, the chemical distributions were observed to be irregular due to a pulsatory convective flow of sea water into the mat. This transport was responsible for a broadening of the O_2-H_2S interface compared to areas without flow and is hypothesized to be the primary reason that Guaymas *Beggiatoa* mats are extremely massive (Gundersen *et al.*, 1992).

It is noteworthy that the measurements made with the chemical sensors in the Sturgeon Bay and Guaymas basin examples were interpreted in light of parallel temperature measurements (see Figures 11.4 and 11.5 and section 3.1 for further discussion). Only in hydrothermal settings, or very shallow coastal areas that are subject to changing sunlight and currents, are temperature variations likely to occur during benthic sensor measurements. Side-by-side pH and pCO_2 microelectrode measurements can be paired and used with the equilibrium relationships of the carbonate system to calculate fine-scale total-CO_2 profiles. These profiles usually agree with coarser-scale pore-water total-CO_2 profiles determined by pore-water extraction and direct measurement of total-CO_2 (Figure 11.6, Komada *et al.*, submitted). However, cases where the carbonate system is apparently out of equilibrium because of very rapid CO_2 production or consumption have led to unrealistic total-CO_2 estimates (de Beer *et al.*, 1997; Komada *et al.*, 1998). It can also be difficult to precisely match measurements of pH and pCO_2 made with separate microelectrodes because sediments are characteristically heterogeneous on small scales (Cai and Reimers, 1993). This last problem can only be avoided with sensors that measure multiple species at the same time.

11.2.3 Voltammetric Microelectrode Analyses

In principle, voltammetric microsensor analysis is similar to amperometry except that multiple electroactive species may be detected simultaneously as a range

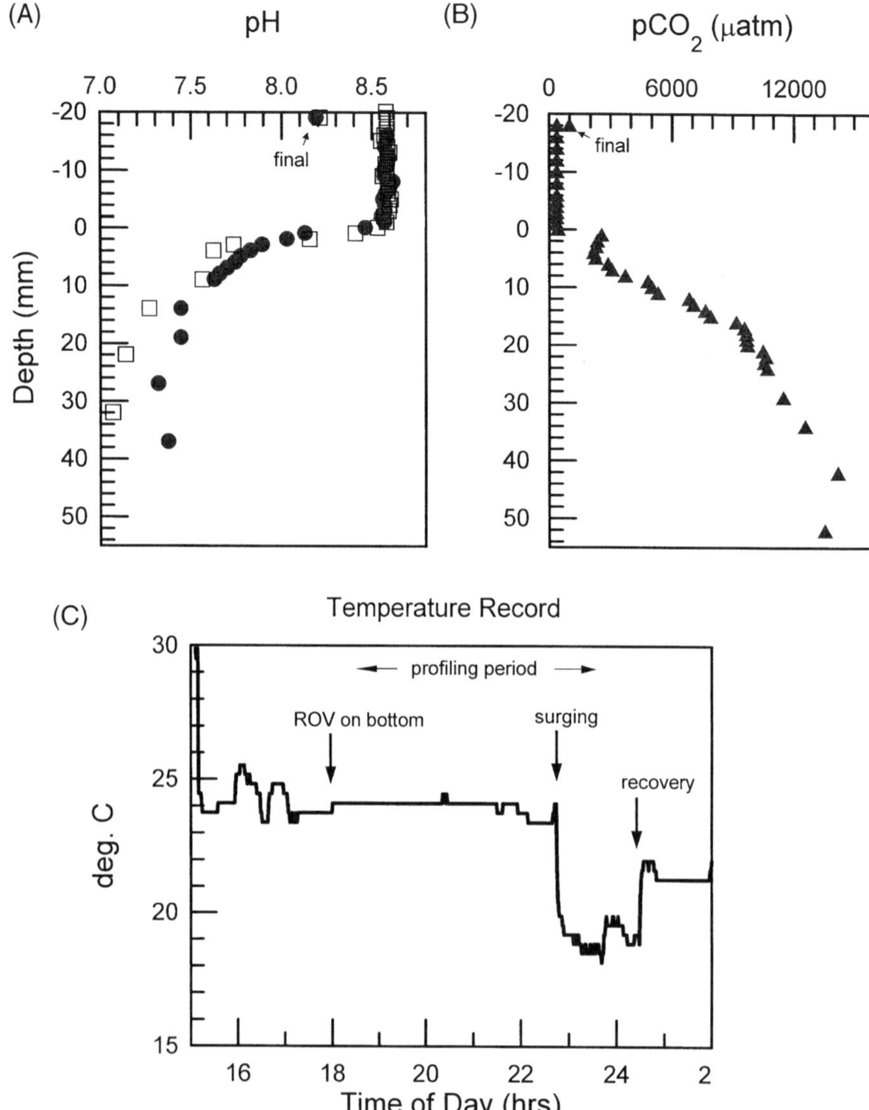

Figure 11.4 (A) and (B) Two pH and a pCO_2 profile(s) measured with potentiometric microelectrodes at a 6 m site in an embayment in Lake Michigan, USA. The microelectrodes were held at each depth in the sediment for periods up to 25 minutes in order for the pCO_2 microelectrodes to respond fully. Final readings taken after profiling differ from water column readings before profiling because of water exchange over the site. The water exchange also lowered the temperature by 4 degrees C. (C) Temperature recorded with a thermistor positioned above the sediment during the course of the deployment. These measurements were recorded in real-time from a Remote Operated Vehicle that was also equipped with benthic chambers (Klump *et al.*, 1997).

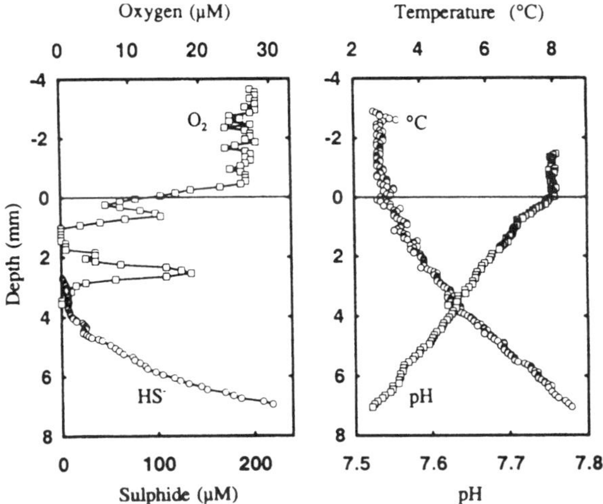

Figure 11.5 Measurements made with a Clark-type O_2 microelectrode, a Ag/AgS type sulfide microelectrode, a glass pH microelectrode and a thermistor probe through a thick *Beggiatoa* mat and underlying sediment in the hydrothermal fields of Guaymas Basin. The microsensors were mounted on a microprofiler that was deployed by the submersible *Alvin*. (Reproduced with permission from Gundersen *et al.*, 1992).

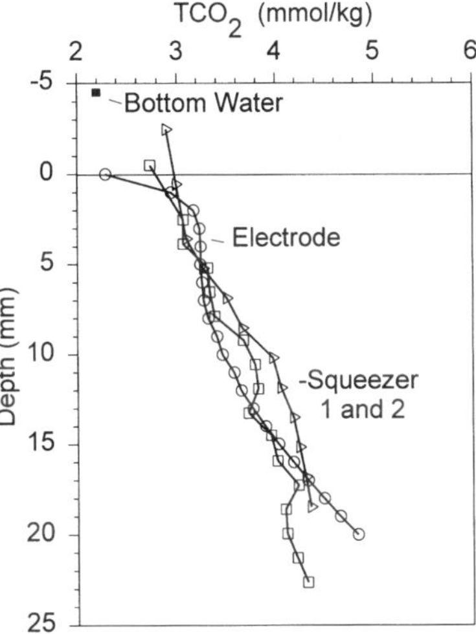

Figure 11.6 A comparison of pore water profiles of total CO_2 determined by (1) equilibrium calculations after pairing points from curves fit to the uppermost 20 mm of the pH (filled circles only) and pCO_2 profiles in Figure 11.4 and (2) direct measurement after separating pore water by the whole-core squeezing method (Bender *et al.*, 1987). The comparison suggests the microelectrode measurements provide the truer representation of the pore water gradient across the sediment-water interface. (Data from work by Reimers, Klump, Boehme and Waples, unpublished).

of potentials is applied (either as a linear or pulsed voltage ramp) between a working Hg-coated microelectrode and a reference electrode (Turner and Whitfield, 1981; Brendel, 1995). A third Pt counter electrode is used to monitor the signal current flowing at the working electrode because this design minimizes feedback problems through the reference. Current-voltage curves are interpreted after suitable (temperature dependent) calibrations to yield concentrations.

Brendel (1995), Brendel and Luther (1995) and Luther *et al.* (1997) pioneered this technique for the study of the pore water distributions of O_2, Mn^{2+}, Fe^{2+}, and total S^{2-}, four principal redox species (or groups of species) involved in early diagenesis. Their microsensors are made typically with tip diameters of $\sim$100 μm, and therefore signal currents are several nanoamps (Table 11.2). One limitation associated with voltammetric techniques in general, is that it is usually necessary to "condition" the sensor tip in between successive measurements by applying a potential that will remove previously deposited electroactive species from the mercury thin-film. These conditioning potentials add to the total analysis time so the sensor can not be used to study processes that may alter concentrations on time scales much shorter than one minute. However, replicate measurements suggest that conditioning procedures do not affect sediment microenvironments around the sensor, and voltammetric microelectrodes have been used very effectively to characterize quasi-steady state one-dimensional pore water distributions and reaction kinetics in laboratory cultures of bacteria. As an example of a pore water study, Luther *et al.* (1997) found evidence that O_2 and Mn^{2+} profiles frequently do not overlap, and they have used this and other pore water information with thermodynamic considerations to suggest Mn^{2+} may be oxidized by reaction with NO_{3-} after Mn^{2+} is produced by the reduction of reactive MnO_2 by NH_3 (both processes producing N_2, Luther *et al.*, 1997). They also

Table 11.2 Half-cell reactions that may occur at a Au/Hg electrode versus a standard calomel electrode, and the slopes of calibration curves run at 25 °C with a 100 μm diameter electrode (Area = 7.85×10^{-3} mm^2) (from Brendel, 1995). O_2 and H_2O_2 data were collected by Linear Sweep Voltammetry; all others were collected by Square Wave Voltammetry. (MDL-Minimum detection limit)* obtained with AIS, Inc. DLK-100. E_p is the peak potential where current from each reaction reaches a maximum.

	E_p (V)	MDL* (M)	calibration slope (nA/μM)
$O_2 + 2H^+ + 2e^- \rightarrow H_2O_2$	–0.30	5	0.152
$H_2O_2 + 2H^+ + 2e^- \rightarrow 2H_2O$	–1.30	5	0.152
$HS^- + Hg \leftrightarrow HgS + H^+ + 2e^-$	–0.62	< 0.2	2.2
$Fe^{2+} + Hg + 2e^- \leftrightarrow Fe(Hg)$	–1.43	10	0.025
$Mn^{2+} + Hg + 2e^- \leftrightarrow Mn(Hg)$	–1.55	3–5	0.070
$2I^- + 2Hg \leftrightarrow Hg_2I_2 + 2e^-$	–0.30	< 0.2	3.2
$2S_2O_3^{2-} + Hg \leftrightarrow Hg(S_2O_3)_2^{2-} + 2e^-$	–0.15	16	0.111
$Fe^{3+} + e^- \leftrightarrow Fe^{2+}$ species	–0.25 to –0.9		colloidal

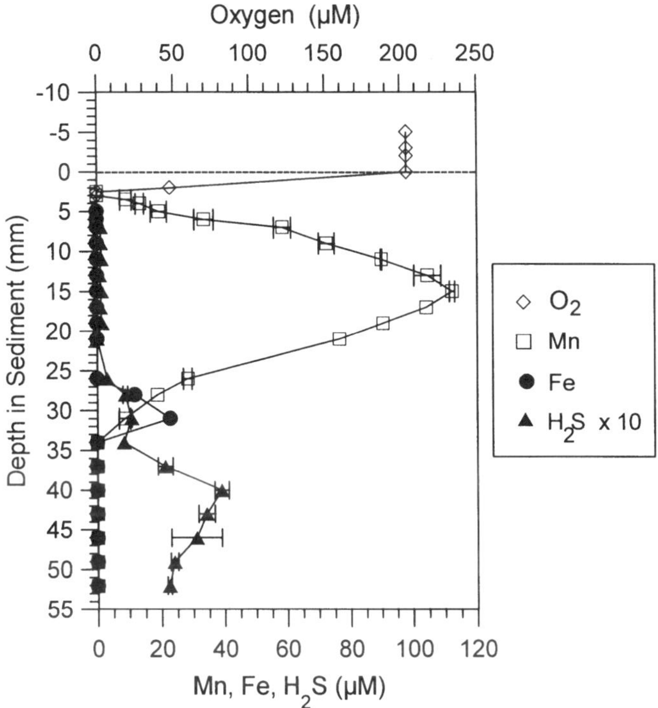

Figure 11.7 Profiles of four redox species measured by a single voltammetric microelectrode *in situ* in Raritan Bay, New Jersey, USA. Data from work in progress by G. Luther and C. Reimers.

verified that H_2O_2, I^-, $S_2O_3^{2-}$, and an Fe(III)-complex may be measured by this technique in pore waters according to half-cell reactions summarized in Table 11.2.

The first use of a solid state Au/Hg voltammetric microelectrode to measure microprofiles *in situ* took place in June 1997 with profiling instrumentation mounted on a Remote Operated Vehicle (Reimers *et al.*, 1997; Figure 11.7). The microelectrode was linked to a shipboard voltammetric analyzer with a 30 m cable that had receiver-transmitter transducers on each end to preserve the signal quality along the cable (constructed by D. Nuzzio, AIS, Inc.). The sensor was thus controlled and monitored in real-time from the research vessel anchored at the shallow study site. Two other profile sets measured at the same site showed large variations in the magnitude of the Mn^{2+} peak, but the O_2 profiles were nearly identical and consistent with profiles measured with Clark-style microelectrodes. It is expected that similar measurements will soon be completed with voltammetric microsensors in deeper environments with the sensors tied to a remotely deployed analyzer, enclosed in a pressure housing.

11.2.4 *In situ* Electrochemical Sensor Instrumentation

As illustrated by the recent development of *in situ* voltammetry, our ability to make *in situ* electrochemical microsensor measurements at the sea floor has grown and continues to grow because of the dual development of new sensors and improved instrumentation. Many components of the sensor equipment of microprofiler and benthic chamber instruments (Figure 11.1) evolved from pre-existing instrumentation, but the merging of sensor equipment into these packages has led to a few unique components and strategies for maintaining sensors under high pressure. For example, Figure 11.8 illustrates the basic design of holders that support microelectrodes at full ocean depths. Each holder is machined from durable polycarbonate[1] rod, and fitted at one end with sleeves (made of Delrin) that help compress o-rings as seals around the microsensors. The other end of the holder acts to join the microsensor with an electrical connector that may penetrate directly to circuitry within an underwater housing (Figure 11.8a and b) or form a cabled connection (Figure 11.8c) (See Colour Plate Section for 11.8b and c). When the sensor is positioned at a distance from the main electronics housing, it can be advantageous to accommodate a small circuit board supporting the first stage of signal amplification within the holder (Figure 11.8c). Pressure compensation is accomplished through a collapsible bulb or tube extending off the side of the holder. These parts, the holder itself, and each microsensor are filled with silicone oil or a non-conducting electronic fluid (Fluorinert 77, 3M Company). The advantage of the electronic fluid is it has a low viscosity and evaporates if spilled, so it does not leave the person setting up the sensors with an oily film on everything he or she handles. The disadvantage of Fluorinert 77 is it is denser than water, so it can not be used to pressure compensate sensors that contain an internal electrolyte solution.

If machined and filled carefully, electrode holders work well to isolate electrodes and connectors from outside sea water. However problems can arise, especially with sensitive potentiometric measurements, if insulation between the connector and housing deteriorates or if the main housing itself is not totally isolated from the electronics system (Archer, 1990). In Figure 11.9, a schematic diagram of the electronics system used to acquire data with a microprofiler fitted with Clark style O_2 microelectrodes and a resistivity sensor[2] is shown as an example of the basic components of instruments designed for *in situ* measurements with electrochemical sensors. As is discussed below, this (or any other similar) design can be easily modified to accept other types of sensors and supporting equipment that produce a standard analog signal. When electrochemical sensors are used in

[1] We have also made the holders from Plexiglas, but polycarbonate has superior strength and abrasion resistance.

[2] The ratio of the resistivity in the sediment to the resistivity of the pore fluid alone is called the Formation Factor. This parameter is routinely computed from resistivity profiles to gage the effects of sediment tortuosity on solute diffusion coefficients (Reimers *et al.*, 1992).

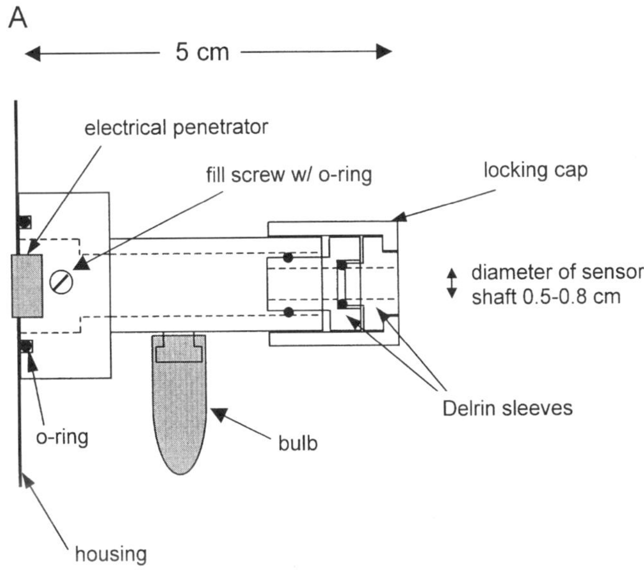

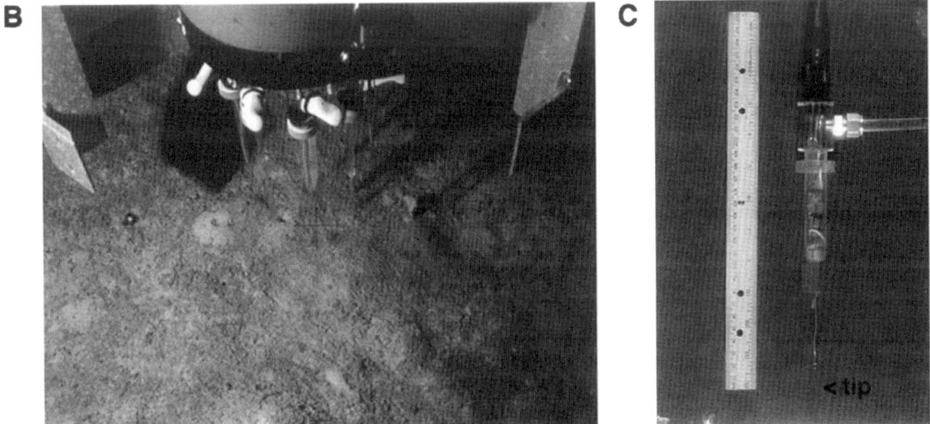

Figure 11.8 (A) Line-drawing of a holder for mounting microsensors on deep-sea instrument packages. This design is screwed directly to the pressure housing and pictured in (B). The design pictured in (C) allows for a cabled connection to an underwater electrical connector/penetrator. Both these designs have been mated to *in situ* microprofiler instruments by C. Reimers. The photo in (B) was taken in 1990 at 4100 meters on the continental rise off California and shows a resistivity sensor and needle-style O_2 and pH microelectrodes in use. Pressure has collapsed the compensating bulbs on the holders. The photo in (C) shows a Clark style O_2 microelectrode assembled with an external preamplifier within the holder. This design has been attached to an underwater motorized manipulator operated apart from the main Microprofiler electronics housing on a Remote Operated Vehicle. *See* Color Plate 11.

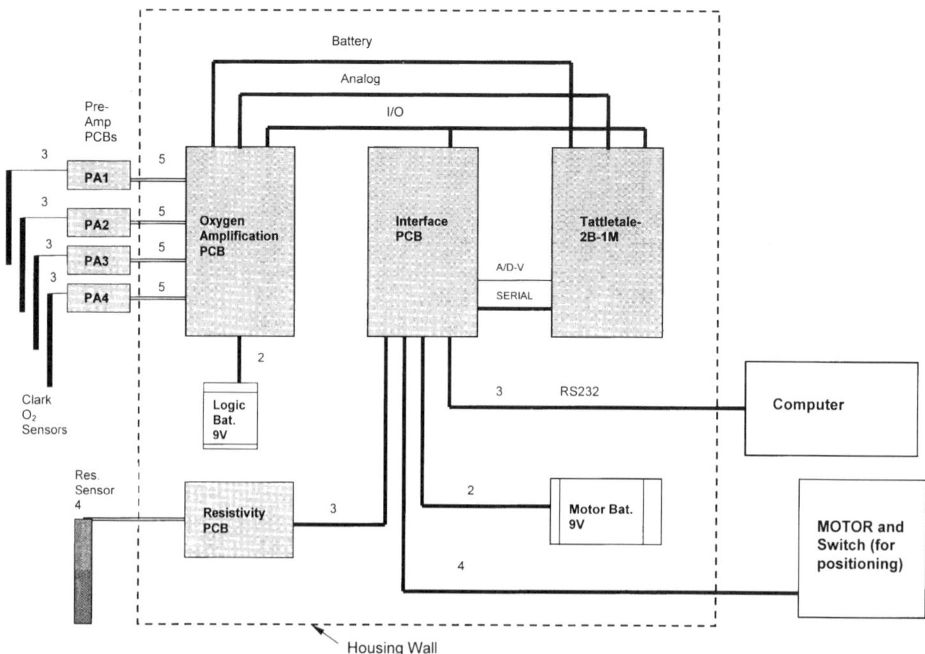

Figure 11.9 Block diagram of the data acquisition components a microprofiler instrument in use by C. Reimers. This design represents a modification of earlier designs developed by Reimers (1987) and Archer (1990). PCB=Printed Circuit Board, numbers indicate the number of wire connections between points, and a mateable underwater penetrator must be at every point where wires cross the pressure housing wall.

conjunction with benthic chambers, the main system difference is all positioning components are omitted or replaced with links to the motorized stirrer that mixes the overlying water within the chamber. The Model 2B Tattletale (Onset Computer Corporation) was chosen as the system controller and data logger for the design portrayed in Figure 11.9 for its compact size, data capacity (1 M), and straight-forward application software (TXBasic). Sensor data can be collected in a pre-programmed remote mode or in real-time through an RS-232 communications port and cable. Thus the system is highly versatile and has been deployable in many configurations by oceanographic landers, submersibles, and ROVs (Reimers *et al.*, 1999).

11.2.5 Microopt(r)odes and Fiber Optical Applications

The third and most recent type of chemical sensor that has been used *in situ* but that is not an electrochemical sensor is a microsensor constructed from an optical fiber. Fiber optical microsensors were introduced to aquatic biology by Klimant *et al.* (1995) and Klimant *et al.* (1997b) who miniaturized a fiber optical O_2 sensor. The measuring principle is based on the dynamic quenching effect of O_2 on the intensity of fluorescence emitted by an indicator (Ruthenium(II) tris-4,7–diphenyl-1,10-phenanthroline perchlorate) (Kautsky, 1939). The sensor is prepared by immobilizing the indicator in a hydrophobic polymer (polystyrene) that is dip coated onto the $\sim 10\,\mu m$ tip of a tapered optical fiber (Klimant *et al.*, 1995; 1997b). An additional layer of black silicone is coated on the sensor tip to shield out both ambient light and emitted light that may be back-scattered from sedimentary particles. The most common optical set up for the sensor is in the reflective mode whereby changes in the intensity of fluorescence emitted by the Ru fluorophor, at a wave length of 610 nm, are detected in response to an excitation wavelength of 450 nm. More details about this and other recent advances within the field of fiber optical microsensors are described in chapter 7 of this volume (Holst *et al.*, 1999). Below we concentrate on the very recent adaptation of such sensors to a benthic lander system and the future potential of opt(r)odes for other *in situ* applications.

The advantages of microopt(r)odes compared to most electrochemical sensors and from the perspective of *in situ* deployment are that they are robust, stable, and very easy to fabricate (Klimant *et al.*, 1995). The much simpler manufacturing procedure of opt(r)odes has the additional benefit of reducing the cost of equipping benthic instruments (especially microprofilers) with microsensors. Testing has shown that despite a pressure sensitive calibration curve, the intensity based $Ru(diph)_3$-O_2 sensor is functional at full ocean depth (Glud *et al.*, 1999) (Figure 11.10). Furthermore, the calibration curve of a microopt(r)ode may be described adequately by only two calibration points provided one is the sensor

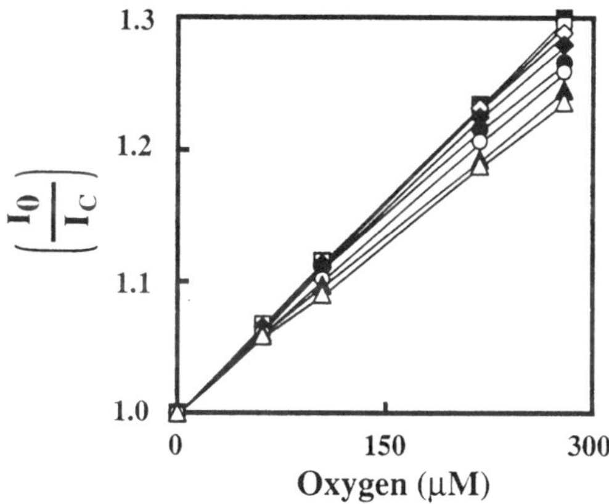

Figure 11.10 Calibration curves for a the $Ru(diph)_3$-Polystyrene O_2 microopt(r)ode at eight different hydrostatic pressures: 1 atm. (■–■); 50 atm. (□–□); 100 atm. (◇–◇); 200 atm. (◆–◆); 300 atm. (●–●); 400 atm. (○–○); 500 atm. (▲–▲); 600 atm. (△–△) where I_0 is the intensity of the sensor signal in the absence of oxygen and I_C is the intensity at the oxygen concentration indicated on the horizontal axis (from Glud *et al.*, 1999).

intensity in the absence of O_2 (i.e. I_0 in Figure 11.10). These qualities have made this sensor well suited for *in situ* work.

The first deployments of fiber optical O_2 microsensors by researchers at the Max Planck Institute for Marine Microbiology (Bremen, Germany) on landers have demonstrated that microopt(r)odes can successfully be applied *in situ* for measuring both microprofiles and total benthic exchange rates of O_2 within a chamber (Figure 11.11; Glud *et al.*, 1999). The opt(r)odes were mated to fiber optical penetrators and mounted in oil filled adapters similar to the holders used for electrochemical sensors (section 11.2.4). A modified and miniaturized version

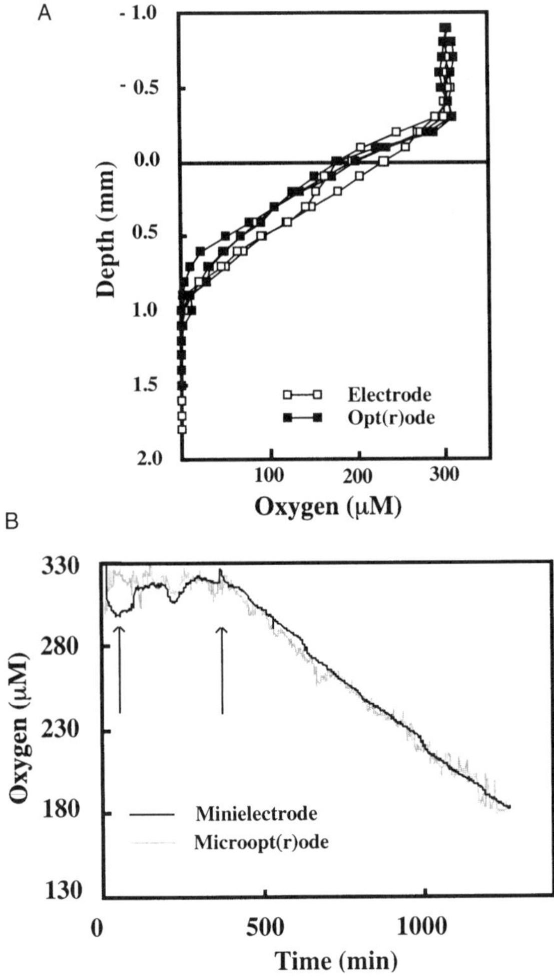

Figure 11.11 Microprofiles and a benthic chamber trace of oxygen concentrations measured with microot(r)odes and microelectrodes *in situ* at 10 m water depth (from Glud *et al.*, 1999).

of the opto-electronics system of an opt(r)ode (i.e. the LED excitation light source, optical filters and photomultiplier tube detector; Holst *et al.*, 1999) was incorporated into the underwater electronics housings of both a microprofiler and a benthic chamber system, side-by-side with the electronics boards for these system's polarographic oxygen microelectrodes. As part of the opto-electronics design, the LED excitation intensity was regulated via a system tied through one side of a fiber coupler that monitors the output intensity of the LED. The other side of the fiber coupler was connected to a two-channel optical switch linked to individual sensors and controlled by a computer processor. The complete size of the 2 channel system is $3 \times 7 \times 7$ cm (Glud *et al.*, 1999). The number of channels can be increased in future deployments by incorporating a multi-channel optical switch.

Any distrust in microopt(r)odes stems from our still rather limited experience with this type of sensor compared to O_2 microelectrodes which we have applied for more than 10 years. Both the optical sensors (including fluorophors and matrix materials) and their supporting instrumentation need further optimization and diversification so that other analytes (e.g. H^+, pCO_2) may be measured *in situ*. Such work is in progress, but already it is fair to say that intensity-based O_2 opt(r)odes represent a realistic alternative to polarographic electrodes for use on benthic platforms (Glud *et al.*, 1999).

11.3 EXPERIMENTAL CONCERNS

Throughout this review we have stressed the advantages of microsensors as compared to macrosensors in benthic research. Due to their small size and fairly simple support systems, microsensors take up very limited space, consume little energy, and the disturbance of the environment in which they operate is kept at a minimum. The small transport distance from the environment to the sensing parts of such sensors ensures that response times and detection limits are generally low. However, the main advantage of microsensors is that they operate at a temporal and spatial scale that is relevant for resolving solute dynamics at benthic interfaces and within microenvironments created by microorganisms.

Despite the many advantages of microsensors, unexpected results, physical disturbances or other sensor related shortcomings occasionally have to be accounted for during evaluation of *in situ* microsensor data. We review below a number of these shortcomings with the caveats that not all are relevant in every environment where microsensors are deployed, and the importance of artifacts related to microsensor measurements is dependent upon the scientific question that is addressed by the data.

11.3.1 Problems Associated with Calibration and Drift

As we have discussed, all chemical sensors that have a history of *in situ* use are sensitive either directly or indirectly to temperature and pressure. It is therefore crucial to perform sensor calibrations under conditions identical to the application conditions, and the simplest approach is to perform the sensor calibration *in situ*. For microelectrodes and microoptodes used on microprofilers or within benthic chambers an important calibration point is most often the sensor reading in bottom waters. In most areas of the deep ocean (one exception being in the vicinity of hydrothermal vents), bottom waters are easy to characterize because they have uniform properties over depth scales of tens of meters, and their chemistry is invariant on time scales that are much longer than the duration of benthic measurements. However, in shallow environments this may not be the case, complicating the task of calibration. Take for example the pH and pCO_2 microprofiles reported in Figure 11.4. Because the pCO_2 sensor used when making these measurements had a relatively long response time, it took over 4 hours to complete the profiles. When the sensors were retracted to the bottom waters they did not return to the same reading each had recorded prior to entering the sediment. This magnitude of change would normally be considered as evidence the sensors had drifted during the measurement period. However, continuous temperature readings in the bottom waters (Figure 11.4c) and simultaneous video observations of water surging at the bottom indicated the changes were in fact real. The site studied was in a shallow bay, which as the wind direction shifted during the evening, was flushed by cooler and more acidic water from outside the bay.

Instances of actual sensor drift may be avoided by testing sensors for at least 24 hours in the laboratory prior to deployments and eliminating unstable sensors. A warm up period *in situ* is also recommended. For example, if sensors on a microprofiler are turned on and allowed to read in the bottom water for 30 minutes before the start of profiling, their drift rate can be evaluated.

11.3.2 Hydrodynamic Interferences

Another concern when performing *in situ* chemical sensor measurements at the sea floor is how to introduce the instrument packages that carry the sensors and the sensors themselves without any disturbance. Typically, benthic landers are free-falling and their descent is accompanied by a bow wave. Video recordings have shown that even the gentlest landing can resuspend fine sediment particles. Such disturbances can potentially destroy the structure of active surface layers and alter solute fluxes and distributions within the uppermost millimeters of the impacted sediments. However, the situation is not hopeless, if the researcher can delay the start of measurements. Reimers *et al.* (1999) have shown that after a perturbation, surface gradients will readjust to their initial condition over time periods that may be only a few hours. The required times depend on the character of the initial solute distribution(s) and the magnitude of the imposed change. Mobile vehicles (e.g. submersibles, ROVs and autonomous Rovers) that can transverse the sea floor can also minimize the imposed change by performing measurements at locations removed from landing sites (Smith *et al.*, 1997; Reimers *et al.*, 1999).

The physical effects of sensors themselves have been shown to be important only in a thin layer of the water column adjacent to the sediment surface, called "the Diffusive Boundary Layer" – DBL (Boudreau and Guinasso, 1982). Although the existence of the DBL had been known for many years, the physical reality and significance first became apparent though a number of recent microelectrode studies (Jørgensen and Revsbech, 1985; Archer *et al.*, 1989b; Gundersen and Jørgensen, 1990). The 0.3–1.5 mm thick DBL is characterized by a linear concentration gradient (Figure 11.2; Archer *et al.*, 1989b; Gundersen and Jørgensen, 1990; Glud *et al.*, 1994a). Recent measurements have shown that a microsensor approaching the sediment surface compresses the DBL by approximately 25–45% just below the tip (Glud *et al.*, 1994b).

The artifact is most likely caused by pressure variations due to acceleration and deceleration of water flow around the shaft of the microsensor. The effect is therefore a function of sensor dimensions, the free flow velocity of the overlying water, and the micro topography of the sediment surface. It is important to consider this effect when evaluating data describing the structure or solute dynamics within the DBL. Compression of the DBL will steepen the diffusive gradient and lead to an overestimation of diffusive fluxes calculated from the segments of sensor profiles through the DBL (Glud *et al.*, 1994b; Lorenzen *et al.*, 1995).

How great the overestimation of computed fluxes may be can be evaluated for oxygen by assuming that an elimination of the DBL would increase the diffusive oxygen flux by a factor of $(C_w/C_0)^{0.5}$. This factor is derived assuming that respiration is constant with depth and O_2 consumption is proportional to the square root of the oxygen concentration at the sediment surface (Boudin, 1968). C_w and C_0 are the oxygen concentration in the turbulent water phase and at the sediment surface, respectively (Boudreau and Guinasso, 1982). Figure 11.12

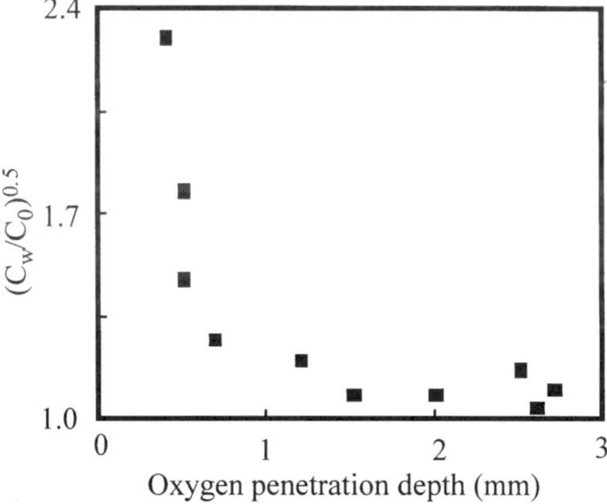

Figure 11.12 The estimated enhancement of O_2 fluxes caused by sensor effects on the DBL as a function of the oxygen penetration depth for a number of selected microprofiles.

shows the estimated DBL-factor as a function of the oxygen penetration depth for a number of profiles selected from recent publications (Gundersen and Jørgensen, 1990; 1991; Kühl *et al.*, 1996; Rasmussen and Jørgensen, 1992; Revsbech and Jørgensen, 1986). DBL-compression is likely to become significant only when the oxygen penetration depth is less than approximately 1.5 mm, conditions found in biofilms, microbial mats or very active sediments. In most coastal and deep-sea sediments (oxygen penetration > 1.5 mm), the artifact will increase the predicted flux by less than 5–10%. To what extent the presence of a profiling instrument mounted on a lander affects the local hydrodynamics and the DBL thickness at the measuring location has not yet been investigated.

11.3.3 Stirring Sensitivity, Sediment Effects and Response Time

The signal of microsensors consuming their analyte (e.g. amperometric and voltammetric electrodes) is dependent on the analyte supply rate (Gust *et al.*, 1987; Revsbech, 1989). Using the oxygen microelectrode described in Figure 11.3 as a model, O_2 has to diffuse from the surrounding medium through the silicone membrane and the electrolyte towards the Au-coated cathode. However, the tip of the microsensor will be surrounded by a DBL, which will impede the oxygen supply to the cathode. The thickness of this DBL, and therefore the sensor signal, will be dependent upon the flow rate of the surrounding medium – the so-called "stirring-effect" (Hale, 1983). Moving the sensor tip from a well mixing water phase into the interstitial waters of a sediment may thus result in a signal decrease which is caused solely by the decreased water flow experienced by the microsensor (Figure 11.13).

Oxygen microelectrodes with stirring effects as high as 50% of readings in aerated sea water have been applied in the laboratory (Gust *et al.*, 1987), but usually the stirring effects of microelectrodes are much less (Revsbech, 1989). It is important that as a routine procedure, sensors used during *in situ* deployments be carefully checked to insure they have stirring effects < 1–2% (Reimers, 1987; Glud *et al.*, 1994a). In coastal environments where the O_2 concentration decrease in the DBL is typically 10–30% of the bottom water concentration, a 1–2% stirring effect is a relatively small fraction of a large apparent concentration gradient (and thereby has little effect on the calculated diffusive exchange rate). However, for deep-sea measurements where the concentration change over the DBL may be only a few percent, a simultaneous decrease of 1–2% caused by a change in flow velocity around the sensor may invalidate the measurements in the DBL and the surficial sediment. In such situations, it is important to apply sensors with the lowest possible (i.e. ~0; Revsbech, 1989) stirring sensitivities.

Potentiometric pH and pCO_2 microelectrodes have also been tested for EMF differences caused solely by introducing the sensor from an overlying water solution into a sediment. When introduced from a buffered solution to an equally buffered sediment slurry in the laboratory, measured responses from pH microelectrodes have been < 1.5 mV (Cai and Reimers, 1993), and for pH and

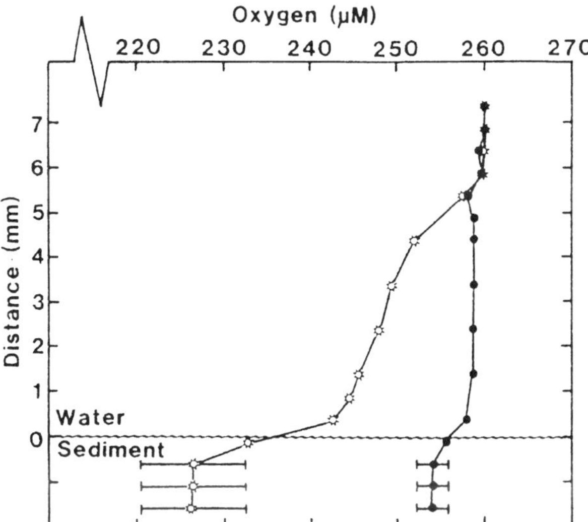

Figure 11.13 Apparent O_2 profiles through a well-mixed water phase into an inactive (no oxygen consumption) artificial sediment consisting of glassbeads. The profiles were measured by two different microsensors with 6% (open symbols) and 1% "stirring effect", respectively (from Revsbech, 1989).

$pCO_2 < 1$ mV and 0 mV, respectively (Komada and Reimers, unpublished data). Thus, any sediment artifact in pH measurements near the sediment-water interface is likely to be no greater than 0.01 pH units. The situation where this artifact (if indeed it exists) may be significant is in studies of deep-sea sediment early diagenesis. For example, Hales *et al.* (1996) have measured *in situ* pH profiles in sediment from the Ontong-Java Plateau in the western Pacific Ocean and observed changes of < 0.03 pH units over the entire first cm below the sediment-water interface. These gentle gradients are interpreted as due to the combined effects of metabolic CO_2 production and $CaCO_3$ dissolution.

Opt(r)odes have no stirring or sediment effects as long as they are coated with black silicone to block out any back-scattering of their emitted light from sediment particles. However, the response time of O_2 microopt(r)odes is ~2 s which is not as fast as most O_2 microelectrodes. Response time must be a consideration in programming microprofilers to measure *in situ* chemical distributions especially if relatively slow responding potentiometric sensors are in use (Table 11.1). By recording multiple readings with a delay time between readings, it is possible to check that sensor readings have stabilized at a given depth.

11.3.4 Scaling and Heterogeneity

The last problems we will address that are associated with applying microsensors *in situ* are identifying the sediment-water interface as a point within recorded microprofiles, one-dimensional scaling, and adequately describing spatial heterogeneity. In the case of O_2, pH and pCO_2 profiles, the first problem is minor since reaction and impeded diffusion in the sediment matrix usually will result in a change in slope of the profile right at the interface (Figure 11.2a). However, for solutes such as H_2S that normally do not show a gradient at the interface, aligning a profile can be very problematic. A solution to this problem has been presented by Klimant *et al.* (1997a), at least for laboratory studies. They cemented a simple optical fiber to another microsensor and measured the amount of light that was reflected back into the fiber and towards the measuring system during profiling. Right at the sediment surface the reflection increased significantly and the position of the sediment surface relative to the obtained concentration profile was resolved to within ± 0.05 mm. Similarly, Reimers *et al.* (1992) used close-up, time-lapse photography *in situ* to identify when bands placed around the shafts of microelectrodes disappeared into the sediment. By knowing the distance from the bands to the sensor tips, and the time each photo and sensor reading was recorded, the sensor reading at the sediment-water interface was determined to within ± 0.5 mm.

The uncertainties quoted above were partly a function of sediment microtopography and partly a function of the minimum resolution the researchers required during their investigations. Observing the benthic interface at the scale of a microsensor tip, the sediment surface is no longer flat but is typically dominated by extended relief. In an example portrayed in Figure 11.14, the surface area of a sediment and a covering DBL can be seen to be 31% and 14% larger, respectively, than the base area (Jørgensen and Des Marais, 1990). For

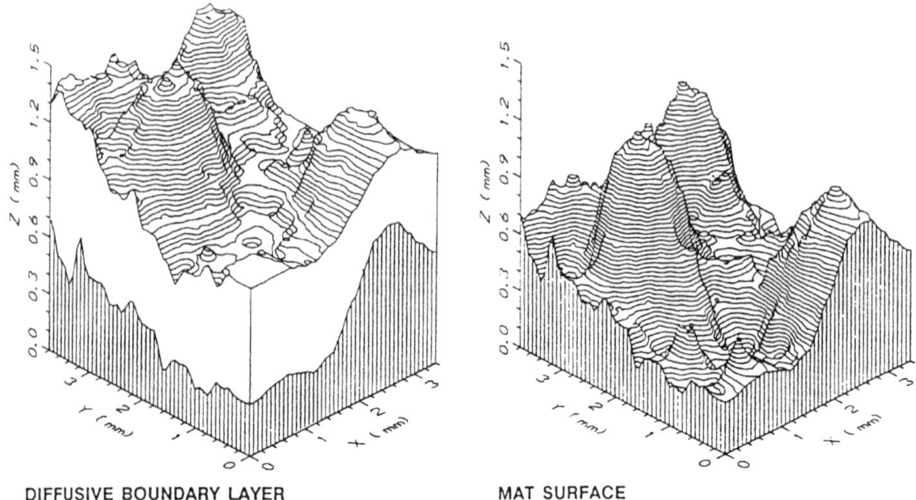

Figure 11.14 An illustration of the microtopography of a sediment and its overlying diffusive boundary layer as determined with oxygen microelectrodes (from Jørgensen and Des Marais, (1990).

simple reasons of geometry, such surface unevenness can cause sediment benthic exchange rates computed from one-dimensional profiles through the DBL (or through the very surface sediment) to be underestimates. Two studies of this effect, using a microbial mat and a coastal sediment, estimated that the total diffusive O_2 uptake was higher by 49% and 150% than what was calculated by a simple one-dimensional microprofile approach (Jørgensen and Des Marais, 1990; Gundersen and Jørgensen, 1990). These two examples probably reflect more intense microtopography than would be expected from the majority of marine sediments, but they illustrate that when deriving fluxes from a few one-dimensional profiles, topography has to be considered.

The other extrapolation problem that is difficult to overcome with a realistic number of one-dimensional *in situ* microprofiles is the full characterization of lateral heterogeneity related to uneven sedimentation, advective pore water flow, or faunal activity. Many sediments are characterized by extensive variations in solute distributions in time and space. Flux chambers have the advantage that they integrate these variations on the spatial scale of the chamber area and the temporal scale of a deployment. However, chamber data can not provide complete insight into the dynamics and governing reactions affecting benthic solutes. Therefore, a combined use of benthic chambers and new non-traditional techniques that may resolve 2-dimensional solute dynamics is expected to become the most information-rich approach for characterizing benthic biogeochemical processes.

11.4 2D CHEMICAL DISTRIBUTIONS ACROSS THE SEDIMENT-WATER INTERFACE

To learn more about the workings of the sediment-water interface in heterogeneous environments, the next logical step is the *in situ* application of a new technique for resolving O_2 distributions in two-dimensions at very high spatial and temporal resolution (Glud *et al.*, 1996a). The technique utilizes the same measuring principle as the $Ru(diph)_3$-polystyrene microsensor described above, however, instead of fixing the immobilized fluorophore at the tip of a fiber, it is coated upon a 175 μm transparent polyester support foil and subsequently covered by a 20 μm black silicone layer. The 3-layered O_2 sensitive foil is thus a "planar opt(r)ode" with a total thickness of approximately 200 μm that may be cut into pieces, inserted into sediment and used with a CCD (Changed Coupled Device) camera to resolve and record two dimensional oxygen distributions (Figure 11.15).

In initial laboratory experiments, the planar opt(r)ode has been fixed to the inside of an aquarium, the two-dimensional O_2 distribution in a tidal flat sediment resolved at a spatial resolution of 26×26 μm, and the O_2 saturation expressed on a linear 8 bit gray scale (Figure 11.16: See colour plate section). If a larger area of coverage was needed, the same imaging could have been run with lenses of longer focal length. However, this change would be accompanied by a corresponding coarser spatial resolution.

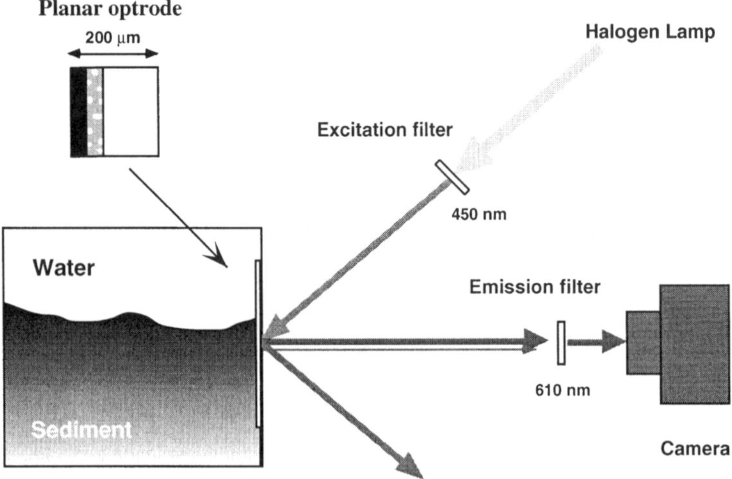

Figure 11.15 The basic set-up for planar opt(r)ode measurements in the laboratory. A halogen lamp equipped with a blue glass filter is used for excitation of the planar opt(r)ode, while the emitted light after filtering is collected by a photometrics CCD camera equipped with a KAF 1400 chip (1317 × 1035 pixels) and macro Nikon lens (redrawn from Glud *et al.*, 1996a).

The planar opt(r)ode technique promises to resolve the spatial heterogeneity across the benthic interface much better than what can be achieved by traditional microsensor techniques especially in heavily bioirrigated sediments (the image in Figure 11.16 is equivalent to more than 650 vertical O_2 microprofiles). In the

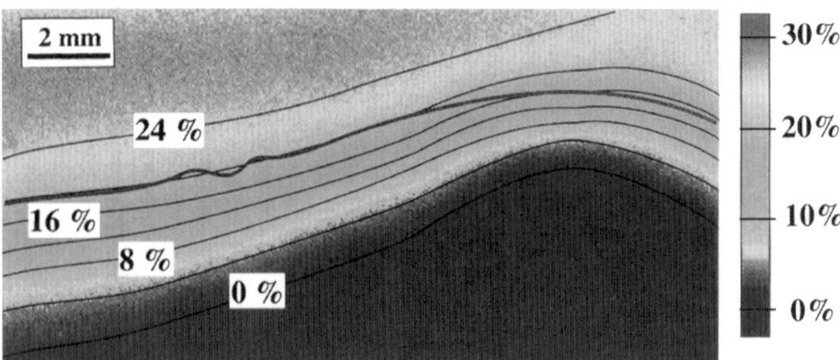

Figure 11.16 The O_2 distribution at the benthic interface of a tidal flat sediment overlain by water equilibrated with a gas mixture of 24 % O_2. Isolines of 0, 4, 8, 12, 16, 20 and 24% O_2 are indicated together with the position of the sediment surface (thick line). The O_2 partial pressure is expressed on a linear scale bar with 256 colors (scale bar shown) for visualization purposes. The O_2 penetration and the calculated diffusive O_2 uptake vary from 1.8–3.5 mm and 10–25 mmol m^{-2} d^{-1}, respectively. Flow direction from left to right (from Glud *et al.*, 1996a). *See* Color Plate 12.

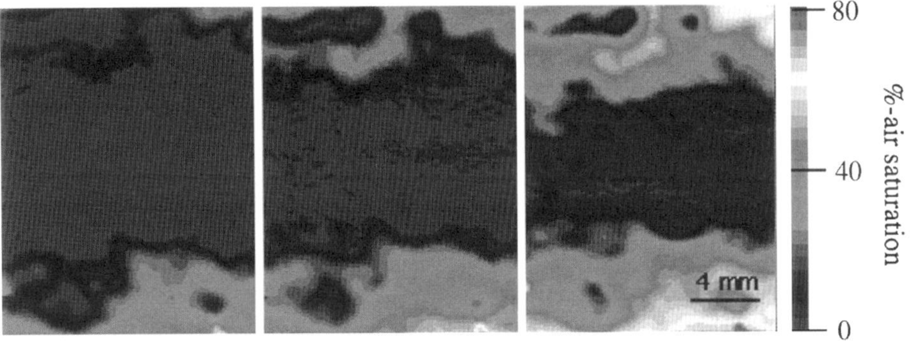

Figure 11.17 The O_2 distribution (expressed on an 8 bit scale) at the base of a biofilm as the flow velocity is increased to 20.5 (at the very left), 26.9 and 35.1 cm s^{-1}, respectively. A scale bar is included in the lower right corner (from Glud *et al.*, 1998). *See* Color Plate 13.

present configuration, images can be obtained every 2 sec which also allows the dynamics of solute exchange to be followed at a very high temporal resolution. Figure 11.17 (See Colour Plate Section) shows how the dynamics of O_2 at the base of a 400 μm biofilm was studied non-intrusively as a function of the free flow velocity of the overlying water (among other things) (Glud *et al.*, 1998). It can be seen how a complex channel structure within the biofilm ensured ventilation to the base of the film and that this ventilation increased as the free flow velocity increased. The images also reflect the extreme heterogeneity of many biofilms, which is nearly impossible to resolve by traditional microsensor techniques.

In summary, the technique of mapping chemical distributions in two dimensions with planar opt(r)odes has only been used in the laboratory and needs further optimization before being applied *in situ*. The main difficulties at this time are the calibration routine is rather cumbersome and the long-term stability is still not satisfactory. However, in combination with lifetime-based signal processing techniques (see chapter 7), planar opt(r)odes will be a valuable tool for resolving the dynamics of O_2 (and potentially other solutes) in heterogeneous benthic systems at high spatial and temporal resolution. For example, we imagine a great advantage of mapping pH and pCO_2 distributions optically so that total-CO_2 distributions may be predicted with much more certainty than is possible with two microelectrodes (see section 11.2.2). Deep-sea camera systems, based on an inverted periscope principle, have for some years been applied for obtaining vertical images of the top 5–20 cm of sediment structures and the borrow systems of infauna (Rhoads and Germano, 1982; Grehan *et al.*, 1992). Equipping such systems with optimized planar opt(r)odes should be possible in the coming years.

References

Archer, D.E. (1990). The dissolution of calcite in deep sea sediments: An *in situ* microelectrode study. *Ph.D. Thesis*, USA: University of Washington, 300.

Archer, D., Emerson, S. and Reimers, C. (1989a). Dissolution of calcite in deep-sea sediments: pH and O_2 microelectrode results, *Geochimica Cosmochimica Acta*, 53, 2831–2845.

Archer, D., Emerson, S. and Smith, C.R. (1989b). Direct measurements of the diffusive sublayer at the deep sea floor using oxygen microelectrodes, *Nature*, 340, 623–626.

Baumgärtl, H., and Lübbers, D.W. (1973). Platinum needle electrodes for polarographic measurement of oxygen. In *Oxygen Supply*, edited by Kessler, M., Bruley, D.F., Clark, L.C., Lübbers, D.W., Silver, A. and Sträuss, J., pp. 130–136. München: Urban & Schwarzenberg.

Baumgärtl, H., and Lübbers, D.W. (1983). Microcoaxial needle sensor for polarographic measurement of local O_2 pressure in the cellular range of living tissue. Its construction and properties. In: *Polarographic Oxygen Sensors Aquatic and Physiological Applications*. Gnaiger, E. and Forstner, H. (eds), pp. 37–65. Berlin: Springer-Verlag.

Berelson, W.M., Hammond, D.E., Smith, K.L. Jr., Jahnke, R.A., Devol, A.H., Hinga, K.R., Rowe, G.T. and Sayles, F. (1987). *In situ* benthic flux measurement devices: Bottom lander technology, *Marine Technology Society Journal*, 21, 26–32.

Berner, R.A. (1963). Electrode studies of hydrogen sulfide in marine sediments, *Geochimica et Cosmochimica Acta*, 27, 563–575.

Boudreau, B.P. (1991). Modelling the sulfide-oxygen reaction and associated pH gradients in porewaters, *Geochimica et Cosmochimica Acta*, 55, 145–159.

Boudreau, B.P. and Guinasso, Jr. N.L. (1982). The influence of a diffusive sublayer on accretion, dissolution, and diagenesis at the sea floor. In: Fanning, K.A., Manheim, F.T. (eds) *The Dynamic Environment at the Ocean Floor*, pp. 115–145, Lexington: Lexington Books.

Bouldin, D.R. (1968). Models for describing the diffusion of oxygen and other mobile constituents across the mud-water interface, *Journal of Ecology*, 56, 77–87.

Brendel, P.J. (1995). Development of a mercury thin film voltammetric microelectrode for the determination of biogeochemically important redox species in porewaters of marine and freshwater sediments, *Ph.D Thesis*, University of Delaware, 141.

Brendel, P.J. and G.W. Luther, III. (1995). Development of a gold amalgam voltammetric microelectrode for the determination of dissolved Fe, Mn, O_2 and S(-II) in porewaters of marine and freshwater sediments, *Environmental Science and Technology*, 29, 751-761.

Cai, W.-J. and Reimers, C.E. (1993). The development of pH and pCO_2 microelectrodes for studying the carbonate chemistry of pore waters near the sediment-water interface, *Limnology and Oceanography*, 38, 1762–1773.

Cai, W.-J., Reimers, C.E. and Shaw, T. (1995). Microelectrode studies of organic carbon degradation and calcite dissolution at a California continental rise site, *Geochimica et Cosmochimica Acta*, 59, 497–511.

Clayton, T.D. and Byrne, R.H. (1993). Spectrophotometer seawater pH measurements: total hydrogen ion concentration scale calibration of m-cresol purple and at-sea results, *Deep-Sea Research*, 40, 2115–2129.

de Beer, D., Glud, A., Epping, E., and Kühl, M. (1997). A fast responding CO_2 microelectrode for profiling sediments, microbial mats and biofilms, *Limnology and Oceanography*, 42, 1590–1600.

de Jong, S.A., Hofman, P.A.G. and Sandee, A.J.J. (1988). Construction and calibration of a rapidly responding pH mini-electrode: application to intertidal sediments, *Marine Ecology Progress Series*, 45, 187–192.

Dickson, A.G. (1992). pH buffers for sea water media based on the total hydrogen ion concentration scale, *Deep-Sea Research*, 40, 107–118.

Distéche, A. (1959). pH measurements with a glass electrode withstanding 1500 kg/cm^2 hydrostatic pressure, *The Review of Scientific Instruments*, 30, 474–478.

Glud, R.N., Gundersen, J.K., Jørgensen, B.B., Revsbech, N.P. and Schulz, H.D. (1994a). Diffusive and total oxygen uptake of deep-sea sediments in the eastern South Atlantic Ocean: *in situ* and laboratory measurements, *Deep-Sea Research*, 41, 1767–1788.

Glud, R.N., Gundersen, J.K., Revsbech, N.P. and Jørgensen, B.B. (1994b). Effects on the benthic diffusive boundary layer imposed by microelectrodes, *Limnology and Oceanography*, 39, 462–467.

Glud, R.N., Gundersen, J.K., Revsbech, N.P., Jørgensen, B.B. and Hüttel, M. (1995). Calibration and performance of the stirred flux chamber from the benthic lander Elinor, *Deep-Sea Research*, 42, 1029–1042.

Glud, R.N., Ramsing, N.B., Gundersen and J.K., Klimant, I. (1996a). Planar optrodes, a new tool for fine scale measurements of two dimensional O_2 distribution in benthic communities, *Marine Ecology Progress Series*, 140, 217–226.

Glud, R.N., Klimant, I., Holst, G., Kohls, O., Meyer, V., Kühl, M. and Gundersen, J.K. (1999). Adaptation, test and *in situ* measurements with O_2 microoptodes on benthic landers, *Deep-Sea Research*, 46, 171–183.

Glud, R.N., Santegoeds, C.M., de Beer, D., Kohls, O. and Ramsing, N.B. (1998). Oxygen dynamics at the base of a biofilm studied with planar optrodes, *Aquatic Microbial Biology*, 14, 223–233.

Grasshoff, K. (1983). Determination of pH. In: *Methods of Seawater Analysis*. Grasshoff, K., Ehrhardt, M. and Kremling, K. (eds) 2nd edition, Verlag Chemie, pp. 85–98.

Grehan, A.J., Keegan, B.F., Bhaud, M. and Guille, A. (1992). Sediment profile imaging of soft substrates in the western Mediterranean: the extent and importance of faunal reworking, *Academic Science Paris (III)*, 309–315.

Gundersen, J.K. and Jørgensen, B.B. (1990). Microstructure of diffusive boundary layers and the oxygen uptake of the sea floor, *Nature*, 345, 604–607.

Gundersen, J.K. and Jørgensen, B.B. (1991). Fine-scale *in situ* measurements of oxygen distribution in marine sediments, *Kieler Meeresforshung Sonderheft*, 8, 376–380.

Gundersen, J.K., Jørgensen, B.B., Larsen, E. and Jannasch, H.W. (1992). Mats of giant sulfur bacteria on deep-sea sediments due to fluctuating hydrothermal flow, *Nature*, 360, 454–455.

Gundersen, J.K., Ramsing, N.B. and Glud, R.N. (1998). Predicting the signal of oxygen microelectrodes from sensor dimensions, temperature, salinity and oxygen concentrations: Development and experimental verification, *Limnology and Oceanography*, 43, 1932–1937.

Gust, G., Booij, K., Helder, W. and Sundby, B. (1987). On the velocity sensitivity (stirring effect) of polarographic oxygen microelectrodes, *Netherlands Journal of Sea Research*, 21, 255–263.

Hale, J.M. (1983). Factors influencing the stability of polarographic oxygen sensors: *Aquatic and Physiological Applications*, pp. 3–17. Heidelberg: Springer Verlag.

Hales, B. and Emerson, S. (1996). Calcite dissolution in sediments of the Ontong-Java Plateau: *in situ* measurements of pore water O_2 and pH. *Global Biogeochemical Cycles*, 10, 527–541.

Holst, G., Kühl, M., Kohls, O., Klimant, I. and Wolfbeiss, O.S. (this volume) Optochemical microsensors.

Jahnke, R.A. and Christiansen, M.B. (1989). A free-vehicle benthic chamber instrument for sea floor studies, *Deep-Sea Research*, 36, 625–637.

Jørgensen, B.B. and Revsbech, N.P. (1985). Diffusive boundary layers and the oxygen uptake of sediments and detritus, *Limnology and Oceanography*, 30, 111–122.

Jørgensen, B.B. and Des Marais, D. (1990). The diffusive boundary layer of sediments: Oxygen microgradients over a microbial mat, *Limnology and Oceanography*, 35, 1353–1355.

Kautsky, H. (1939). Quenching of luminescence by oxygen, *Transactions Faraday Society*, 35, 216–219.

Klimant, I., Meyer, V. and Kühl, M. (1995). Fiber-optic oxygen microsensors, a new tool in aquatic biology, *Limnology and Oceanography*, 40, 1159–165.

Klimant, I., Holst, G. and Kühl, M. (1997a). A simple fiberoptic sensor to detect the penetration of microsensors into sediments and other biogeochemical systems, *Limnology and Oceanography*, 42, 1638–1643.

Klimant, I., Kühl, M., Glud, R.N. and Holst, G. (1997b). Optical measurements of oxygen and other environmental parameters in microscale: Strategies and biological applications, *Sensors and Actuators* B, 38–39, 29–37.

Klump, J.V., Paddock, R., Lovalvo, D., Doroodchi, M., Reimers, C., Waples, J., MacKenzie, R., Wroczynski, S. and Vande Slundt, D. (1997). Real-time studies of benthic-pelagic interactions using BESS: An ROV driven benthic shuttle system, *Marine Technology Society Journal*, 31, 21–33.

Komada, T., Reimers, C.E. and Boehme, S.E. (1998). Dissolved inorganic carbon profiles and fluxes determined using pH and pCO_2 microelectrodes, *Limnology and Oceanography*, 43, 769–781.

Kühl, M., Glud, R.N., Ploug, H. and Ramsing N.B. (1996). Microenvironmental control of photosynthesis and photosynthesis-coupled respiration in an epilithic cyanobacterial biofilm, *Journal of Phycology*, 32, 799–812.

Kühl, M. and Revsbech, N.P. (1999). Microsensors for the study of interfacial biogeochemical processes. In: *The Benthic Boundary Layer Transport and Biogeochemical Processes*, Boudreau, B. and Jørgensen, B.B. (eds), Oxford University Press.

Kühl, M., Steuckart, C., Eickert, G. and Jeroschewski, P. (1998). A H_2S microsensor for profiling biofilms and sediments: application in an acidic lake sediment, *Aquatic Microbial Ecology*, 15, 201–209.

Lorenzen, J., Glud, R. and Revsbech, N.P. (1995). Impact of microsensor-caused changes in diffusive boundary layer thickness on O_2 profiles and photosynthetic rates in benthic communities of microorganisms, *Marine Ecology Progress Series*, 119, 237–241.

Luther, G.W. III, Sundby, B., Lewis, B.L., Brendel, P.J. and Silverberg, N. (1997). Interactions of manganese with the nitrogen cycle: alternative pathways to dinitrogen, *Geochimica et Cosmochimica Acta*, 61, 4043–4052.

Millero, F.J. (1986). The thermodynamics and kinetics of the hydrogen sulfide system in natural waters, *Marine Chemistry*, 18, 121–147.

Rasmussen, H. and Jørgensen, B.B. (1992). Microelectrode studies of seasonal oxygen uptake in a coastal sediment: role of molecular diffusion, *Marine Ecology Progress Series*, 81, 289–303.

Reimers, C.E. (1987). An *in situ* microprofiling instrument for measuring interfacial pore water gradients: methods and oxygen profiles from the North Pacific Ocean, *Deep-Sea Research*, 34, 2017–2035.

Reimers, C.E., Fischer, K.M., Merewether, R., Smith, K.L., Jr. and Jahnke, R.A. (1986). Oxygen microprofiles measured *in situ* in deep ocean sediments, *Nature*, 320, 741–744.

Reimers, C.E., Jahnke, R.A. and McCorkle, D.C. (1992). Carbon fluxes and burial rates over the continental slope and rise off central California with implications for the global carbon cycle, *Global Biogeochemical Cycles*, 6, 199–224.

Reimers, C.E., Jahnke, R.A. and Thomsen, L. (1999). *In situ* boundary layer measurements. In: *The Benthic Boundary Layer Transport and Biogeochemical Processes*, Boudreau, B. and Jørgensen, B.B. (eds), Oxford University Press.

Reimers, C.E., Luther, G.W., III, Lovalvo, D. and Nuzzio, D. (1997). Real-time measurement of pore water redox species and pH using voltammetric, potentiometric and amperometric microelectrodes from an ROV. In: *Marine Analytical Chemistry for Monitoring and Oceanographic Research Proceedings*, Blain, S. *et al.* (eds) Brest France.

Reimers, C.E., Ruttenberg, K.C., Canfield, D.E., Christiansen, M.B. and Martin, J.B. (1996). Porewater pH and authigenic phases formed in the uppermost sediments of the Santa Barbara Basin, *Geochimica et Cosmochimica Acta*, 60, 4037–4057.

Revsbech, N.P. (1983). *In situ* measurement of oxygen profiles of sediments by use of oxygen microelectrodes. In *Polarographic Oxygen Sensors Aquatic and Physiological Applications*, edited by Gnaiger, E. and Forstner, H., pp. 265–273, Berlin: Springer-Verlag.

Revsbech, N.P. (1989). An oxygen microelectrode with a guard cathode, *Limnology and Oceanography*, 34, 474–478.

Revsbech, N.P. and Jørgensen, B.B. (1986). Microelectrodes and their use in microbial ecology. In *Advances in Microbial Ecology*, 9, edited by Marshall, K.C., pp. 293–352. New York: Plenum.

Revsbech, N.P., Jørgensen, B.B. and Blackburn, T.H. (1983). Microelectrode studies of the photosynthesis and O_2, H_2S, and pH profiles of a microbial mat, *Limnology and Oceanography*, 28, 1062–1074.

Revsbech, N.P. and Ward, D.M. (1983). Oxygen microelectrode that is insensitive to medium chemical composition: Use in an acid microbial mat dominated by *Cyanidium caldarium*. *Applied and Environmental Microbiology*, 45, 755–759.

Rhoads, D.C. and Germano, J.D. (1982). Characterization of organism-sediment relations using sediment profile imaging: An efficient method of remote ecological monitoring of the seafloor (Remots™-system), *Marine Ecology Progress Series*, 8, 115–128.

Sayles, F.L. and Dickinson, W.H. (1991). The ROLAI^{2}D lander: a benthic lander for study of exchange across the sediment-water interface, *Deep-Sea Research*, 38, 527–545.

Smith, K.L. Jr., Glatts, R.C., Baldwin, S.E., Beaulieu, R.J., Uhlman, A.H., Horn, R.C. and Reimers, C.E. (1997). An autonomous, bottom-transecting vehicle for making long time-series measurements of sediment community oxygen consumption to abyssal depths, *Limnology and Oceanography*, 42, 1601–1612.

Smith, R.M. and Martell, A.E. (1976). *Critical Stability Constants*, Vol. 4, Inorganic ligands. Plenum.

Tengberg, A., de Bovée, F., Hall, P., Berelson, W., Chadwick, B., Ciceri, G., Crassous, P., Devol, A., Emerson, S., Gage, J., Glud, R.N, Graziottin, F., Gundersen, J.K., Hammond, D., Helder, W., Hinga, K., Holby, O., Jahnke, R., Khripounoff, A., Lieberman, H., Nuppenau, V., Pfannkuche, O., Reimers, C., Rowe, G., Sahami, A., Sayles, F., Schlüter, M., Smallman, D., Wehrli, B. and de Wilde, P. (1995). Benthic chamber and profiling landers in oceanography – A review of design, technical solutions and functioning, *Progress in Oceanography*, 35, 253–294.

Turner, D.R. and Whitfield, M. (1981). The classification of electroanalytical techniques. In *Marine Electrochemistry A Practical Introduction*, Whitfield, M. and Jagner, D. (eds), pp. 67–98, Chichester: Wiley.

Visscher, P.T., Beukema, J. and van Gemerden, H. (1991). *In situ* characterization of sediments: measurements of oxygen and sulfide profiles with a novel combined needle electrode, *Limnology and Oceanography*, 36, 1476–1480.

Wallman, K., Linke, P., Suess, E., Bohrmann, G., Sahling, H., Schlüter, M., Dählmann, A., Lammers, S., Greinert, J. and von Mirbach, N. (1997). Quantifying fluid flow, solute mixing, and biogeochemical turnover at cold vents of the eastern Aleutian subduction zone, *Geochimica et Cosmochimica Acta*, 61, 5209–5219.

Wilson, T.R.S., McPhail, S.D., Braithwaite, A.C., Koch, B., Dogan, A. and Disteche, A. (1989). An instrument to measure pH and formation factor *in situ* across the seawater-sediment interface, *Deep-Sea Research*, 36, 315–321.

Zhao, P. and Cai, W.-J. (1997). An improved potentiometric pCO_2 microelectrode, *Analytical Chemistry*, 69, 5052–5058.

12. *IN SITU* MEASUREMENT OF LABILE SPECIES IN WATER AND SEDIMENTS USING DGT

WILLIAM DAVISON and HAO ZHANG

Institute of Environmental and Natural Sciences, Environmental Science Division, Lancaster University, Lancaster LA1 4YQ, UK

12.1 INTRODUCTION

The first publication on the technique of diffusive gradients in thin-films (DGT) appeared only recently (Davison and Zhang, 1994). Since then DGT has been used to measure (i) kinetically labile trace metals in seawater and freshwater (Zhang and Davison, 1995; Zhang *et al.*, 1996; Davison and Hutchinson, 1997), (ii) caesium, phosphate and sulfide in solution (Chang *et al.*, 1998; Zhang *et al.*, 1998a; Teasdale *et al.*, 1999) (iii) fluxes and concentrations of metals, phosphate and sulfide in sediment porewaters at a spatial resolution down to 100 μm (Zhang *et al.*, 1995a; Davison *et al.*, 1997; Soares, 1998; Zhang *et al.*, 1998a; Teasdale, 1999) and (iv) fluxes and concentrations of metals in soils (Zhang *et al.*, 1998b; Hooda *et al.*, 1999). This rapid development and application is due to the simplicity of DGT which comprises a layer of hydrogel overlying a layer of immobilised binding agent. Solutes freely diffuse through the hydrogel layer before being fixed by the binding agent. The device is deployed for a fixed period of time and then the mass of accumulated solute in the binding layer is measured. Because the diffusion layer thickness during the *in situ* deployment is determined by the known thickness of the hydrogel layer, the mean concentration in solution during deployment can be calculated using Fick's first law of diffusion. Thus, with DGT the chemical separation is made *in situ*, but the subsequent analysis is performed later under controlled laboratory conditions.

DGT developed from the sister technique of diffusive equilibration in thin-films (DET) which has been used to measure solutes in porewater at mm resolution (Davison *et al.*, 1991). In DET a layer of hydrogel without a layer of binding agent is held in a plastic assembly and inserted into the sediment. Solutes equilibrate with the 95% water of the hydrogel in a similar fashion to dialysis. Because a thin-gel is used (0.5–1 mm), complete equilibration is usually achieved in hours (Harper *et al.*, 1997). On removal the solutes are either fixed by chemical treatment or the gels are rapidly sliced at the required resolution. Solutes are either measured in the dried gel using a beam technique or in solution after being back-equilibrated with the gel slices. DET has been used to make porewater measurements of Fe and Mn (Davison, 1991; Davison *et al.*, 1994; Zhang *et al.*, 1995), NH_4^+, Cl^-, SO_4^{2-} and NO_3^- (Krom *et al.*, 1994; Zhang *et al.*, 1999), total CO_2 (Haraldsson *et al.*, 1997) and Ca, Mg, K and Na (Zhang *et al.*, 1999).

Without constraining the gel into compartments it can only accurately reproduce true porewater gradients and maxima at a spatial resolution greater than 1 mm (Harper *et al.*, 1997). Furthermore, the small volumes of gel limit sensitivity, preventing its use for trace components. These restrictions led to the development of DGT which automatically pre-concentrates and fixes solutes. While DET relies on equilibrium being established, DGT continuously removes solutes. Consequently, it is a kinetic procedure which only responds to labile species. This kinetic capability of DGT gives it a much wider application than high resolution pore water measurements, and enables its use as an *in situ* speciation tool in waters and for direct flux measurements in sediments.

12.2 USE IN WATER

12.2.1 Principles

Figure 12.1 provides a conceptual view of a DGT device operating at steady state while it is placed in solution. A binding agent such as a resin, selective to one of the ions in solution, is immobilised in a thin layer of hydrogel. It is separated from solution by an ion permeable hydrogel layer of thickness Δg. When there is convection in the bulk solution a thin layer of solution of thickness δ remains in contact with the gel surface. Transport of ions through this layer, known as the

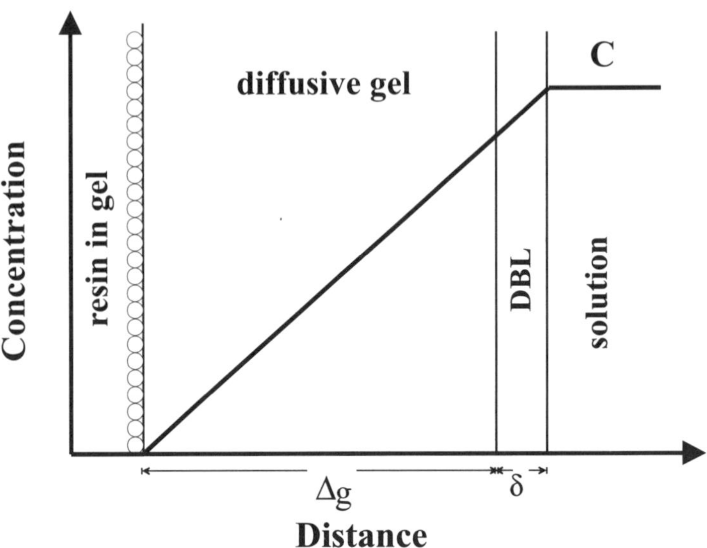

Figure 12.1 Conceptual view of the steady state concentration gradient of a solute through a DGT device deployed in a well stirred solution with solute concentration, C.

diffusive boundary layer (DBL), is solely by molecular diffusion. A steady state flux, F, of an ion is given by Fick's first law of diffusion (equation 12.1),

$$F = D\frac{dC}{dx}, \tag{12.1}$$

where D is the diffusion coefficient and dC/dx is the concentration gradient. If diffusion coefficients of ions in the gel are the same as in water, the flux is given by equation 12.2,

$$F = D(C - C')/(\Delta g + \delta), \tag{12.2}$$

where C is the bulk concentration of an ion and C' its concentration at the boundary between the binding layer (known as the resin-gel) and the ion permeable layer (known as the diffusive gel). If the free metal ions are in rapid equilibrium with the resin, with a large binding constant, C' is effectively zero providing the resin is not saturated. In well stirred solutions the boundary layer thickness, δ, is negligibly small compared to the thickness of the diffusive layer, Δg. Equation (12.2) then simplifies to equation (12.3).

$$F = DC/\Delta g. \tag{12.3}$$

In practice the DGT device is deployed for a fixed time, t. On retrieval the resin-gel layer is peeled off and the mass of the accumulated ions in this layer is measured. The mass can be measured directly in the resin-gel layer by drying it and using a beam technique such as proton induced x-ray emissions (PIXE) (Zhang *et al.*, 1995b; Davison *et al.*, 1997) or in the case of radionuclides by direct counting (Chang, 1998). More commonly, ions in the resin-layer are eluted with a known volume, Ve, of solution (1M HNO_3 in the case of metals bound to Chelex) (Davison *et al.*, 1994; Zhang *et al.*, 1995a; Zhang and Davison, 1995). The concentration of ions in the eluent, Ce, are then measured by any suitable technique. As the elution is in batch mode only a fraction of the bound ions are retrieved. The ratio of the eluted to bound metal is known as the elution factor, fe. Values of fe of 0.8 have been reported for Zn, Cd, Cu, Ni and Mn and 0.7 for Fe when using 1 or 2 M HNO_3 (Davison and Zhang, 1995). Ideally values of fe should be measured by individual research groups, especially if conditions such as acid strength are varied. Taking the elution factor into account the accumulated mass of ions in the resin-layer can be calculated from equation (12.4)

$$M = Ce(Vg + Ve)/fe, \tag{12.4}$$

where Vg, the volume of gel in the resin-layer, is often negligibly small. The measured mass, M, can be used to calculate the flux through the known area of the device, A (equation 12.5).

$$F = M/At. \tag{12.5}$$

Equating (12.3) and (12.5) and rearranging (equation 12.6), demonstrates that the concentration in the bulk solution

$$C = M\Delta g/DtA \quad (12.6)$$

can be calculated from the known values of Δg, D and A, the measured deployment time, t, and accumulated mass, M. This feature of DGT whereby concentration is calculated from defined quantities make it ideal for *in situ* use as there is no requirement for calibration.

12.3 METHODOLOGY

12.3.1 Gel Preparation

The DGT technique relies on diffusion through the gel layer being reproducible. Depending on the proportions of cross-linker used and its composition, hydrogels based on acrylamide can have a wide range of pore sizes which determine effective diffusion coefficients. The hydrogel most commonly used for DGT is physically robust and permits free diffusion of simple solutes with diffusion coefficients for metal ions being indistinguishable from those measured in water (Zhang and Davison, 1999). It uses a patented cross linker derived from agarose. The preparation procedures for DGT using a Chelex resin suitable for trace metals are given below.

Chemicals: Acrylamide solution (30%). (**Caution!** Acrylamide solution is a suspected human carcinogen. Handle with gloves.) DGT gel cross-linker solution (2%) (from DGT Research Ltd., Skelmorlie, Bay Horse Road, Quernmore, Lancaster LA2 0QJ, UK). Ammonium persulphate (10%, 1 g in 10 g of water) (prepared daily). N,N,N′N′-Tetramethylethylenediamine (TEMED, 99%). Ion-exchange resin (Chelex-100, 200–400 mesh).

Gel solution preparation and casting: Mix 25 ml of acrylamide solution, 17.5 ml of deionised water and 7.5 ml of DGT gel cross-linker solution (measured by weighing 7.5 g) together in a clean plastic container. Make sure the gel solution is well mixed by shaking or stirring. It can be stored in a refrigerator (4 °C) for at least three months. Pipette 10 ml of gel solution into a container and add 70 μl of ammonium persulphate solution followed by 25 μl of TEMED solution. Mix them well and cast between two glass plates. Hydrate the gel in MQ water (or deionised water) for one day (changing the water 2–3 times), then store the gel in $NaNO_3$ solution (0.01–0.1 M).

To cast gels: Acid wash and rinse glass plates and plastic (typically 0.25 or 0.4 mm thick) spacers. Wipe dry with clean tissue papers rather than air-dry. Place a spacer round three edges and clip glass plates together. To make pipetting of the gel solution easier offset the glass plates slightly, leaving a 1 mm overlap on the edge without spacer. Pipette the solution in a smooth controlled fashion. If air bubbles appear tilt plates to remove them before continuing pipetting. Maintain the assembly at about 42 °C for 45–60 minutes until the gel is completely set (no remaining liquid). Prize open the assembly after the gel has set completely.

Resin gel preparation: First soak the resin in MQ (or deionised water) and then remove excessive water with clean tissues. Pipette 10 ml of gel solution into a container. Add 2 g (wet weight) of Chelex resin and mix well. Sufficient resin should be used to ensure that the resin density on the gel surface is maximal without affecting casting and setting of the gel. Then add 60 µl of ammonium persulphate solution and 20 µl of TEMED. Follow the casting procedures as indicated above, making sure the resin particles are suspended in the gel solution before pipetting by mixing well. Resin will settle on one side of the gel by gravity if laid flat during setting. Hydrate the Chelex gel in MQ water (or deionised water) and store it in MQ water (or deionised water).

The gel assembly must be deployed in a holder which ensures that only the diffusive gel contacts solution. One design based on a simple tight-fitting piston (Figure 12.2) (Zhang and Davison, 1995) is commercially available (DGT Research Ltd., UK). It consists of a backing cylinder and front cap with a 2.0 cm diameter window. A layer of resin-gel is placed on top of the piston face, ensuring that the side with the settled resin faces upwards. A layer of diffusive gel is placed on top of it. For field applications a wet 100 µm thick, 0.45 µm Millipore cellulose nitrate filter or Gelman polysulphone filter is placed on top of the diffusive gel. The cap is then placed on top and pressed tight so that the layers are in good contact with one another and there is a good seal between filter and cap. The filter has been shown to behave as an extension of the diffusive gel layer, there being no measurable difference in diffusion coefficients through a gel or a gel plus filter combination (Zhang and Davison, 1999; Chang *et al.*, 1998). The filter minimises adhesion of particles to the surface of the device. A typical resin-gel thickness is 0.4 mm and typical diffusive gel thicknesses are 0.4–0.8 mm.

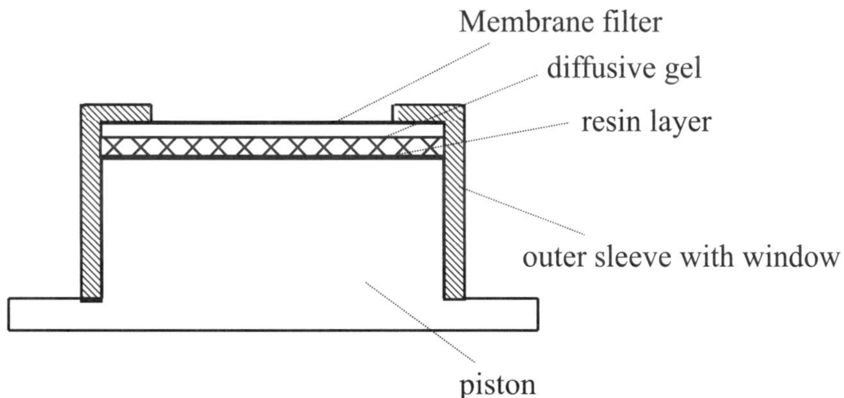

Figure 12.2 Schematic representation of a plastic DGT holder based on a simple piston design.

12.3.2 Verification of Principles

The basic principles of DGT have been verified in the laboratory. A typical routine to test the performance of DGT assemblies prior to field deployment is as follows.

Into a 3 L plastic container, mix 2 L of MQ water (or deionised water) and 20 ml of 1 M $NaNO_3$ solution. Spike an appropriate amount of Cd standard solution (or mixed metal standard) to make up a 10 ppb solution (make sure that the pH of the solution is above 5). Now you have created an immersion solution containing 0.01 M $NaNO_3$ with 10 ppb of Cd (or mixed metals). You can also use synthetic lake water or synthetic seawater rather than $NaNO_3$. Place three DGT units (0.8 mm thick diffusive gels) in the immersion solution. Make sure the solution is well stirred but not cavitating. Take an aliquot of the immersion solution for analysis at the beginning of the experiment. After about 4 hours, sample the immersion solution once again for analysis. Take the DGT units out of the solution and rinse the surface with MQ water. Record the deployment time and the solution temperature during exposure. Remove the Chelex gel and place it in a clean sample tube. Add 1 ml of 1 M $NaNO_3$ solution and leave it at least overnight before analysis. The Cd concentration in the eluent acid should be about 20 ppb if you have followed the above procedures at 25 °C.

12.3.3 Theoretical Response

When gel assemblies with a single diffusive gel layer thickness were deployed in stirred 1 ppb $CdCl_2$ solutions for various times, their measured mass increased linearly with time (Figure 12.3). Similarly when different diffusive layer thicknesses were used and deployment time was fixed, the measured mass was inversely proportional to gel layer thickness (Figure 12.3). In both cases the experimental data points agreed with the theoretical response calculated from the known concentration in solution, C, using equation 12.6. The diffusive layer thickness (gel layer plus filter), Δg, the exposed surface area, A, and the deployment time, t, were measured directly. The diffusion coefficient, D, was taken to be the same as the value in water. Direct measurements using a diffusion cell with simple metal salt solutions have confirmed that the diffusion coefficients of metal ions in gel and filter assemblies are indistinguishable from values in water, indicating that there is no reaction between metal ions and the gel (Davison *et al.*, 1994; Zhang and Davison, 1999). By contrast, the diffusion of phosphate is slightly impeded by the gel (Zhang *et al.*, 1998a). Typical values for the diffusion coefficients of metal ions in water are given in Table 12.1. They have been calculated from values in Li and Gregory (1974) using the Stokes-Einstein equation linking the temperature dependence to viscosity and the known temperature dependence of viscosity, as detailed in Zhang and Davison (1995). Clearly, as D changes with temperature, the DGT response is temperature dependent. If the structure, chemical reactivity and dimensions of the gel do not vary with temperature, the DGT temperature response should solely depend

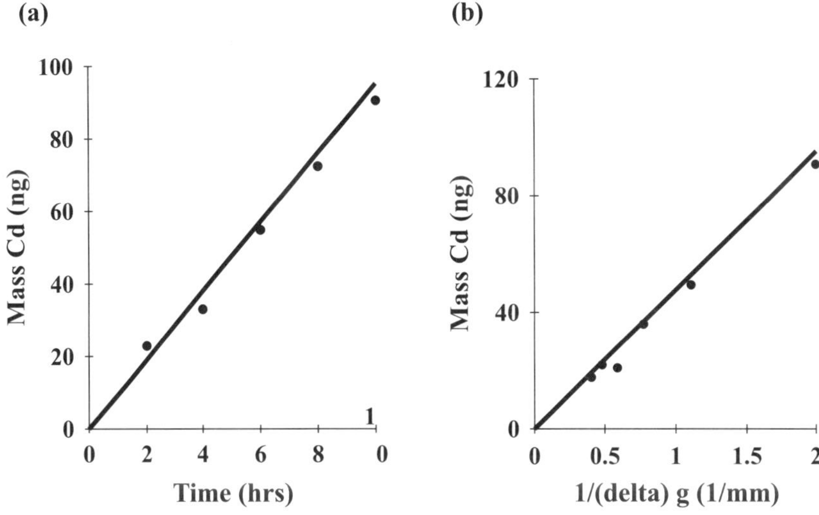

Figure 12.3 Measured mass of Cd in the resin layer (a) for gel assemblies with a 0.5 mm diffusion layer immersed in a stirred $CdCl_2$ solution (1 μg l^{-1} Cd) for different times; (b) for gel assemblies with different gel layer thicknesses, delta g, exposed to $CdCl_2$ solution (1 μg l^{-1} Cd) for 10 hours. The solid lines are predicted by equation (6).

Table 12.1 Diffusion coefficients ($\times 10^{-6}$ cm^2 sec^{-1}) of metal ions in water at various temperatures.

T (°C)	*Cd*	*Co*	*Cu*	*Fe*	*Mn*	*Ni*	*Pb*	*Zn*
4	3.74	3.64	3.82	3.75	3.58	3.54	4.92	3.73
6	4.01	3.91	4.10	4.02	3.85	3.80	5.29	4.00
8	4.30	4.19	4.39	4.31	4.12	4.07	5.66	4.28
10	4.59	4.48	4.69	4.60	4.41	4.35	6.05	4.58
12	4.90	4.78	5.01	4.91	4.70	4.64	6.46	4.89
14	5.22	5.09	5.33	5.23	5.01	4.94	6.88	5.20
16	5.55	5.41	5.67	5.56	5.32	5.25	7.31	5.53
18	5.89	5.74	6.02	5.90	5.65	5.58	7.76	5.87
20	6.24	6.08	6.38	6.26	5.99	5.91	8.22	6.22
22	6.60	6.44	6.75	6.62	6.34	6.25	8.70	6.59
24	6.98	6.80	7.13	7.00	6.70	6.61	9.20	6.96
26	7.36	7.18	7.53	7.38	7.07	6.97	9.71	7.34
28	7.76	7.57	7.93	7.78	7.45	7.35	10.23	7.74
30	8.17	7.96	8.35	8.19	7.84	7.74	10.77	8.15
32	8.59	8.37	8.78	8.61	8.24	8.13	11.32	8.56
34	9.02	8.79	9.22	9.04	8.65	8.54	11.89	8.99

on D. This has found to be the case. The measured accumulated mass was theoretically predicted using equation 12.6 from 5 to 35 °C with values of D taken from Table 12.1 (Zhang and Davison, 1995).

12.3.4 Ionic Strength and pH

Initial measurements made using DGT of 3 ppm Cd in solutions of various concentrations of sodium nitrate from 10 nM to 1 M were constant (Zhang and Davison, 1995). This independence of ionic strength was supported by the theoretical response being obtained when measurements were made in sea water with known concentrations of labile Zn (Figure 12.4). These observations are consistent with our theoretical understanding of diffusional fluxes being proportional to a concentration gradient. They are thus independent of activity and hence ionic strength. Diffusion coefficients in seawater have been reported to be at most 8% smaller than in water (Li and Gregory, 1994).

The DGT response to pH depends on the properties of the binding agent. Using Chelex to bind cadmium it is virtually independent of pH between pH 5 and 9. At pH 4 the response is still 90% of theoretical, but at pH 3 or lower binding to Chelex is ineffective, resulting in a very poor response.

12.3.5 Solution Flow and Biofouling

In practice, diffusion to the resin-gel of a DGT assembly will depend on the total thickness of the diffusive gel layer and the diffusive boundary layer (DBL) in solution. In well stirred solutions the DBL must be negligibly small because the theoretical response is obtained. If there is no stirring in the solution a lower response is obtained (Zhang and Davison, 1995; Zhang *et al.*, 1998a). When DGT devices were placed in a flume with their faces parallel to the flow the response was virtually independent of flow for all measurable flow velocities (0.02–0.26 m s^{-1}) (Davison and Hutchinson, 1997). It is expected, then, that the DBL will be negligibly small in rivers, estuaries and the well-mixed waters of seas and lakes. Deployment of devices with two or more different thicknesses permits calculation of concentration irrespective of flow and hence DBL thickness. When the DBL thickness, δ, is significant, equation 12.7 applies (Zhang and Davison, 1995).

$$\frac{1}{M} = \frac{\Delta g}{DCtA} + \frac{\delta}{DCtA}. \tag{12.7}$$

If measurements are made using several different gel layer thicknesses a plot of $1/M$ versus Δg should be linear with a slope given by $1/DCtA$ and an intercept by $\delta/DCtA$, enabling C and δ to be calculated. With two gel layer thicknesses, Δg_1 and

Δg_2, C can be calculated from the measured values of M_1 and M_2 (equation 12.8) (Zhang *et al.*, 1998a).

$$\frac{1}{M_1} - \frac{1}{M_2} = \frac{\Delta g_1 - \Delta g_2}{DCtA}. \tag{12.8}$$

Deployment of two or more devices, with different gel layer thicknesses, simultaneously may also allow correction for biofouling. If the effects of biofouling are restricted to physically altering the surface they can be likened to a change in δ. Providing this surface modification is the same on all devices, it will not affect the calculation of concentration.

12.3.6 Capacity

DGT has been shown to behave theoretically provided the capacity of the resin to bind ions is not exceeded. If a non selective binding agent is used, other more common ions will quickly accumulate and saturate the resin. When a general cation exchange resin was used to measure caesium and strontium the performance was good in synthetic solutions, but in natural waters a linear response between measured mass and time was only found for a few hours (Chang *et al.*, 1998). After that time the resin became saturated with major cations such as calcium. With a selective binding agent, however, very long deployment times can be used. Chelex is very selective for trace metals. DGT devices for trace metals are estimated to have maximum deployment times in the ocean of about 2 years and several months in more contaminated coastal waters (Davison and Zhang, 1995). The accumulation of cadmium from synthetic lake water in the laboratory has been shown to increase linearly with time for at least one month (Davison and Hutchinson, 1997). When radioactive caesium was measured by deploying DGT with a selective binding agent *in situ* in a lake for 2 months, the concentrations agreed well with direct lake-water measurements (Chang, 1998). Over these times the major problem is expected to be biofilm growth.

12.3.7 Speciation

In principle two factors determine which species are measured by DGT, the binding and the pore size of the gel. Chelex has very strong binding groups, at high effective concentration, which will out-compete most other ligands for metal ions, especially at the low concentrations of organic components present in most natural waters. Metal is continuously removed from solution to the resin. If metal-ligand complexes rapidly dissociate they will contribute to this flux, but if they are inert they will not. The time available for this dissociation is roughly the time taken for a metal to diffuse through the gel layer which is determined by the gel layer thickness (Zhang and Davison, 1995). For a typically 0.4 mm thick gel it is about 2 minutes. Consequently only labile (dissociation time < 1 min) metal complexes are measured, in an analogous fashion to anodic stripping voltammetry (Buffle, 1988). Any metal complex measured by DGT has to diffuse through the pores of the gel. If the complex is too large it will clearly be retarded. The agarose-derived cross linked (DGT Research Ltd., UK) polyacrylamide gel composition used in the preliminary development of DGT has a very open structure which allows the unretarded diffusion of simple metal ions; the diffusion coefficient in the gel is generally indistinquishable from that in water depending on cross linker batches (Zhang *et al.*, 1995a). Zhang and Davison (1999) have shown that humic and fulvic substances extracted from both waters and soils diffuse through the original composition gel. However, as their molecular diffusion coefficient is appreciably less than that of simple metal ions, the transport of metal humic species to the resin layer in DGT will be retarded compared to simple metal ions. Therefore if DGT with an open gel as the diffusion layer was used to measure metal ions in an unknown natural water, the proportional contribution of any humic complexed metal would not be known. By varying the composition and cross-linker of polyacrylamide gel, it is possible to restrict the pore size so that while simple metal ions (tested for Zn^{2+} and Cu^{2+}) still diffuse freely, the diffusion of fulvic and humic species is so slow that their contribution to the DGT accumulated metal would be negligible (Zhang and Davison, 1999). Therefore a DGT device which uses such a constrained gel (DGT Research Ltd) should measure only small labile species in solution. Although simple organic species, such as acetate complexes, would be included, in practise this fraction is likely to be dominated by simple inorganic species. By the simultaneous use of DGT with open and constrained gels, it should be possible to measure labile inorganic metal and total labile metal, thereby obtaining labile organically complexed metal by difference.

12.3.8 Field Applications

Because DGT is virtually unique in being able to measure labile species *in situ* it is difficult to verify its performance in the field. When Zn was determined in a seawater sample using anodic stripping voltammetry without any sample pre treatment there was good agreement with independent DGT measurements (Figure 12.4) (Davison and Zhang, 1994). Linear responses for the measured mass with respect to time and $1/\Delta g$ have also been obtained from DGT deployments made directly in coastal water using different deployment times and/or gel layer thicknesses (Davison and Zhang, 1994). *In situ* DGT measurements have also been made in estuaries where anodic stripping voltammetry (ASV) and cathodic stripping voltammetry (CSV) was used on samples collected from the same location (Twiss and Moffett, 1998). Good agreement was obtained between free metal ion concentrations measured by CSV or by DGT, assuming DGT measures only labile inorganic species.

DGT devices with different thicknesses has been deployed *in situ* in surface North Atlantic water and at 500 m depth (Zhang *et al.*, 1996). In both cases the

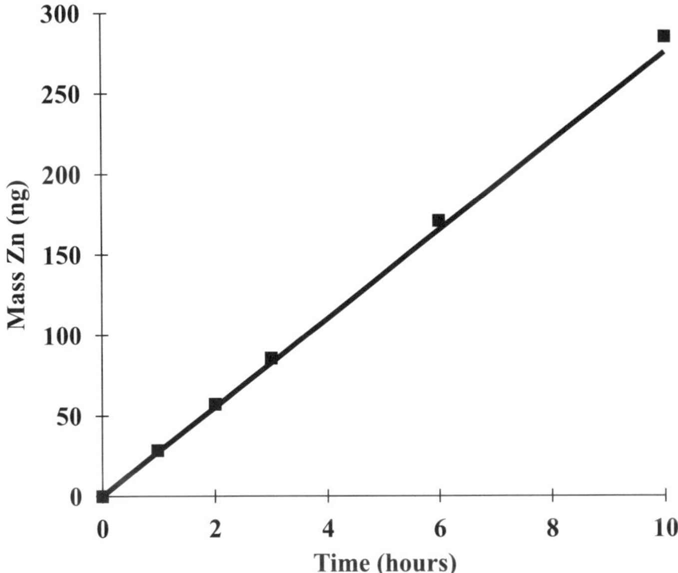

Figure 12.4 Measured mass of zinc in the resin layer (solid squares) for gel assemblies immersed for different times in a stirred solution of natural seawater (pH 7.8) in the laboratory (28 °C). The concentration of labile Zn was measured directly by anodic stripping voltammetry as 31 nM and was used with equation (6) to calculate the straight line.

measured masses of Cu, Zn, Ni, Co, Pb and Mn increased linearly with $1/\Delta g$. Concentrations were consistent with other measurements of ocean metals. Metals have also been measured *in situ* in lakes and rivers using DGT (Davison and Hutchinson, 1997). An attractive feature of DGT is its ability to pre-concentrate *in situ* which greatly aids the measurement of trace metals by overcoming contamination problems associated with sampling and handling. With the standard gel holder (Figure 12.2), the minimum practical elution volume is 0.4 ml. When a 1 mm thick diffusion layer is used this results in a concentration in the eluent 20 times that in the solution being measured for a 24 hour deployment. If electrothermal atomic absorption spectroscopy, with a typical detection limit of 1 nmol l^{-1}, is used, the DGT detection limit will be about 50 pmol l^{-1}.

12.4 USE IN SEDIMENTS AND SOILS

12.4.1 Principle of Bulk Measurements

Organisms present in soils and sediments and most *in situ* measurement devices alter local concentration gradients. To appreciate the information obtained when making measurements with DGT in sediments and soils it is important to regard

DGT as an *in situ* perturbation experiment. DGT perturbs the environment in which it is placed by providing a continuous sink for ions. Pore water concentrations are locally lowered and a flux is established to the device. For a particular DGT geometry there is a limiting maximum flux determined by the diffusive gel layer thickness and the initial concentration in the pore waters. This maximum flux is the flux which would be measured if DGT was deployed in a stirred solution. DGT provides a measure of the mean flux to the device during its deployment, designated here by F_{DGT}. Three situations may arise when DGT is used in soils and sediments (Figure 12.5).

(i) **Fully sustained:** Ions removed from the soil solution or pore waters by the DGT device are rapidly re-supplied from solid phases adjacent to the device; the concentration in solution is effectively buffered to a constant value. If the mass of accumulated ions in the resin layer is measured after a known deployment time, the local *in situ* concentration in the soil solution can be calculated using equation (12.6). The DGT flux, F_{DGT}, can be obtained using equation (12.5). This flux is dictated by the diffusive layer thickness and the concentration in the pore waters. Clearly the sediment or soil could potentially supply a greater flux. The potential maximum possible flux, F_{pm}, is greater than F_{DGT}.

(ii) **Unsustained:** There is no resupply from the solid phase to solution. The supply of ions to the DGT device is solely by diffusion from the pore waters

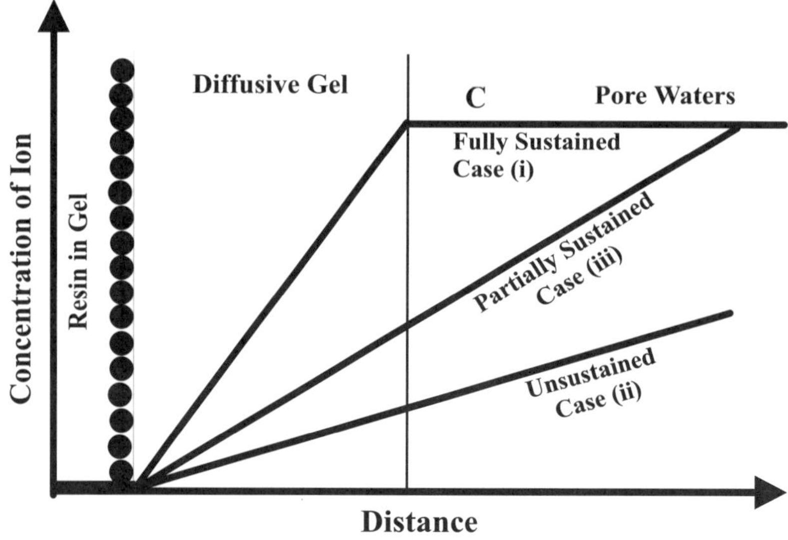

Figure 12.5 Schematic representation of concentration gradients established in a DGT gel assembly in contact with pore waters of sediments or soils for three cases when the pore water concentrations are fully or partially sustained or unsustained by resupply from the solid phase.

or soil solution which become progressively depleted, first in the vicinity of the DGT device and then extending into the sediment or soil. The DGT flux consequently declines with deployment time. However, the initial concentration in solution can be calculated from the DGT measured mass using a numerical solution of the time-dependent diffusion equations (Zhang *et al.*, 1995).

(iii) **Partially sustained:** There is some resupply of ions from solid phase to solution, but it is insufficient to sustain the initial pore water concentration and to satisfy fully the DGT demands. The flux measured by the DGT device can be equated to the *in situ* local flux from solid phase to solution induced by DGT and is close to the potential flux of the soil ($F_{DGT} \sim F_{pm}$). F_{pm} will be approached if measurements are made using DGT devices with progressively thinner diffusive layers.

The above three cases can be simply identified if pore water concentrations, C_{soln}, are measured by an independent method. Equation 12.6 is assumed to apply; pore water concentrations are calculated from the DGT deployment and designated C_{DGT}. The ratio, R, (equation 12.9) has a value close to one for

$$R = C_{DGT}/C_{soln} \tag{12.9}$$

case (i) when the concentration in solution adjacent to the device is maintained constant by rapid resupply from the solid phase. For cases (ii) and (iii) R is dependent on the diffusive gel layer thickness and the deployment time. A 0.4 mm thick diffusive gel layer and 24 hours deployment, case (ii), where supply is solely by diffusion, results in an R value of about 0.1 (Harper *et al.*, 1998). Measured R values between 0.1 and 1, or more practically between 0.2 and 0.95, imply that there is some resupply from solid phase to solution, but that it cannot fully sustain the DGT demand (case (iii)).

The response of a sediment or saturated soil to a DGT device has been modelled assuming firstly that there is a labile pool of ions associated with the solid phase of concentration C_{sp} that are in reversible equilibrium with ions in the solution (Harper *et al.*, 1998). The partition can be described by a constant, K_d,

$$K_d = C_{sp}/C_{soln}. \tag{12.10}$$

Secondly the resupply from solid phase to solution is assumed to obey first order kinetics; that is the rate of resupply from solid phase to solution is proportional to the concentration of the labile solid phase pool

$$\text{Rate of resupply} = k_{-1}C_{sp}, \tag{12.11}$$

where k_{-1} is a rate constant. Clearly the ability of the soil or sediment to sustain the DGT demand will depend on this rate of resupply and therefore on the first order rate constant and K_d. A large K_d implies that there is a large solid phase reservoir capable of resupplying local porewater depletion. This resupply will only be effective, however, if the rate constant, k_{-1}, is large. Harper *et al.* (1998) have quantitatively related K_d and k_{-1} to R. If there is an independent measure of K_d, the measured value of R can be used to calculate a value for the resupply rate constant, k_{-1}.

Interpretation of R in terms of k_{-1} and K_d is particular to the diffusive gel layer thickness and deployment time used. Usually a reasonable pseudo steady state is obtained for a 24 hour deployment with typical 0.4 and 0.8 mm diffusive gel layer thicknesses. Measurements made for times less than 2 hours can still be influenced appreciably by the initial response prior to pseudo steady state being achieved. Significant depletion of the local pool of labile solid phase ions is likely to occur for one week or longer deployments, reducing the measured value of R.

When independent measurements of pore water concentrations are unavailable the different degrees of sustainability may sometimes be distinguished if several deployments are made using different gel layer thicknesses. A thicker diffusion layer places a lower demand for resupply from the solid phase so that the fully sustained case is more likely to apply. When the well sustained case applies at all gel layer thicknesses, a plot of the measured flux against $1/\Delta g$ is a straight line. (Zhang *et al.*, 1995a). Similarly C_{DGT} should be constant, independent of Δg.

12.4.2 Applications

Few systematic data are so far available for soils and sediments. Trace metals were measured in a freshwater sediment at 1.25 and 2.5 mm spatial resolution (Zhang *et al.*, 1995a) (see next section). They were compared to pore water measurements made at 1 cm intervals using a syringe extraction procedure. For Cd and Zn the R value exceeded 0.95, indicating that the pore waters were well buffered by rapid resupply from the solid phase. Ni, Cu and Fe had R values in the range 0.3–0.4 indicating that porewaters were partially sustained by resupply from the solid phase. In these cases the DGT measurements provided directly the mean *in situ* flux from solid phase to porewaters during the 24 hour deployment. Values of R of 0.04–0.08 for Mn indicated diffusional only resupply with no contribution from the solid phase.

DGT measurements were made in soils variously treated with sewage sludge with and without metal amendments (Zhang *et al.*, 1998b). A thick soil slurry was obtained by raising the moisture content to 120%. After allowing the soil water mixture to equilibrate for 24 hours the same DGT holders used for solution work (Figure 12.2) were pushed into the slurry. In untreated soils the R value for Zn and Cd was indistinguishable from one, suggesting, as for the sediment example, that the soil solution is well buffered by rapid resupply from the solid phase. For Zn and Cd in soils with higher treatments of sludge and for Cu and Ni in all soils, the resupply from soil to solution was unable to sustain fully the DGT demand

(R~0.3). The DGT measured flux therefore provided an estimate of the *in situ* flux from soil to solution. Not surprisingly the *in situ* flux for Zn, Cd, Ni and Cu increased with increasing sludge application due to a large pool of labile metal. The change in R value for Zn and Cd on adding sludge suggests that there are two separate pools of metal associated with the solid phase. Zn and Cd appear to have higher resupply rate constants from the solid phase in untreated soils. However, the actual rate of resupply is higher in the treated soils due to the much higher concentration associated with the solid phase.

The effect of soil moisture content on the DGT response has been systematically investigated (Hooda *et al.*, 1999). It is predictable and consistent with known soil transport properties.

12.5 HIGH RESOLUTION MEASUREMENTS

12.5.1 Principles

Although the first reported DGT measurements were in solution (Davison and Zhang, 1994), DGT developed conceptually from its sister technique of DET which had been used for measuring pore waters at high spatial resolution (Davison *et al.*, 1991; Davison *et al.*, 1994; Krom *et al.*, 1994). DET uses a diffusive gel only. Pore waters equilibrate with the 95% water of the gel. On removal the gels are sliced or the metals are fixed prior to measurement by a beam technique. Until the gels are sliced or chemically fixed re-equilibration of solutes occurs. This and other effects limits the spatial resolution of high fidelity DET measurements to at best 1–2 mm (Harper *et al.*, 1997). Recently resolution has been improved by constraining the gel in 200 μm wide slits, analogous to a miniature version of dialysis peepers (Fones *et al.*, 1998).

With DGT, metal ions are effectively fixed spatially as soon as they bind to the resin. In a typical deployment assembly (Figure 12.6) 15 × 5 cm sheets of resin gel and diffusive-gel, along with a filter, are held between two plastic plates which are clipped or screwed together. Contact with the solution is made via a 10 × 1 cm window in the top plate. Firm pressure between the plates is essential to avoid ingress of solution through the edges of the assembly, although excessive pressure causing deformation of the gels must be avoided. The total thickness of the assembly is 3–5 mm. Twin assemblies have been deployed (Shuttleworth, 1999). In these, two gel sandwiches are arranged back to back on either side of a plastic sheet.

The DGT assembly is smoothly inserted into the sediment and left typically for 24 hours. Shorter deployments are more likely to exhibit artifacts associated with the initial sediment disturbance. After retrieval and washing, the filter and diffusive gel are peeled off and the resin gel underlying the window sliced at the desired resolution. Slices as thin as 1 mm can be cut by hand with a teflon-coated blade or by using a guillotine attached to a vernier (Zhang *et al.*, 1995a; Shuttleworth, 1999). A multi-cutter system has been used to slice at mm intervals

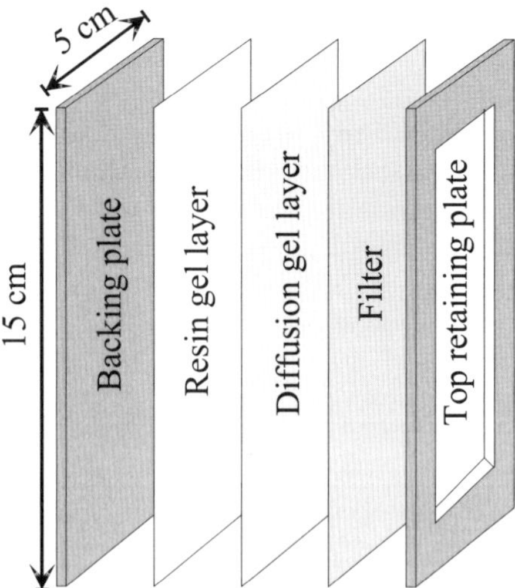

Figure 12.6 Schematic representation of the components of a DGT assembly used for deployment in sediments.

(Soares, 1998). Slicing more thinly is likely to introduce errors due to the non-uniform distribution of the 75–150 µm diameter beads of Chelex 100. The gel slices are eluted with acid prior to measurement in a similar fashion to the solution measurements. However, the very small area of the resin-gel results in a low accumulated mass. Consequently for metals present at very low concentrations, elution volumes may have to be as small as 100–300 µl.

Spatial resolution has been improved further by using a beam technique to measure metals directly in the resin-gel without slicing (Davison *et al.*, 1997). A chelating resin with a 0.2 µm bead size was used to ensure a uniform distribution of bound metal. The whole assembly, which used ceramic rather than plastic supporting plates, was ~1 mm thick. After deployment the resin-gel was dried onto a membrane filter and metals were measured directly on the filter by rastering a 1 µm proton beam using PIXE (proton induced x-ray emissions). 100 µm spatial resolution was achieved in the vertical plane when the beam was used to average the concentration in 100 × 500 µm rectangles. Alternatively 2.5 × 2.5 mm image intensity maps of pore water concentration were measured. The spatial fidelity of the measurement would be compromised if smaller than 100 µm averages were used, due to diffusional spreading of features in the pore water concentration profile during transport through the diffusion layer.

It is important to remember when using high resolution DGT that only a flux is directly measured. The resulting profiles against depth show spatial variations in the flux to the DGT device. Only if additional information is available (pore water concentrations, measurements with different gel layer thicknesses) can the measured flux profile be interpreted as concentration versus depth.

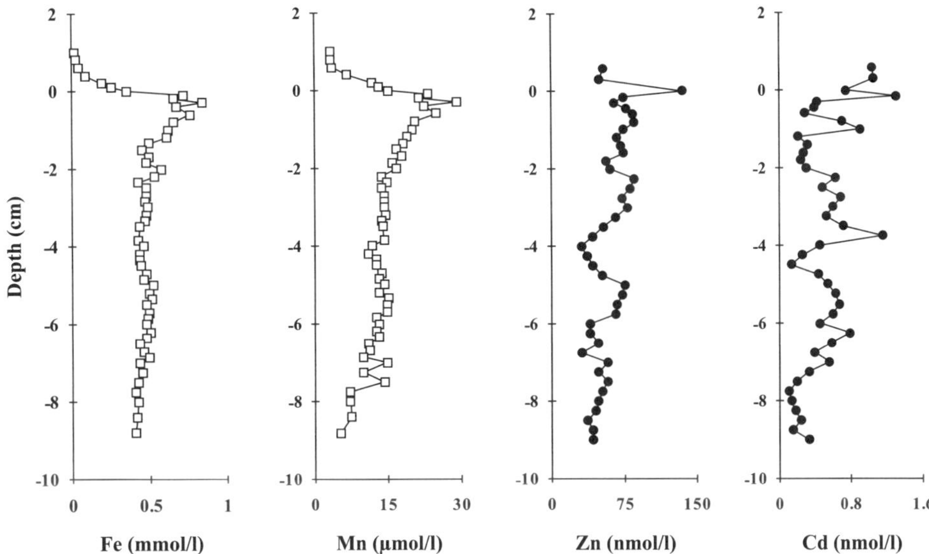

Figure 12.7 Concentration profiles of Fe and Mn measured by DET and Zn and Cd measured by DGT in pore waters in Esthwaite Water in July 1993. Gel assemblies were inserted *in situ*.

12.5.2 Applications

DGT has been used to show that there are pronounced maxima of trace metals at the sediment water interface of a productive lake, Esthwaite Water, (Zhang *et al.*, 1995a) and a productive contaminated stream, Sankey Brook (Davison *et al.*, 1997), (Figure 12.7). The sharp concentration gradient from surface sediments to overlying waters suggests that there may be a considerable remobilization flux of trace metals which has been overlooked, due to the inadequate spatial resolution of previous measurements. Consequently benthic organisms may be exposed to higher concentrations of trace metals than previously thought. Initial measurements at 1 mm resolution indicate that the maxima of trace metals in Esthwaite Water are 1.5–3.0 mm above the Fe and Mn maximum, suggesting that reductive remobilization may not be the source. Moreover, the Zn maxima in Esthwaite Water was separated by 1.5 mm from the Cd, Ni and Cu maxima and in Sankey Brook sediment pore-water maxima of Zn, Mn and As were separated by 1–3 mm. Clearly mechanistic differences in supply and removal are being resolved by DGT. In Esthwaite Water trace metals from the 1–2 mm wide trace metal maxima were completely removed above the maxima of Fe(II) and Mn(II). As it is unlikely that there will be any sulphide present above the Fe(II)/Mn(II) maxima, the results indicate that metal sulfide formation is probably not the removal mechanism for trace metals. The steepness of the metal maxima gives information about the rate of removal of metal (Davison and Zhang, 1996). The resulting half life of < 3 minutes is consistent with removal by binding to surfaces. This overall picture indicated mobilization from organic material and rapid binding to oxyhydroxides.

Two dimensional images of pore water concentrations obtained using 2.5×2.5 mm PIXE rastors have shown horizontal as well as vertical gradients (Davison *et al.*, 1997) and revealed microniche structure. Its impact on the calculation of remobilization fluxes will depend on whether it is on a small enough scale to be effectively averaged by the horizontal averaging inherent in most pore water measurements, including DGT.

When DGT was used to measure trace metals at 100 µm resolution at a sediment-water interface overlain by a microbial mat dominated by *Oscillatoria* there was clear evidence of remobilization of Zn and Mn and possibly Fe at the surface of the biofilm. Such measurements have only been made in one system where the precise sources and sinks of the metals have yet to be determined. In particular it is not yet known whether the microbial mat is acting as a relatively passive collector of sedimentary material, which may be remobilized once it is physically trapped, or if the metals are being released due to direct physiological involvement of the microbial mat.

By using alternative binding agents, DGT has been used to measure other solutes in pore waters. Phosphate has been measured at 1 mm spatial resolution in freshwater sediments by immobilizing ferri-hydrite in the backing layer of gel (Zhang *et al.*, 1998a). The slices were eluted with sulphuric acid prior to colorimetric measurement. Silver iodide was used to bind sulfide (Teasdale *et al.*, 1999). The gel strips were treated with acid to liberate H_2S which was trapped in NaOH and then measured colorimetrically. Reaction of the pale yellow AgI to form Ag_2S produced various shades of grey, depending on the amount of Ag_2S formed. This colour change was used to provide a quantitative estimate of concentration using a colour density scanner. Spatial resolution to 250 µm could be achieved.

12.5.3 The Future

In the short time since its invention DGT has proved to be a very versatile technique. Although development has been rapid, it has so far mainly been undertaken in one laboratory. Wider acceptance in the scientific community clearly requires corroboration of these initial findings and applications to diverse environments. The work so far has largely served to open the door on new areas of science which require more thorough investigation. The potential of DGT as an *in situ* speciation and monitoring tool will only be fully realised when the species measured are carefully defined under controlled laboratory conditions and systematic comparative measurements with other techniques are made in the field. Better means of deploying DGT in demanding environments, such as the deep sea, are clearly required. Sound experimental work is required to understand more fully what DGT measures in sediments and soils. It is vital in these situations to ally DGT to alternative techniques which provide information on pore-water speciation. Theoretical treatments can be tested by establishing whether resupply rate constants vary when conditions are changed. Understanding of the DGT response, along with more sophisticated modelling, will lead to a better appreciation of solid-solute interactions. High resolution measurements have revealed previously unknown pore-water structure and mechanistic differences in solute mobilization and binding. A better understanding of this micro-world requires measurements on many different systems, allied to carefully controlled laboratory investigations. Improved deployment systems which minimize artifacts, such as those associated with disturbance, should be designed. Finally the principles of DGT are universal and may be applied to any solute providing there is a selective binding agent available. There is clearly a bright future for DGT as understanding of its measurements increase and new determinands emerge.

References

Buffle, J. (1988). Complexation Reactions in Aquatic Systems; *Ellis-Horwood*: Chichester, UK, 1988.

Chang, L. (1998). Development of DGT for the measurement of radioactive and stable caesium and strontium in natural waters, *PhD Thesis*, Lancaster University.

Chang, L., Davison, W., Zhang, H. and Kelly, M. (1998). Performance characteristics for the measurement of Cs and Sr by diffusive gradients in thin-films (DGT), *Anal. Chim. Acta*, 368, 243–253.

Davison, W., Fones, G.R. and Grime, G.W. (1997). Dissolved metals in surface sediment and a microbial mat at 100 μm resolution, *Nature*, 387, 885–888.

Davison, W., Grime, G.W., Morgan, J.A.W. and Clarke, K. (1991). Distribution of dissolved iron in sediment pore waters at submillimetre resolution, *Nature*, 352, 323–325.

Davison, W. and Hutchinson, W. (1997). An assessment of the feasibility of using DGT procedures to measure trace metals and radionuclides in rivers, *Environment Agency* R&D Technical Report, 92.

Davison, W. and Zhang, H. (1994). *In situ* speciation measurements of trace components in natural waters using thin-film gels, *Nature*, 367, 545–548.

Davison, W. and Zhang, H. (1996). Lacustrine chemistry at the micron scale. In *International Symposium on the Geochemistry of the Earth's surface 1996* (Ed. Bottrell, S.H. University of Leeds, pp. 337–341.

Davison, W., Zhang, H. and Grime, G.W. (1994). Performance characteristics of gel probes used for measuring the chemistry of pore waters, *Env. Sci. Technol.*, 28, 1623–1632.

Fones, G.R., Davison, W. and Grime, G.W. (1998). Development of constrained DET for measurement of dissolved iron in surface sediments at sub-mm resolution, *Science Tot. Environ.*, 221, 127–137.

Haraldsson, C., Anderson, L.G., Hassellov, M., Hulth, S. and Ohlsson, K. (1997). Rapid, high-precision potentiometric titration of alkalinity in ocean and sediment pore waters, *Deep Sea Res.*, 44, 2031–2044.

Harper, M.P., Davison, W. and Tych, W. (1997). Temporal, spatial and resolution constraints for porewater sampling devices using diffusional equilibration: dialysis and DET, *Env. Sci. Technol.*, 31, 3110–3119.

Harper, M., Davison, W., Zhang, H. and Tych, W. (1998). Solid phase to solution kinetics in sediments and soils interpreted from DGT measured fluxes, *Geochim. Cosmochim. Acta.*, 62, 2757–2770.

Hooda, P.S., Zhang, H., Davison, W. and Edwards, A.C. (1999). Measuring bioavailable metals by diffusive gradients in thin films (DGT): soil moisture effects on its performance in soils, *European Journal Soil, Sci.*, 50, 285–294.

Krom, M.D., Davison, P., Zhang, H. and Davison, W. (1994). High resolution pore water sampling using a gel sampler: an innovation technique, *Limnol. Oceanogr.*, 39, 1967–1972.

Li, Y.H. and Gregory, S. (1974). Diffusion of ions in sea water and in deep sea sediments, *Geochim. Cosmochim, Acta*, 38, 703–714.

Shuttleworth, S. (1999). Dynamics of iron and manganese and trace metals in lacustrine sediments, *PhD Thesis*, University of Lancaster.

Soares, A. (1998). An investigation of early diagenetic processes in marine coastal sediments by the DGT technique, *PhD Thesis*, University of Southampton.

Teasdale, P., Hayward, S. and Davison, W. (1999). Measurement of sulfide in waters and porewaters by DGT. *Anal. Chem.*, 71, 2186–2191.

Twiss, M.R. and Moffett, J. (1998). *pers. comm.*

Zhang, H. and Davison, W. (1999). Diffusion characteristics of hydrogels used in DGT and DET techniques, *Anal. Chim. Acta*, in press.

Zhang, H. and Davison, W. (1995). Performance characteristics of diffusion gradients in thin films for the *in situ* measurement of trace metals in aqueous solution, *Anal. Chem.*, 67, 3391–3400.

Zhang, H., Davison, W., Gadi, R. and Kobayashi, T. (1998a). *In situ* measurement of phosphate in natural waters using DGT, *Anal. Chim. Acta.*, 370, 29–38.

Zhang, H., Davison, W. and Grime, G.W. (1995). New *In situ* procedures for measuring trace metals in pore waters. Dredging, remediation, and containment of contaminated sediments, *ASTM STP 1293*, 170–181.

Zhang, H., Davison, W., Knight, B. and McGrath, W. (1998b). *In situ* measurements of solution concentrations and fluxes of metals in soils using DGT, *Environ. Sci. Technol.*, 32, 704–710.

Zhang, H., Davison, W., Miller, S. and Tych, W. (1995a). *In situ* high resolution measurements of fluxes of Ni, Cu, Fe, and Mn and concentrations of Zn and Cd in porewaters by DGT, *Geochim. Cosmochim. Acta.*, 59, 4184–4192.

Zhang, H., Davison, W. and Ottley, C. (1999). Remobilisation of major ions in freshly deposited lacustrine sediments at overturn, *Aquatic Sci.*, in press.

Zhang, H., Davison, W. and Statham, P. (1996). *In situ* measurements of trace metals in seawater using diffusive gradients in thin-films (DGT). Proceedings of the "Fourth International Symposium on the Geochemistry of the Earth's Surface", Bottrell, S.H. (editor in Chief) 138–142.

13. CHEMICAL SENSOR TECHNOLOGY: CURRENT AND FUTURE APPLICATIONS

JOHN M. TOKAR and TOMMY D. DICKEY

13.1 INTRODUCTION

Over the past decade, population growth has increased the sense of urgency to better understand human impact on the oceans and atmosphere. It is vital to improve measurements of critical variables if we are to be able to distinguish natural from anthropogenic changes. New chemical measurement techniques are continually being sought to improve detection limits, accuracy, reliability and to obtain *in situ* and real-time data in the global ocean. An important aspect of understanding complex oceanic processes is being able to detect and quantify the true or *in situ* value of a chemical component, rather than the altered value, that may have resulted following a period of degradation or change that may occur after a sample is removed from its natural environment (Johnson, 1992).

Traditional measurement methodologies currently used in chemical oceanography such as gas chromatography, high performance liquid chromatography (Yi, 1992) and infrared spectrometry are primarily utilized in the laboratory and are quite sophisticated (NRC Report, 1993). Instruments are calibrated and results are compared using Standard Reference Materials, or SRMs (Trahey, 1995; Cantillio, 1995). Many field analysis techniques are often based on proven laboratory methodologies using instrumentation originally designed for the controlled laboratory environment. These devices do not always respond favorably when installed aboard research vessels. Methods using wet chemical techniques often require that samples be collected, stored, preserved, and transported from the field into the laboratory before chemical analysis can occur. Further, solvent extractions and separations are usually required before actual chemical analysis can begin. This approach is problematic in that biological, chemical, and physical processes cause changes in the chemical measurement of interest. The ability to identify and quantify trace levels of metals, synthetic organics, and other toxic substances in the field, lags behind the ability to identify and quantify these same compounds in the laboratory. Low signal to noise ratios and sample contamination are problematic as well.

In addition, the high costs associated with sophisticated laboratory analysis of environmental samples preclude their widespread use in many monitoring programs. *In situ* measurements of chemical parameters (i.e. dissolved oxygen, pH, pCO_2, etc.) are currently limited and include *in situ* seawater probes based primarily on polarographic and other electro-chemical principles. These and other improved sensor systems can provide valuable information on the interaction of biogeochemical, and physical processes in marine systems. Scientific areas of particular concern include the greenhouse effect (Sarmiento, 1996), ocean pollution (NOAA Strategic Plan, 1997), and hydrothermal vents (Edmond *et al.*, 1982). This review will focus on *in situ* measurement technologies and how they can address these problems. We will also highlight some of the important advances in chemical sensors and suggest future applications for new and emerging technologies that should be beneficial for *in situ* measurement and monitoring of chemical components in the marine environment.

13.2 PROBLEMS OF INTEREST

13.2.1 Greenhouse Gases

The problem of identifying and quantifying the oceanic chemical processes that may contribute to global climate change is difficult and challenging. The documented increase of "greenhouse" gases in the atmosphere, especially chloroflorocarbons and carbon dioxide, is also an ocean research area that has received much attention in recent years (Rodhe, 1990; Karl *et al.*, 1991). Atmospheric CO_2 has been measured continuously since the 1950's to determine global concentrations with results showing increasing concentrations (Keeling, 1994). However, the oceans have not been sampled to the degree necessary to accurately determine global oceanic CO_2 concentrations and fluxes. Oceanic sources and sinks (Tans, 1990), and seasonal and geographic variability (Maier-Reimer, 1987) are known to be important for the global CO_2 budget, but have not been accurately determined (Landrum, *et al.*, 1996).

Current estimates of oceanic CO_2 uptake using isotopic carbon methods have been reported to be $2.1 +/- 0.75$ Gton C yr^{-1} (Quay, 1992), with significant inter-annual variability (Francey *et al.*, 1995). Global CO_2 estimates have also been determined by extensive ocean observations using CO_2 partial pressure differences between the marine atmosphere and surface water, where the Atlantic Ocean is likely the major sink (60% uptake or 0.60–1.34 Gt Cyr^{-1}) and the equatorial Pacific Ocean is apparently a source (Takahashi *et al.*, 1997). Considerable uncertainty and controversy presently surrounds the carbon budget in many regions such as the Southern Ocean.

The interaction and exchange of greenhouse gases at the air/sea interface and the flux of these chemicals to the sea floor is governed by complex processes (Liss, 1983; 1986) and are of high priority within ocean research communities. Progress in understanding and modeling of these processes has been hampered largely because of inadequate methodologies and sampling. Measurements of key chemical species (e.g. CO_2) are primarily derived from discrete samples obtained during ship operations (Bradshaw, 1988). The high cost and great time associated with collection of these data from ships has resulted in infrequent and non-synoptic data sets at limited geographic sites. Current progress in developing new sensor technologies suggests that many critical chemical measurements may be obtained in the future from remotely deployed moorings, drifters, autonomous underwater vehicles (AUVs), and offshore platforms as well as ships of opportunity.

13.2.2 Ocean Pollution

The effects of ocean pollutants in estuaries and in near-coastal waters and their transport into offshore waters, has remained high on the public and political agenda since the early 1970's (Gross, 1976). Great progress has been made in understanding the effects and underlying mechanisms associated with a number of toxic substances, including polychlorinatedbiphenyl (PCBs), chlorinated pesticides in marine mammals (Shantz *et al.*, 1995), toxic metals, (Saager *et al.*, 1997) and other toxic organics (Saliot, 1997) found in the marine environment. Advances in analytical instrumentation have made possible the identification and analysis of chemical pollutants at very low (picomolar) concentrations. Future progress in understanding of marine and estuarine waters as coherent functioning systems and our ability to predict natural compounds and contaminant input and their effects, will depend on development of new, accurate and reliable field measurement techniques. Sampling equipment needs improvement because of the need for lower detection limits and the greater probability of sample contamination. New instrumentation is needed that can provide *in situ* measurements at trace levels of selected suites of chemical components of seawater, at relatively low expense, with low maintenance, and minimal human interaction.

Over the last few decades, a number of oxygen depletion (< 1 ppm) events have occurred in U.S. coastal and estuarine waters that have resulted in mass mortalities of shellfish, bottom fish and other commercially important marine species. Specific areas impacted by these events included the New York Bight (Swanson, 1979) and Chesapeake Bay (Mackiernan, 1987). These events were linked to toxic inputs and nutrient enrichment that resulted in massive algal or other dinoflagellate blooms, followed by eutrophication (Burkholder, 1997).

Previous work (Eccles, 1987, Hennet, 1988; Klainer *et al.*, 1988) has demonstrated the potential of a variety of chemical sensors for measuring toxic substances in fresh and salt water. Future development and deployment of strategically placed observational systems could effectively monitor "at risk" marine and estuarine areas. These sensor systems could send data to a central collection and analysis facility to provide real-time information for directing mitigating operations. In addition, these data could be used to establish compliance or non-compliance to environmental regulations (Tokar, 1991).

13.2.3 Hydrothermal Vents

The discovery of hydrothermal vents (Corliss *et al.*, 1979) has opened a new scientific field of research bearing on viability of life in extreme conditions.

Hydrothermal venting areas are often located at depths of more than 4000 meters with water temperatures approaching 350 °C, making it a very hostile environment for both human exploration and instruments. There are presently a number of deep ocean measurement programs designed to measure and study hydrothermal vent fluid at slow-spreading centers and the effect of venting fluids on the ocean environment. RIDGE (Ridge Inter-Disciplinary Global Experiments) is an international program supported by the National Science Foundation and international agencies (RIDGE Program Announcement, 1995) and a sub program called VENTS, which was established in 1984 (Hammond *et al.*, 1991). These programs conduct research on oceanic impacts and consequences of submarine volcanoes and hydrothermal centers on selected ocean ridge/rift subduction zones. Program goals also include quantifying geochemical signatures by measuring near bottom temperatures, venting fluids, and minerals in the water column and on the sea floor.

13.3 INTERDISCIPLINARY SAMPLING CONSIDERATIONS – TIME AND SPACE DOMAINS

It should be emphasized at the outset that virtually all-important environmental problems require interdisciplinary approaches and necessarily chemical, physical, biological, and optical data sets. Optimally, these data should be collected simultaneously and span time and space scales relevant to the processes of interest (Figure 13.1: after Dickey, 1991; Time-Space Diagram). For global problems, this means that variability over ten orders of magnitude in space and time is encompassed. Present capabilities for obtaining requisite atmospheric and physical oceanographic data are quite advanced in contrast to those for their chemical, biological, and optical counterparts. This is not surprising, considering the greater complexity and nonconservative nature of the chemistry and biology of the oceans. Nonetheless, remarkable advances are being made in the areas of *in situ* and remote sensing of the bio-optical properties of the upper ocean (Dickey, 1991; Dickey *et al.*, 1997; 1998). In fact, many bio-optical variables can now be made at the same time and space scales as physical variables. It is imperative that chemical and additional biological measurement capabilities reach this point as well.

13.4 SAMPLING STRATEGIES AND DATA UTILIZATION

Clearly, detection limits, precision and accuracy of ocean measurements are important. However, the oceans are naturally dynamic with large amplitude periodic and episodic variability. Monitoring long-term change presents challenges beyond those faced by laboratory scientists in that large volumes of data obtained in an uncontrolled and typically harsh environment are just as important as precision and accuracy. In addition, limited numbers of variables may be

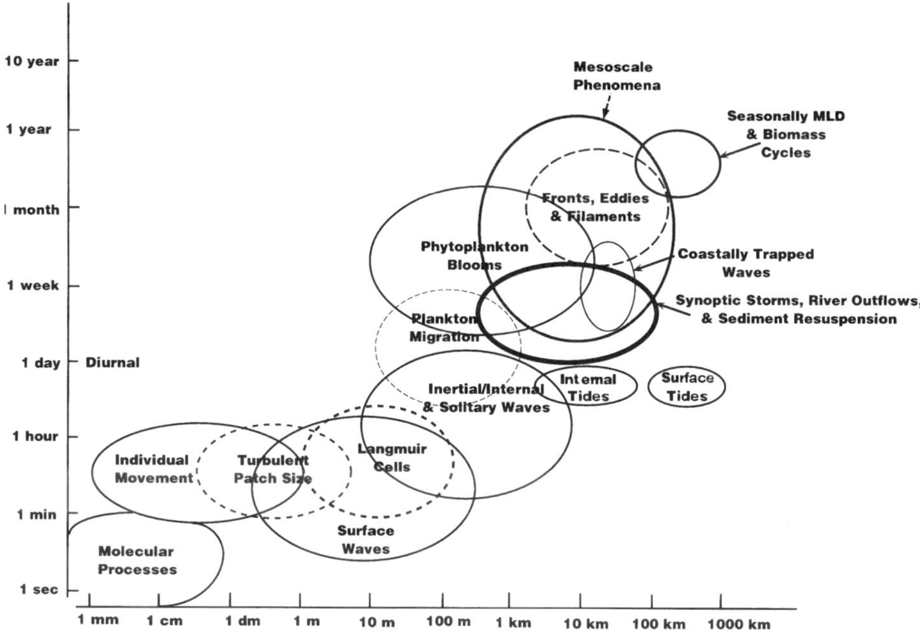

Figure 13.1 Time-space diagram indicating various physical and biological processes relevant to chemical distributions and processes.

directly measured in the marine environment, so it is important to define the most critical. Because of the sparseness of data, interdisciplinary numerical models capturing the primary physical variability over broad time and space scales will be needed for chemical as well as biological and ecological studies and predictions. Some major interdisciplinary oceanographic programs have adopted multi-platform approaches as conceptualized in Figure 13.2 (after Dickey, 1991). These include the Climate and Global Change (C&GC) Program (Dickey, 1997), Joint Global Ocean Flux Study (JGOFS) (Dickey, 1993; Dickey and Siegel, 1993), and the Global Ocean Ecosystem Dynamics (GLOBEC) program (Dickey, 1993; 1998). It is likely that the Global Ocean Observing System (GOOS) (US GOOS Report, 1992; US GOOS Report, 1997) will also follow this trend. Further, numerical modeling of physical and biological dynamics is central to these collective programs. Success depends upon accurate, reliable, and cost effective chemical measurement technologies.

13.5 KEY CHEMICAL MEASUREMENTS

Different oceanographic problems require specific suites of chemical measurements. A few examples are presented here for illustration. Some of the critical

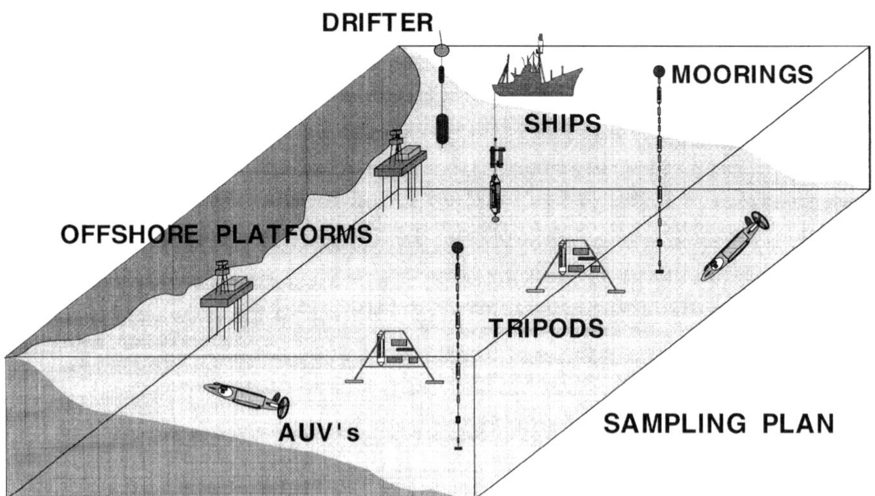

Figure 13.2 Schematic indicating a variety of sampling platforms which may be used for chemical, physical, and biological observations.

chemical variables with relevance to global climate change include: partial pressures of carbon dioxide (pCO_2) and other greenhouse gases, total CO_2 (TCO_2), pH, dissolved oxygen (DO), macronutrients such as nitrates (NO_3), ammonium (NH_4), silicates (SiO_4), and phosphates (PO_4), and micronutrients such as iron (Fe^{3+}) and cobalt (Co^{2+}). Open ocean measurements of trace elements such as lead (Pb) are also of great interest (Wu, 1997). A myriad of chemical species are important for coastal marine pollution (ref.; e.g.: PCBs, DDT, toxic metals, etc.) and for hydrothermal vent emissions (ref.: e.g. DO, Fe^{2+}, Mn^{2+}, pH, H_2S). Dissolved oxygen (DO) is necessary to support marine life and is an important parameter required in water quality assessment programs. Early detection of potential anoxia events would be useful in predicting conditions that might lead to fish mortality. Real-time DO information is useful in monitoring low oxygen events and, in some cases, permit action to prevent such conditions from progressing. Similarly, nutrient measurements can be used for monitoring adverse eutrophication.

As mentioned previously, chemical measurements on the same time scales as the physical and biological processes of interest are needed for many applications in order to understand and model the processes which are causing environmental variability, thus observations down to scales of minutes for several months should be targeted for each *in situ* deployment (e.g. Dickey, 1991; Johnson, 1994). To date, only a few research groups have successfully measured a limited number of the variables mentioned above (e.g. NO_3, DO, pCO_2, and pH) from autonomous platforms (Jannasch, 1994; Degrandpre *et al.*, 1995; 1997; Dickey *et al.*, 1998; Langdon, 1984; Friedrich *et al.*, 1995). Biofouling of sensors and analyzers remains as a problem for data quality and duration of sampling. There are no known reliable *in situ* sensors or analyzers for several important compounds such as PCBs and many other toxic organics in the marine environment.

It is interesting to note that generally speaking, electrical power for sensors and data acquisition and telemetry are no longer serious limiting factors (Dickey *et al.*, 1993; 1998). Considering the increasing importance of oceanic chemical measurements at appropriate time and space scales, it is evident that chemical sensor and analyzer technologies, which can be used for ocean applications, should receive considerable attention and support.

13.6 A MULTI-PLATFORM APPROACH

Remote sensing of the physical and certain aspects of the biological variability of the upper ocean via satellites has stimulated new insights concerning processes of the upper ocean. This technique is increasingly used as a quantitative tool to diagnose and predict the physical and biological states of the upper layer as well. Unfortunately, remote sensing of chemical species is far more difficult and at this point virtually intractable. Further, acquisition of subsurface chemical (e.g. biological) data from space would be even more difficult. Thus, *in situ* observations remain extremely important for chemical oceanography. Ships have served our community well; however, their limitations in terms of cost, availability, poor synoptic sampling, sample degradation and contamination, etc., have forced utilization of other platforms as well. The schematic shown in Figure 13.2 illustrates a variety of platforms, several of which can now utilize bio-optical and chemical sensors or systems as well as physical measurement devices. The time-space diagram shown in Figure 13.3 (after Dickey, 1991) provides a means of estimating the utility of different platforms in space (horizontal aspect depicted in figure) and re-emphasizes the need for deploying sensors from *in situ* platforms. The previous advances in bio-optical systems and a few chemical sensors, which are deployed from moorings and drifters (Dickey, 1991; Dickey *et al.*, 1998), provides impetus for deployment of chemical sensors from moorings, bottom tripods, drifters, floats, autonomous underwater vehicles (AUVs), and offshore platforms. Considerations for chemical sensors and systems include response time, drift characteristics, size, power requirements, data storage and telemetry, durability, reliability, stability/drift, and susceptibility to biofouling. It should be noted that analyzers have been successfully deployed from moorings and drifters, however, chemical sensors will likely be preferable, if not required, for some platforms such as towed bodies and AUVs. Below, brief summaries of representative activities using different platforms are presented.

13.6.1 Moorings

Interdisciplinary moored measurement systems and sensors are being used primarily by the research community to study environmental changes in the

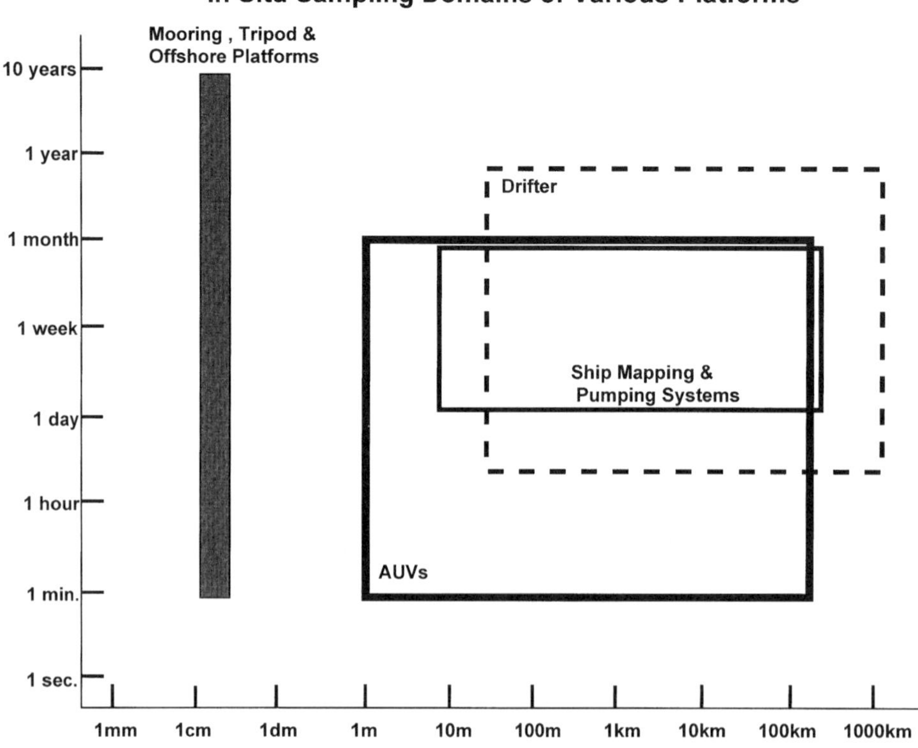

Figure 13.3 Time-space diagram showing the horizontal scale and temporal sampling domains of platforms shown in Figure 13.2.

ocean on time scales from minutes to years (Dickey, 1991; Dickey, 1996; Chavez *et al.*, 1997; Dickey *et al.*, 1998; DeGrandpre *et al.*, 1997). This work has led to discoveries of new processes (e.g. primary production variability associated with ENSO and equatorial long waves, sediment resuspension through internal solitary waves, cloud-induced and diel fluctuations in phytoplankton biomass, phytoplankton blooms associated with incipient stratification and frontal- and eddy-trapped inertial waves, etc.).

A relatively large number of bio-optical parameters are being measured from moorings at present (Dickey *et al.*, 1998). However, only a few chemical measurements are being made. Nonetheless, measurements of nitrate, pCO_2, and DO have enabled remarkable new insights into primary and new production and gas exchange across the air-sea interface (DeGrandpre *et al.*, 1995; 1997; Friederich *et al.*, 1995; Jannasch *et al.*, 1994). Interestingly, several diverse and often adverse oceanic regions have been studied using interdisciplinary moored systems. These include: the equatorial and central Pacific, eastern North Atlantic south of Iceland and western North Atlantic near Bermuda, Mediterranean Sea, Arabian Sea, and several coastal regions of the U.S., England, Germany, and Norway.

In several cases, data have been telemetered in near real-time (Dickey *et al.*, 1993; 1998; Chavez *et al.*, 1997). Mooring hardware is in place at a few locations (e.g. TOGA-TAO array, Foley *et al.*, 1998; Bermuda Testbed Mooring, Dickey *et al.*, 1998; Monterey Bay Research Institute (MBARI), Chavez *et al.*, 1997; Hawaii Ocean Time-series (HOT) mooring; Karl, personal communication). These arrays are also being used to test a variety of new chemical, as well as biological measurement systems. Because of biofouling, useful data from moorings is usually limited to a few months in the open ocean; however work is underway to mitigate this problem. Moored systems have proven their value in the research realm and need to be deployed in critical regions for studies of seasonal through decadal variability and long-term monitoring purposes. High temporal resolution will continue to be needed to minimize sampling induced uncertainties (via undersampling and aliasing).

13.6.2 Bottom Tripods

Benthic processes may be studied and monitored using instrumentation deployed on bottom tripods (e.g. Jahnke, 1990; Tengberg *et al.*, 1998). Bottom tripods and their instrumentation may be placed in virtually the same environments as moorings provided systems are designed to withstand great pressures in the deep sea. Essentially the same suite of sensors and samplers deployable from moorings can be used on bottom tripods. The chemical species of interest will vary depending on the type of environment (e.g. harbor, coastal, or open ocean) and problems of interest.

13.6.3 Offshore Platforms

Offshore oil production platforms provide a unique opportunity for conducting oceanic research (Cooper *et al.*, 1997; Busch, 1997). Recent discussions with the major oil companies have paved the way for developing effective partnerships between the ocean science community and industry. These large and very stable platforms/facilities often have space for manned research laboratories and are equipped with adequate power and other needed service ideal for oceanographic studies. They offer several advantages over shipboard platforms, including absolute stability in high sea states, suitability for time series measurements, and can be manned continuously or indefinitely. It should be possible to launch AUVs from these platforms for spatial sampling as well. Active platforms would not be preferable for all types of measurements because of possible chemical contamination that may result from drilling operations.

13.6.4 Drifters and Floats

Whereas moorings and bottom tripods can be used to provide high temporal resolution, long-term measurements at fixed locations (Eulerian), drifters and floats may be used to provide spatial data by effectively following water parcels (Lagrangian). Physical oceanographers have utilized these methodologies to great advantage for several decades (Dickey, 1998b). Although several thousand drifters have been utilized for circulation studies, only a few have been equipped with optical and/or chemical sensors to date. Within the past decade, bio-optical oceanographers have begun to deploy optical sensors from drifters and floats (e.g. Hitchcock *et al.*, 1989; Abbott *et al.*, 1990; Honjo, 1990; Dickey, 1991; Chavez *et al.*, 1997; Foley *et al.*, 1998) Recent drifter studies utilizing chemical sensors (e.g. nitrate and oxygen) were conducted during an iron fertilization experiment (IRONEX II) in the equatorial Pacific, which was designed to test the hypothesis that iron limits phytoplankton growth in high nutrient-low chlorophyll waters. Recently, Merlivat and Brault 1995 developed a buoy called CARIOCA for CARbon Interface Ocean Atmosphere, which is used to deploy a pCO_2 sensor, a thermistor, fluorometer, and a wind anemometer. The system is designed to act as a drifter (with a holeysock drogue); however, it can also be moored. It has been recently tethered to the Bermuda Testbed Mooring buoy for open ocean testing and studies.

13.6.5 Underway Shipboard Sampling

Research vessels equipped with onboard uncontaminating seawater pumping systems have become valuable research platforms that have provided spatial maps of near surface chemical and biological as well as physical variables (Feely *et al.*, 1987). In earlier work (Tokar, 1981) a gas lift system was designed and constructed to collect large volumes of sea water while on station, to perform structure analysis of marine humic and fulvic acids (Harvey *et al.*, 1983; Tokar 1983; Carter *et al.*, 1989) in the open ocean. More recently, shipboard pumping systems have helped to bridge gaps in oceanographic CO_2 data sets (Feely *et al.*, 1994; 1995). They pump water from an inlet port in the bow, located a few meters below the surface to the ship's laboratory, making seawater readily available for chemical analysis. This configuration allows continuous and near real-time measurements while the vessel is underway at normal cruising speed, and has resulted in efficient mapping of surface concentrations. A number oceanographic research cruises conducted in the equatorial Pacific have benefited from the use of onboard sea water pumping systems to measure total inorganic carbon, total alkalinity, and pH (Millero *et al.*, 1993; Feely *et al.*, 1998).

CO_2 in surface seawater is usually reported as the fugacity (fCO_2) (Chen *et al.*, 1995) or partial pressure (pCO_2) (Feely *et al.*, 1987), and is measured by infrared gas analyzers and gas chromatographs (Weiss, 1981). The water temperature and surface wind speed must be known and correction factors applied, before accurate fCO_2 values are determined. Underway shipboard fCO_2 measurements have been accomplished using both discrete and continuous sampling methods (Wanninkhof, 1993; Wanninkhof *et al.*, 1996). They include an underway system that measures the mixing ratio of ambient air and oceanic CO_2 in head space equilibrium with surface seawater which is pumped continuously into a 24 l polycarbonate equilibrator (Butler *et al.*, 1988). Another modified method (Chipman, 1993) uses discrete water sample aliquots equilibrated within a measured headspace, where CO_2 is quantitatively converted to methane and then analyzed. Detection of CO_2 for both systems is accomplished using a non-dispersive infrared analyzer. Results from these studies have shed new light upon open ocean CO_2 real-time analysis and mapping.

The importance of a new line of *in situ* chemical sensors for determining the spatial and temporal variability of fCO_2 cannot be overemphasized. Sensors and sensor systems that provide high resolution and accuracy are critical to understanding the seasonal and geographical variability of fluxes in the global ocean. New sensors will be needed for use in discrete and underway sampling from ships, as well as for measuring oceanic and atmospheric CO_2 from moored systems.

13.6.6 Manned Submersibles, ROVs, and AUVs

Manned submersibles have been in use for ocean exploration since the 1960's (Busby, 1976; 1990). Equipping these crafts with *in situ* sensor packages has played a major role in increasing our understanding of chemical, physical and biological processes in the ocean (Kalvaitis, 1997). Sensors deployed from submersibles, remotely operated vehicles (ROVs), and AUVs can provide effective real-time, *in situ* capabilities for locating the source of venting fluids by measuring key analytes (Fornari, 1997) as well as for use in deep ocean exploration (Hui, 1997). In addition, long-term unattended monitoring from bottom-mounted sensor instrument packages could provide useful information on temporal and spatial variations in hydrothermal venting regions of the ocean floor.

As mentioned previously, *in situ* monitoring of key chemical components (i.e. dissolved oxygen, pH, Fe^{2+}, Mn^{2+}, and H_2S) of venting hydrothermal fluids is required to adequately study oceanic vents. These components are normally measured by collecting water at the venting site, using specially designed water samplers operated from manned submersibles or from ROVs. A novel submersible chemical analyzer or "SCANNER" is a manifold sampler system (Coale *et al.*, 1991) which was designed for deployment from a submersible and was developed to meet the need to efficiently obtain samples of hydrothermal venting fluids for subsequent laboratory analysis (Johnson, 1986). With this system, considerable time is required during a dive to locate the vent, collect the sample, and return to the surface, where the water samples are then analyzed by conventional laboratory techniques.

Recently, various new *in situ* chemical detection systems have been developed and used from manned submersibles that have virtually eliminated the need to return to the surface to perform the analysis. One such system, SUAVE or Submersible System Used to Assess Vented Emissions, is an integrated *in situ* measurement system that contains a chemical analyzer which can measure four chemical species, including Mn^{+2}, Fe^{+2}, Fe^{+3}, and H_2S (Massoth *et al.*, 1995). It is also equipped with a CTD and light scattering sensors. The system was developed for use in the VENTS program. In principle, ROVs and AUVs can be used to perform many of the tasks of a manned submersible.

AUVs are rapidly emerging as important platforms for obtaining spatial data sets. As the costs of AUVs decrease, it should be possible to deploy several at a time for specific studies and even as monitoring platforms over particular ocean sections. For pollution studies or monitoring, AUVs could in principle be programmed to follow and track plumes. Chemical sensors will likely be used rather than chemical analyzers because of size, flow and response time constraints.

13.7 FUTURE CHEMICAL SENSOR TECHNOLOGY

As discussed earlier in this chapter, current field analysis techniques are often based on proven methodologies using laboratory instrumentation. Future large-scale oceanographic programs will demand new, reliable, and improved chemical sensor technology for rigorous and repeated use in the ocean. This section will discuss and suggest some innovative and emerging sensor technologies that may prove useful in oceanic measurement and monitoring programs in the future.

13.7.1 Fiber Optic Chemical Sensors

Fiber optic chemical sensors offer a completely new approach from methods currently used by chemical oceanographers (Tokar, 1990). These sensors have a number of advantages over conventional wet chemical sensing techniques. They can be made to be rugged and simple to operate. Since signals are transmitted optically rather than electrically, they are immune to all electrical and electromagnetic interference (Lerner, 1997). Analytical monitoring can be accomplished remotely, and unlike potentiometric sensing, a reference signal is not required. Measurements using a fiber optic sensor system are done *in situ* and in real-time. This section will describe basic sensor operating principles and highlight some of the recent advances in fiber optic sensors for ocean applications.

The optical sensor or modulation device is but one component of an integrated system that also includes an excitation light source, optical fibers, a photodiode detector, and other associated components such as connectors, couplers, and signal processing/data logging equipment (Figure 13.4). Light sources can range from incandescent quartz halogen lamps or LEDs to infrared solid state or gas lasers. The modulated optical parameters of light may include amplitude, phase, color, state of polarization, or a combination of these.

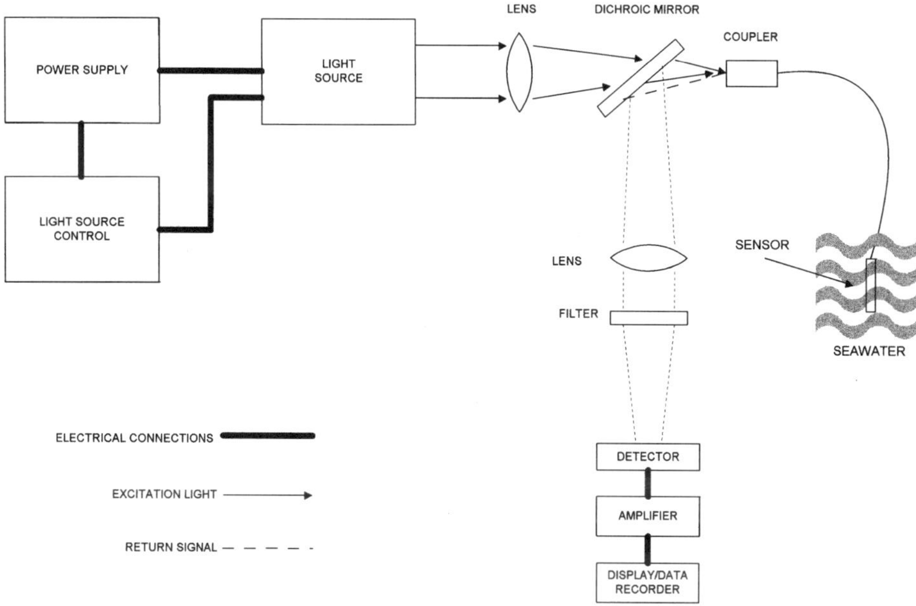

Figure 13.4 Typical fiber optic chemical sensor system showing the configuration of various functional components. Optical fibers in the 50 to 500 μm range are used most often, with an extrinsic chemical sensor located in the sampling region. These type systems make remote, *in situ* sensing possible.

The fiber optic chemical sensor is usually made up of analyte specific sensing reagents immobilized on the side or located at the tip of an optical fiber. To construct an effective sensor, it is necessary to immobilize suitable and sufficient sensing reagents (i.e. fluorescent dyes) to the sensing area of the fiber (Shakhsher, 1994). Surface area limitations due to the small size (10–50 microns) are overcome by using various surface amplification techniques (Seitz, 1984; Seitz *et al.*, 1993), including a unique type of covalent immobilization (Munkholm *et al.*, 1986); imbedding the reagent into a membrane (Duportail, 1983) or polymer (Hsu, 1987) placed at the tip of the fiber. Attaching large surface area porous glass fibers (Heitzmann, 1985) or porous beads to the fiber tip and containing very sensitive reagents within a reservoir cell (Figure 13.5), have also been accomplished (Goswami *et al.*, 1989).

Fiber optic sensors normally fall into two major categories: intrinsic and extrinsic (Tokar, 1990b). Intrinsic sensors use the optical fiber itself, to sense the parameter being measured. This becomes possible when the fiber is altered by the physical or chemical external variable being sought. These external variations cause an alteration of the optical properties (total internal reflection) of the fiber, resulting in measurable light intensity, phase, or polarization changes.

Intrinsic refractive index sensors have been used previously (Kawahara, 1983) to measure hydrocarbons (HC) in water. In this design, the unclad fiber is coated

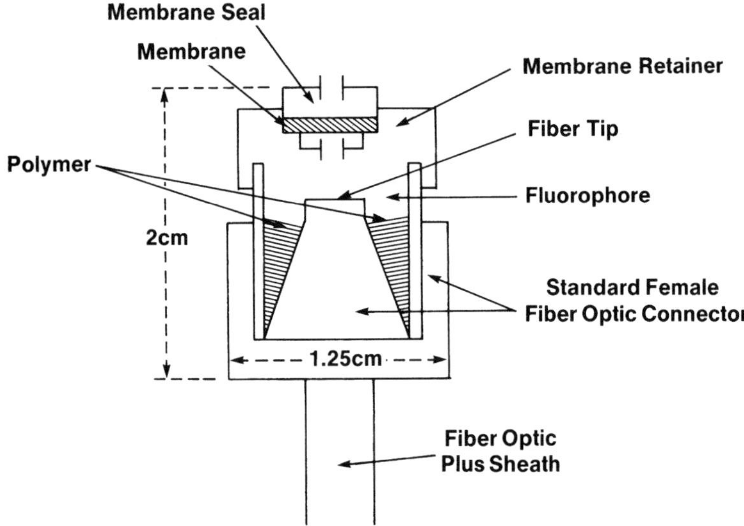

Figure 13.5 Reservoir cell fiber optic chemical sensor is a modified aluminum fiber optic coupler, which contained the fluorescent reagent in a machined 2-μl reservoir. The reagent was held within the cell by a gas permeable membrane and a TFE seal, located at the top portion of the schematic. (reproduced with the kind permission of Compass Publications)

with an organophilic compound that absorbs HC (Figure 13.6). Because the index of refraction of water is lower than the fiber core, total internal reflection

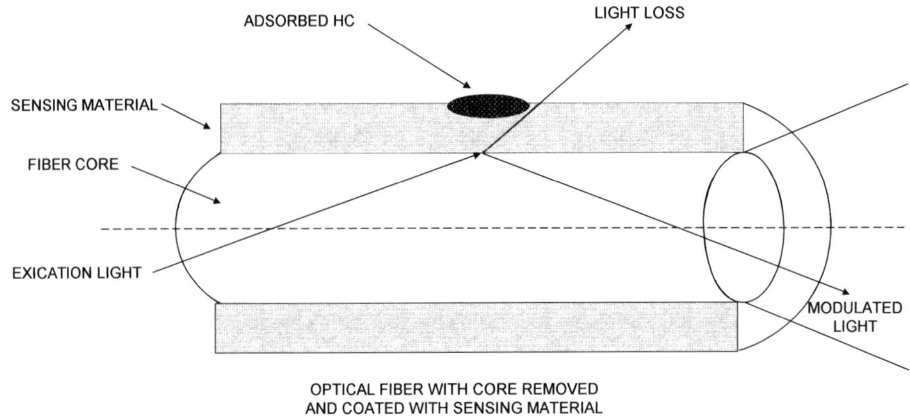

Figure 13.6 Refractive index fiber optic chemical sensor used to detect hydrocarbons (HC) in water. This device modulates the excitation light when a HC coats the sides of the unclad fiber surface. The resultant reduced light intensity at the detector may then be related to the amount of HC present.

occurs with no light loss. When a small quantity of HC comes in contact with the unclad fiber surface, the resultant higher index of refraction will cause the light to refract out of the fiber, thereby reducing the light intensity at the detector. This light loss is then related quantitatively to the amount of the adsorbed contaminant. Recent studies have also been successful using a reactive layer between the core and an exterior clad made of sensing material (Klainer, 1989) and by using species specific thin metal film clad (Klainer, 1992).

In an extrinsic sensor, the optical fiber passively transmits light between the sensor located in the medium to be measured and the photodetector. This type of system may use a single strand of fiber to both convey the excitation light into the medium and return the reflected light to the detector. Alternatively, a second fiber may be used to return the light. A number of extrinsic fiber optic sensors have been designed and tested both in the laboratory and the field to measure selected analytes in water (Chudyk, 1985). One recent example is a pH sensor using optical reflectance coupled to the swelling of a polystrene membrane, which was developed for and tested in water (Zhang *et al.*, 1997). Other examples of successfully measured chemical species using fiber optic sensors include ammonia (Rhines, 1987), and methane (Tai, 1987).

It is possible to measure dissolved carbon dioxide in seawater with fiber optic sensors using pH changes. CO_2 dissolved in water exists in equilibrium with different species as shown below:

$$CO_2(gas) + H_2O \rightleftharpoons CO_2(aq)$$

$$CO_2(aq) + H_2O \rightleftharpoons H_2CO_3$$

$$H_2CO_3 \rightleftharpoons H^+ + HCO_3^-$$

At equilibrium, pH and CO_2 (aq) are equated to one another. The hydrogen ion (H^+), sensitive fluorophore, or sensing reagent is isolated from the environment by a CO_2 permeable membrane. The dissolved CO_2 from the water passes through the membrane, decreases the pH of the dye, and results in a measurable change in fluorescence intensity. An experimental CO_2 reservoir cell design (Figure 13.6) sensor (Goswami *et al.*, 1989) has been developed and tested in seawater and uses modulation in luminescence intensities of the sensing reagent hydroxypryrenetrisulfonate (HPTS) as the basis for the sensor. Laboratory tests and at-sea trials using this sensor connected with a 400 μm plastic clad silica optical fiber to a custom made filter fluorometer resulted in a linear response for dissolved CO_2 in the 0–600 ppm range.

An adsorbance-based fiber-optic sensor was developed for the *in situ* measurement of sediment pore water $CO_{2(aq)}$ (Hales, 1997). This device also relies on pH changes of a dye solution contained within a gas-permeable membrane. It was made rugged and tested in the equatorial Atlantic at depths exceeding 4500 meters with a response time within a few minutes.

Another example of a pCO_2 fiber optic sensor for seawater was developed based upon fluorescence using a combination of dyes (about 10 μl total volume) contained within a silicone membrane bulb attached to the tip of a single optical fiber (Goyet, 1992) and coupled to a commercial fiber optic fluorometer. When tested at sea, it yielded a precision of 3% in the range of 300 to 500 ppm of pCO_2. A renewable-reagent fiber optic sensor (DeGrandpre, 1993; DeGrandpre *et al.*, 1997) also tested at sea, was developed and operated by measuring light intensity at the adsorbing wavelength of colorimetric acid-base indicator. The indicator was continuously delivered to the sensing tip of the fiber using capillary tubing and yielded superior sensitivity and stability.

A reservoir cell sensor has also been successful in measuring oceanic dissolved oxygen (Goswami, 1988; Tokar, 1989). This sensor is based on luminescence quenching of a fluorophore (Ruthenium(II)tris(bipyridine)) by oxygen. This system included a custom built spectrometer containing an Argon ion laser light source, detector and suitable electronics. Unlike the CO_2 sensor, this device used a Teflon membrane with a 1 μm pore size. This sensor was also tested at sea, with linear response from 0–22 ppm, and was intercalibrated using Winkler chemical titration (Dawes, 1988).

The CARIOCA buoy mentioned earlier in this chapter, was successfully instrumented with a pCO_2 sensor to determine CO_2 flux at the air/sea interface. The principle of this sensor is based on measuring the change in optical adsorbance of a diluted dye solution in seawater as a function of pH (Merlivat, 1995). The dye is renewed following each measurement and contained within a optical cell separated from seawater by semi-permeable membrane

A fully autonomous fiber optic sensor system for measuring low-level dissolved CO_2 in seawater uses a fixed reagent fluorescent dye (carboxy-SNAFL-1) immobilized at the end of an optical fiber in a gas permeable membrane (Tabacco *et al.*). The system was deployed on a buoy, is completely reversible, and operated in the dynamic range of 200–1000 ppm pCO_2 for continuous and unattended monitoring. It was tested along with the CARIOCA pCO_2 sensor on the Bermuda Testbed Mooring.

Currently, there is a limited number of fiber optic sensors available for use in seawater. Additional research and development is needed to improve response time, reproducibility, and long-term reliability. With the continued availability of inexpensive, high quality optical fiber, advances in solid-state technology (Saini, 1994), surface amplification technology and analyte specificity, we believe that in time fiber optic chemical sensors will come of age, and be available for routine use in marine chemistry studies.

13.8 MICROELECTROMECHANICAL SYSTEMS (MEMS)

The microelectromechanical system, or MEMS, is a relatively new technology for making and combining miniaturized mechanical and electronic components out of silicon wafers using micro-machining techniques (Payne, 1995). The resultant miniaturized structures (10 s of microns in size) exhibit useful mechanical properties and functionality that can serve as the basis for physical or chemical sensors. These micro-sensors have evolved from the silicon microchip industry and can take full advantage of commercially available complementary metal oxide semiconductor (CMOS) processing (Klaassen *et al.*, 1996), which allows precise and efficient manufacturing of MEMS sensing technology.

Many advantages may be achieved over conventional sensor technologies including auto-calibration, self-testing, digital compensation, small size, and economical production (Wise, 1991). MEMS sensors or "transducers" are micro-structures that can convert mechanical, thermal, magnetic, or chemical inputs into a measurable electrical charge or signal. Typically, small mechanical structures are "machined" using photolithography and etching of silicon substrates, where a particular physical configuration is desired (e.g. a free standing beam isolated from the surrounding substrate, Figure 13.7). When an electric voltage is passed through a beam structure, a particular chemical species (analyte of interest) can be adsorbed to the structure based on its chemical composition. The adsorbed material increases the weight of the beam, resulting in a reduction of the vibrational frequency of the beam. The operation of this type of chemical sensor would utilize measurements of frequency change (Gabriel, 1995).

Scientists at the National Institute of Standards and Technology (NIST) are engaged in research on the development of solid state sensor array technology for the real-time measurement of multiple species in gas mixtures using sensors which are based on specially designed micro-hotplate arrays (Semancik *et al.*, 1994). Each element in an array is a multi-layer micro-structure consisting of a heater, a thermometer/heat distribution plate, and sensing film electrical contacts

Figure 13.7 Free standing beam structure etched in silicon (courtesy of Sandia National Laboratories) demonstrates that a change in a physical parameter (weight of the beam) can be useful in determining the presence of a particular chemical species. This structure is of the order of 100 μm in size.

(Figure 13.8). Typical single element sizes range from 50 to 200 micrometers. The micro-hotplates have low power requirements (tens of milliwatts), a large operating temperature range approaching 800 degrees Celsius, and heating time constants of 1 to 5 milliseconds (Suehle, 1993).

One sensing principle that can be utilized in such devices follows. Sensing films composed of a semiconducting oxide such as SnO_2 or TiO_2, are deposited on selected micro-hotplates using self-lithographic techniques. When a chemical

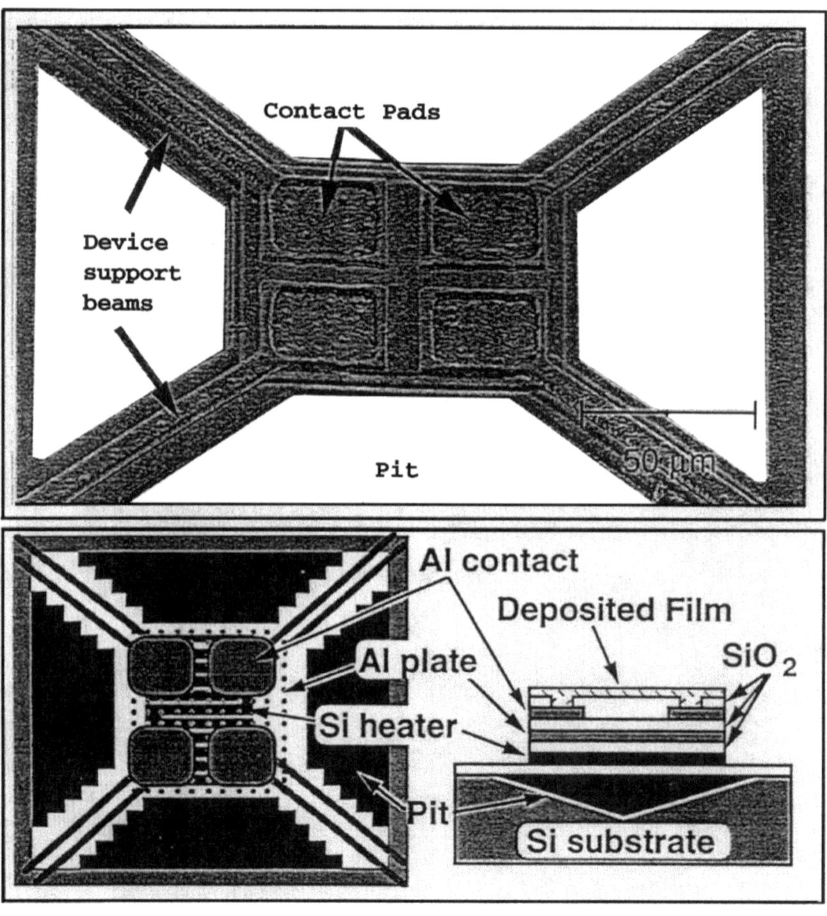

Figure 13.8 Scanning electron micrograph (courtesy of NIST) showing a suspended microhotplate sensing structure with 4 top-surface contact pads for conductance measurements on an active film. Top and cross-sectional end view schematics of the layers (not to scale) in this device are shown in the lower panel.

species is adsorbed on a film surface, a measurable change in its electrical conductance occurs. Catalytic metal surface-dispersed additives alter response sensitivity of the metal oxide sensing films of individual array elements enhancing their selectivity to the analytes of interest. The use of time-varied temperature control for the array elements also allows analyte-specific response signatures to be measured. These capabilities support a novel sensing approach that has excellent potential to discriminate and quantify individual chemical species in multi-component gas mixtures. These sensors provide sensitivity, selectivity, and capability for real-time monitoring. They can be manufactured at low cost with mechanical robustness.

One specific application of MEMS technology presently being examined is for detection and quantification of oxygenated hydrocarbons found in automotive exhaust emissions as well as a range of organic species (Semancik, 1998). Research investigations include the study of the mechanisms governing surface adsorption and desorption processes and use dynamic temperature-programmed sensing techniques that take advantage of the rapid thermal control capabilities of the micro-hotplate sensor structures. Signal processing algorithms capable of species detection and quantification are also being studied.

MEMS chemical sensors have been successfully used in the laboratory for biomedical instruments (Paula, 1996), in the automotive industry (Marshall, 1997) and for other applications (Wise, 1990). Other requirements have prompted research and development in gas sensing in a wide range of applications including, environmental monitoring (indoor building air), industrial process control, and worker safety.

Although there are no documented uses of MEMS technology for ocean instrumentation, current research suggests that MEMS may have application for measuring trace gases at the ocean air-sea interface. Accuracy and precision could be further improved if the individual sensors were polled electronically and simultaneously, and data were averaged. As individual MEMS became exhausted or fouled by sea spray, other sensors could come on line by using integrated electronic circuitry, allowing the system to continue measuring the components of interest. The inherently small size of MEMS would make this type of sensor package extremely sensitive and result in a fast response time. In addition, their low power requirements could make this type of array suitable for installation on a buoy or drifter using battery and solar power.

13.9 SUMMARY

Several societally important environmental problems involve the chemistry of the oceans. Here, we have highlighted the need for chemical sensors for monitoring, studying, and predicting variability induced by greenhouse gases, pollution, and hydrothermal vents. Effective assessment of the chemical state of important compounds in the world oceans requires the application of new and innovative approaches for measurements on time scales ranging to several decades, and over broad oceanic spatial extents. A major goal is to obtain chemical data on the same time and space scales as physics so that measurements and modeling have high fidelity to nature. If this goal is met, we would then gain a better understanding of chemistry, physics, biology, and how to ascertain changes in the environment to make informed predictions and policy decisions. Multiple observational platforms, each with particular sampling attributes, will need to be deployed strategically (e.g. key sites, nested arrays, etc.) Accurate, reliable and cost effective sensors and systems that remain operational for extended periods without human intervention, will allow for the determination of impending conditions that may lead to undesirable long-term trends in marine pollution and changes in the earth's climate. Large numbers of data will be just as important as accuracy and resolution, thus low cost, durability, etc. will be critical factors.

The development and application of new and improved sensor systems for real-time, *in situ* measurements of oceanic chemical compounds have provided cost-effective alternatives to current techniques for some chemical and physical environmental parameters. *In-situ* chemical sensors have the potential for contributing significantly to understanding the complex processes that occur in the world oceans. There has been much progress made in the development of off-the-shelf sensors suitable for use in freshwater systems. Application of these concepts and technologies to chemical analysis of seawater is relatively new, but offers great potential. State-of-the art technologies will provide new and innovative approaches to improved measurement and monitoring capabilities in chemical oceanography.

Here, we have described two emerging chemical sensor technologies: fiber optic chemical sensors and MEMS. The former is being utilized at sea presently while the latter remains primarily a laboratory method, but with potential for ocean application. Fiber optic sensor technology has shown great promise, and is expected to become available for routine oceanic use in the beginning of the next century. Further research is needed to improve sensor long-term stability, reproducibility and resistance to biofouling. MEMS have shown encouraging results for sensing physical parameters, but work is needed to fully realize their full potential for chemical sensing.

It is noteworthy that power, data storage, and telemetry are no longer major limiting factors for many of the emerging chemical sensor technologies. Thus, there is great impetus for accelerating the research and development for new chemical sensors with fast response times, low drift, high durability, and low cost that are capable of measuring increasing numbers of chemical species. In order to develop new chemical technologies, continued laboratory and field-testing will be necessary. Ocean testbed moorings are now available and effectively used for testing and intercomparing results obtained using emerging and relatively well-developed technologies. Once sensors are thoroughly tested, they can be used on other ocean platforms including AUVs.

Finally, data obtained from emerging chemical as well as biological, optical, and physical measurement sensors and systems will need to be optimally utilized in order to provide critical information for the environmental problems as illustrated here. Thus, technologists, observationalists, and modelers will need to form strong interdisciplinary collaborations at the international level.

References

Abbott, M.R., Brink, K.H., Booth, C.R., Blasco, D., Codispoti, L.A., Niiler, P.P. and Ramp, S.R. (1990). Observations of phytoplankton and nutrients from a Lagrangian drifter off northern California, *Journal of Geophysical Research*, 95, 9393–9409.

Bradshaw A.L. and Brewer, P.G. (1988). High precision measurements of alkalinity and total carbon dioxide in seawater by pontentiometric titration – 1. Presence of unknown protolyte(s)?. *Marine Chemistry*, 23, 69–86.

Burkholder, J.M. and Glasgow, H.B. Jr. (1997). Pfiesteria piscidida and other pfiesteria-like dinoflagellates: Behavior, impacts, and environmental controls, *Limnology and Oceanography*, 42, (5, part 2), 1052–1075.

Busby, R.F. (1976). *Manned Submersibles*, Office of the Oceanographer of the Navy.

Busby, R.F. (1990). Undersea Vehicles Directory – 1990–91, 4th. Edn, Busby Associates, Inc.

Busch, W.S. (1997). Government-industry alliance: new strategy for environmental research, *MTS Journal*, 27(2), 5–9.

Butler, J.H., Elkins, J.W., Brunson, K.B., Egan, R.B., Conway, T.J. and Hall, B.D. (1988). Trace gases in and over the west Pacific and east Indian oceans during the El Nino southern oscillation event of 1987, *NOAA Data Report ERL ARL-16*, 104.

Cantillo, A. (1995). Standard and Reference Materials for Environmental Science: *NOAA Technical Memorandum NOS ORCA 94* (Part 1 and 2).

Carter, K.L., Steward, R.G., Harvey, G.R. and Ortner, P.B. (1989). Marine humic and fulvic acids: Their effects on remote sensing of ocean chlorophyll, *Limnology and Oceanography*, 43(1) 68–81.

Chavez, F.P., Pennington, J.T., Herlein, R., Jannasch, H., Thurmond, G. and Friederich, G.F. (1997). Moorings and drifters for real-time interdisciplinary oceanography, *J. Atmos. Ocean. Tech.*, 14, 1199–121.

Chen, H., Wanninkhof, R.H., Feely, R. and Greeley, D. (1995). Measurements of fugacity of carbon dioxide in seawater: An evaluation of a method based on infrared analysis, *NOAA Technical Memorandum ERL AOML-85*, PB95–271029, 54.

Chipman, D.W., Marra, J. and Takahashi, T. (1993). Primary production at 47 °N and 20 °W in the North Atlantic Ocean: a comparison between the ^{14}C incubation method and mixed layer carbon budget observations, *Deep Sea Research*, II, 40, 151–169.

Chudyk, W.A., Carrabba, M.M. and Kenny, J.E. (1985). Remote detection of groundwater contaminates using far-ultraviolet laser-induced fluorescence, *Analytical Chemistry*, 57(7), 1237–1242.

Coale, K.H., Chin, C.S., Massoth, G.J., Johnson, K.S. and Baker, E.T. (1991). *In Situ* chemical mapping of dissolved iron and manganese in hydrothermal plumes, *Nature*, 352, 325–328.

Cooper, C.K., Fooffistall, G.Z., Hamilton, R.C. and Ebbesmeyer, C.C. (1997). Utilization of offshore oil platforms for meteorological and oceanographic measurements, *MTS Journal*, 27(2), 10–23.

Corliss, J.B., Dymond, J., Gordon, L.I., Edmond, J.M., von Herzen, R.P., Ballard, R.D., Green, K., Williams, D., Bainbridge, A., Crane, K. and an Andel, T.H. (1979). Submarine thermal springs on the Galapagos rift, *Science*, 203, (4385), 1073–1083.

Dawes, C.J. (1988). The Winkler procedure for measurement of dissolved oxygen. In *Experimental Phycology: A Laboratory Manual*, edited by Lobban, C.S., Chapman, D.J. and Kemer, B.P., pp. 78–82.

DeGrandpre, M.D., Hammar, T.R., Wallace, W.R. and Wirick, C.D. (1997). Simultaneous mooring-based measurements of seawater CO_2 and O_2 off Cape Hatteras, North Carolina, *Limnology and Oceanography*, 42, 21–28.

DeGrandpre, M.D., Hammar, T.R., Smith, S.O. and Sayles, F.L. (1995). *In situ* measurements of seawater pCO_2, *Limnology and Oceanography*, 40, 969–975.

DeGrandpre, M.D. (1993). Measurement of seawater pCO_2 using a renewable-reagent fiber optic sensor with colorimetric detection, *Analytical Chemistry*, 65, 331–337.

Dickey, T. (1991). The emergence of concurrent high-resolution physical and bio-optical measurements in the upper ocean, *Reviews of Geophysics*, 29, 383–413.

Dickey, T. (1993). Technology and related developments for interdisciplinary global studies, *Sea Technology*, 47–53.

Dickey, T. (1997). Emerging technologies in biological, chemical, optical, and physical sampling of the oceans, in *Proc. of the Int. Workshop on Ocean. Biol., and Chem. Data Management*, NOAA Tech. Rep. NESDIS 87, 115–124.

Dickey, T. and Jones, B.H. (1996). A decade of interdisciplinary process studies, *Ocean Optics*, 8, 254–259.

Dickey, T., Frye, D., Jannasch, H., Boyle, E., Manov, D., Sigurdson, D., McNeil, J., Stramska, M., Michaels, A., Nelson, N., Siegel, D., Chang, G., Wu, J. and Knap, A. (1998). Initial results from the Bermuda Testbed Mooring program, *Deep-Sea Research*, I, 45, 771–794.

Dickey, T., Plueddemann, A. and Weller, R. (1998b). Current and water property measurements in the coastal ocean. In *The Sea*, edited by Robeinson, A. and Brink, K., pp. 367–398.

Dickey, T.D. and Siegel, D.A. (eds.) (1993). Bio-optics in U.S. JGOFS, U.S. Joint Global Ocean Flux Study Planning Report 18, U.S. JGOFS Planning and Coordination Office, Woods Hole, MA, 180.

Dickey, T.D., Douglass, R.H., Manov, D., Bogucki, D., Walker, P.C. and Petrelis, P. (1993b). An experiment in duplex communication with a multi-variable moored system in coastal waters, *Journal of Atmospheric and Oceanic Technology*, 10, 637–644.

Dickey, T.D., Frye, D., Jannasch, H.W., Boyle, E. and Knap, A.H. (1997). Bermuda Sensor System Testbed, *Sea Technology*, 81–86.

Duportail, G. and Weinreb, A. (1983). Photochemical changes of fluorescent probes in membranes and their effect on the observed fluorescence anisotropy values, *Biochemica et Biophysica Acta*, 736(2) 171–177.

Eccles, L.A., Simon, S.J. and Klainer, S.M. (1987). *In situ* Monitoring at Superfund Sites with Fiber Optics, II. Plan for Development, *Technical Report*, EPA/600/X87/415, Las Vegas, NV.

Edmond, J.M., Von Damm, K.L, Mc Duff, R.E. and Measures, C.I. (1982). Chemistry of hot springs on east the east Pacific rise and their effuent dispersal, *Nature*, 297, 187–191.

Feely, R.A., Gammon, R.H., Taft, B.A., Pullen, P.E., Watterman, L.S., Conway, T.J., Gendron, J.F. and Wisegarver, D.P. (1987). Distribution of chemical tracers in the eastern equatorial Pacific during and after the 1982–1983 El Nino/Southern Oscillation event, *Journal of Geophysical Research*, 92(C6), 6545–6558.

Feely, R.A., Wanninkhof, R., Cosca, C.E, Murphy, P.P., Lamb, M.F. and Steckley, M.D. (1995). CO_2 distributions in the equatorial Pacific during the 1991–1992 ENSO event, *Deep-Sea Research, II*, 42(2–3), 365–386.

Feely, R.A., Wanninkhof, R., Cosca, C.E., McPhaden, M.J., Byrne, R.H., Millero, F.J., Chavez, F.P., Clayton, T., Campbell, D.M. and Murphy, P.P. (1994). The effect of tropical instability waves on CO_2 distributions along the equator in the eastern equatorial Pacific during the 1992 ENSO event, *Geophysical Research Letters*, 21(4), 277–280.

Feely, R.A., Wanninkhof, R., Goyet, C., Archer, D.E. and Takahashi, T. (1998). Variability of CO_2 distributions and sea-air fluxes in the central and eastern equatorial Pacific during the 1991–94 El Niño, *Deep-Sea Research*, II, 44, 1851–1867.

Foley, D., Dickey, T., McPhaden, M., Bidigare, R., Lewis, M., Barber, R., Lindley, S., Manov, D. and McNeil, J.D. (1998). Longwaves and primary productivity variations in the equatorial Pacific at 0°, 140°, *Deep-Sea Research*, II, 44, 1801–1826.

Fornari, D.J. and Perfit, M.R. (1997). Sampling and imaging requirements for deep submergence science, *MTS Journal*, 31, 77–79.

Francy, R.J., Tans, P.P., Allison, C.E., Enting, I.G., White, J.W.C. and Trolier, M. (1995). Changes in oceanic and terrestrial carbon uptake since 1982, *Nature*, 373, 326–330.

Friederich, G.E., Brewer, P.G., Herlien, R. and Chavez, F.P. (1995). Measurement of sea surface partial pressure of CO_2 from a moored bouy, *Deep-Sea Research*, 42, 1175–1186.

Gabriel, K.J. (1995). Engineering Microscopic Machines, *Scientific American*, 273(3), 150–153.

GCOS-14, 1995, Global climate observing system plan for the Global Climate Observing System, May 1995, WMO/TD, No. 681, 49.

Goswami, K., Kennedy, K.A., Dange, D.K., Klainer, S.M. and Tokar, J.M. (1989). A fiber optic chemical sensor for carbon dioxide dissolved in sea water, *SPIE proc.*, Boston, MA, 1172, 1172–1126.

Goswami, K., Klainer, S.M. and Tokar, J.M. (1988). Fiber optic chemical sensor for the measurement of partial pressure of oxygen, *SPIE proc.*, Boston, MA, 990–111.

Goyet, C., Walt, D.R. and Brewer, P.G. (1992). Development of a fiber optic sensor for measurement of pCO_2 in sea water: design criteria and sea trials, *Deep-Sea Research*, 39, 1015–1026.

Gross, G.M. (1976). *Middle Atlantic Continental Shelf and the New York Bight*. Special symposium, Volume 2. American Society of Limnology and Oceanography Inc. Lawrence, Kansas: Allen Press, Inc.

Hales, B., Burgess, L. and Emerson, S. (1997). An adsorbance-based fiber optic sensor for $CO_{2(aq.)}$ measurement in porewaters of sea floor sediments, *Marine Chemistry*, 59, 51–92.

Hammond, S., Baker, E., Bernard, E., Massoth, C., Fox, R. Feely, Embley, R., Rona, P. and Cannon, G. (1991). NOAA's VENTS program targets oceanic thermal effects, *Eos*, 72 (59), 561 and 565–566.

Harvey, G.R., Boran, D.A., Chesal, L.A. and Tokar, J.M. (1983). The structure of marine fulvic and humic acids, *Marine Chemistry*, 12, 119–132.

Heitzmann, H.A. (1985). Optical sensor with beads, *U.S. Patent No. 4,557,900*.

Hennet, J.C. and Whelan J. (1988). *In situ* chemical sensors for detecting and exploring ocean floor hydrohthermal vents, *Technical Report WHOI* 88–53, Woods Hole, MA.

Hitchcock, G.L., Lessard, E.J., Dorson, D., Fontaine, J. and Rossby, T. (1989). The IFF: the isopycnal float fluorometer, *Journal of Atmospheric and Oceanic Technology*, 6, 19–25.

Honjo, S., Krishfield, R. and Plueddemann, A. (1990). The Arctic Environmental Drifting Buoy (AEDB) Report of Field Operations and Results: August 1987–April 1988, *Woods Hole Oceanographic Institution Technical Report WHOI 90–02*, 128.

Hui, Li (1997). New robotic vessel extends deep-ocean exploration, *Science*, 278, 1705.

Hsu, L. (1987). Dye containing silicon polymer composition, *U.S. Patent No. 4,712,865*.

Jahnke, R. (1990). Ocean flux studies: a status report, *Reviews of Geophysics*, 28, 381–398.

Jannasch, H.W., Johnson, K.S. and Sakamoto, C.M. (1994). Submersible, osmotically pumped analyzers for continuous determination of nitrate *in situ*, *Analytical Chemistry*, 66, 3352–3361.

Johnson, K.S., Beehler, C.L. and Sakamoto-Arnold, C.M. (1986). A submersible flow analysis system, *Anal. Chim. Acta*, 179, 245–257.

Johnson, K.S. and Jannasch, H.W. (1994). Analytical chemistry under the sea surface: monitoring ocean chemistry *in situ*. Naval Research Reviews, 4–12.

Johnson, K.S., Coale, K.H. and Jannasch, H.W. (1992). Analytical chemistry in oceanography, *Analytical Chemistry*, 64(22), 1065–1075.

Kalvalitis, A.N. (1997). A critical capability-submersible sampling and sensing, *MTS Journal*, 31(3), 68–71.

Karl, T.R., Heim, R.R. Jr. and Quayle, R.G. (1991). The greenhouse effect in North America: If not now, when? *Science*, 251, 1058–1061.

Kawahara, F.K., Fiutem, R.A., Silvus, H.S., Newman, F.M. and Frazar, J.H., (1983). Development of a novel method for monitoring oils in water, *Analytica Chimica Acta*, 151, 316–327.

Keeling, C.D. and Whorf, T.P. (1994). Atmospheric CO_2 records from the sites in the SIO air sampling network, in Trends, 1993: A compendium of data on global change (eds. Boden, T.A., Kaiser, D.P., Sepanski, R.J. and Stoss, F.W.), Carbon Dioxide Information Analysis Center, pp. 16–24.

Klaassen, E.H., Reay, R J. Storment, C., Audy, J., Henry, P., Brokaw, A.P. and Kovacs, T.A. (1996). Micromachined thermally isolated circuits. In *Proc. Solid-State Sensor and Actuator Workshop*, Hilton Head, S.C., 127–131.

Klainer, S., Dandge, D., Goswami, K., Simon, S. and Eccles, L. (1988). EPA, Report #600/X-88/259.

Klainer, S.M. (1989). Fiber optic which is an inherent chemical sensor, *U.S. Patent No. 4,846,548*.

Klainer, S.M. (1992). Planar and other waveguide refactive index sensors using metal cladding, *U.S. Patent No. 5,165,005*.

Landrum, L.L., Gammon, R.H., Feely, R.A., Murphy, P.P., Kelly, K.C., Cosca, C.E. and Weiss, R.F. (1996). North Pacific Ocean CO_2 disequilibrum for spring through summer, 1985–1989, *Journal of Geophysical research*, 101, 28,539–28,555.

Langdon, C. (1984). Dissolved oxygen monitoring system using a pulsed electrode: design, performance, and evaluation, *Deep-Sea Research*, 31, 1357–1367.

Lerner, E.J. (1997). Fiberoptic sensors monitor environmental conditions, *Laser Focus World*, 107–112.

Liss, P and Merlivat, L. (1986). *The Role of Air-Sea Exchange in Geochemical Cycling*, Buat-Menard, Ed. *Adv. Sci. Inst. Ser.* 185, Reidel, Hingham.

Liss, P.S. (1983). Gas transfer: Experiments and geochemical implications, In *Air-Sea Exchange of Gases and Particles*, edited by Liss, P.S. and Slinn, W.G.N., pp. 241–298. D. Reidel Publishing Company.

Mackerieran, G.B. (1987). Dissolved oxygen in the Chesapeake Bay. A Maryland Sea Grant Publication, College Park, Maryland.

Maier-Reimer, E. and Hasselmann, K. (1987). Transport and storage of CO_2 in the ocean – an inorganic oceanic-circulation carbon cycle model, *Climate Dynamics*, 2, 63–90.

Marshall, (1997). MEMS technologies – on the brink of maturity, *Research & Development*, 39(8), p. 20.

Massoth, G.J., Baker, E.T., Feely, R.A., Butterfield, D.A., Embley, R.E., Lupton, J.E., Thompson, R.E. and Cannon, G.A. (1995). Observations of manganese and iron at

Co-Axial seafloor eruption site, Juan de Fuca Ridge, *Geophysical Research Letters*, 22(2), 151–154.

Merlivat, L. and Brault, P. (1995). CARIOCA Buoy: Carbon dioxide monitor, *Sea Technology*, 36(10), 23–30.

Millero, F.J., Byrne, R.H., Wanninkhof, R., Feely, R., Clayton, T., Murphy, P. and Lamb, M F. (1993). The internal consistancy of CO_2 measurements in the equatorial Pacific, *Marine Chemistry*, 44, 269–280.

Munkholm, G., Walt, D., Milanovich, F. and Klainer, S.M. (1986). Polymer modification of fiber optic chemical sensors as a method of enhancing fluorescence signal for pH measurement, *Analytical Chemistry*, 57, 7, 1427.

NOAA Strategic Plan, A Vision for 2005: *NOAA Internal Report*, U.S. GPO 1996–411222/50368.

NRC Report (1993). *Applications of analytical chemistry to oceanic carbon cycle studies*, ISBN 0–309–0428–8: Washington DC, National Academy Press.

Paula, G. (1996). MEMS sensors branch out. *Mechanical Engineering*, 118(10), 64.

Payne, R.S., Sherman, S., Lewis, S. and Howe, R.T. (1995). Surface micromachining: from vision to reality to vision, *Proc. IEEE International Solid-State Circuits Conference*, Paper TA 9.6, 164–165.

Quay, P.D., Tilbrook, B. and Wong, C.S. (1992). Oceanic uptake of fossil fuel CO_2: carbon – 13 evidence, *Science*, 256, 74–79.

Rhines, T.D. and Arnold, M.A. (1987). Simplex optimization of a fiber-optic ammonia sensor based on multiple indicators, *Analytical Chemistry*, 60, 76–81.

RIDGE Program Announcement (1995). NSF 95–132.

Rodhe, H. (1990). A comparison of the contribution of various gases to the greenhouse effect, *Science*, 248, 1217–1219.

Saager, P.M., Baar, H.J.W., de Jong, J.T.M., Nolting, R.F. and Schijf, J. (1997). Hydrography and local sources of dissolved trace metals Mn, Ni, Cu, and Cd in the northeast Atlantic Ocean, *Marine Chemistry*, 57, 195–216.

Saini, D.P. (1994). Chip level waveguide sensor, *U.S. Patent No. 5,439,647*.

Saliot, A. and Martin, J.M. (1997). Fourth international symposium on model esturaies: The biogeochemistry of organic compounds and trace metals in macrotidal estuaries, *Marine Chemistry*, 58, 1–147.

Sarmiento, J.L. and Le Quere, C. (1996). Oceanic carbon dioxide uptake in a model of century-scale global warming, *Science*, 274, 1346–1350.

Schantz, M.M., Koster, B.J., Oakley, L.M., Schiller, S.B. and Wise, S.A. (1995). Certification of polychlorinated biphenyl congeners and chlorinated pesticides in a whale blubber standard reference material, *Analytical Chemistry*, 67, 901–910.

Seitz, W.R., Hassen, K., Conway, V. and Pan, S. (1993). Structure effects on mechanical and swelling properties of amine modified polystyrene beads for pH sensing. In proc. of Symposium on Chemical Sensors, *Electrochem. Soc.*, 74–80.

Seitz, W.R. (1984). Chemical sensors based on fiber optics, *Analytical Chemistry*, 56(1), 16A–34A.

Semancik, S. and Cavicchi, R.E. (1998). Kinetically-controlled chemical sensing using micro machined structures, *Accounts of Chemical Research*, 31, 279–287.

Semancik, S., Cavicchi, R.E., Gaitian, M. and Suehl, J.S. (1994). Temperature-controlled, micromachined arrays for chemical sensor fabrication and operation, *U.S. Patent No. 5,345,213*, Washington, DC.

Shakhsher, Z., Seitz, R.W. and Legg, K.D. (1994). Single fiber optic pH sensor based on changes in reflection accompanying polymer swelling, *Analytical Chemistry*, 66, 1731–1735.

Suehle, J.S., Cavicchi, R.E., Kreider, K.G., Gaitan, M. and Semancik, S. (1993). *IEEE Elec. Device Lett.*, 14, 118–120.

Swanson, R.L. and Sindermann, C.J. (1979). Oxygen depletion and associated benthic mortalities in New York Bight, 1976. *NOAA Professional Paper* 11, 345.

Tai, H., Tanaka, H. and Yoshino, T. (1987). Fiber-optic evanescent-wave methane-gas sensor using optical absorption for the 3.392–μm line of a He-Ne laser, *Optics Letters*, 12(6), 437–439.

Takahashi, T., Feely, R.A., Weiss, R.F., Wanninkhof, R.H., Chipman, D.W., Sutherland, S.C. and Takahashi, T.T. (1997). Global air-sea flux of CO_2: an estimate based on measurements of sea-air pCO_2 difference, *Proc. Natl. Acad. Sci. USA*, 94, 8292–8299.

Tans, P.P., Fung, I.Y. and Takahashi, T. (1990). Observational constraints on the global atmospheric CO_2 Budget, *Science*, 247, 1431–1438.

Tengberg, A., *et al.* (1998). Benthic chamber and profiling landers in oceanography – A review of design, technical solutions and functioning, *Prog. Oceanog.*, 35, 253–294.

Tokar, J.M., Harvey, G.R. and Chesal, L.A. (1981). A gaslift system for large volume water sampling, *Deep-Sea Research*, 28A(11), 1395–1399.

Tokar, J.M., Harvey, G.R. and Chesal, L.A. (1983). A portable system for extraction of organics from thousand liter volumes of seawater, *NOAA Technical Memorandum ERL AOML-53*.

Tokar, J.M., Woodward, W.E. and Goswami, K. (1990a). Fiber optic chemical sensors: exploring the light fantastic, *Sea Technology*, 31(4), 45–49.

Tokar, J.M., Klainer, S.M. and Goswami, K. (1991). Oceanic measurements: fiber optic chemical sensor (FOCS) brings a new dimension, *OTC Proc.*, 6546, 416–422.

Tokar, J.M., Pugh, L. and Goswami, K. (1990b). The use of fiber optic sensors for *in-situ* chemical measurements in the ocean, *ASTM Standard Technical Publication*, 1087, 65–75.

Tokar, J.M., Woodward, W.E. and Goswami, K. (1989). The measurement of oceanic biogeochemical compounds using fiber optic chemical sensors, *OCEANS 89 Proc.*, 5.

Trahey, N.M., editor (1995). *NIST standard reference material catalog, 1995–96 edition*, NIST Spec. Publ. 260.

U.S. GOOS Report (1992). First steps toward a U.S. GOOS, *Report of a Workshop on Priorities for U.S. Contributions to a Global Ocean Observing System*, JOI, Woods Hole, MA, October 14–16, 1992, 48.

U.S. GOOS Report (1997). Joint GCOS GOOS WCRP ocean observations panel for climate (OOPC):GOOS Report No. 41, Baltimore, MD, UNESCO.

Wanninkhof, R., Feely, R.A., Chen, H., Cosca, C. and Murphy, P.P. (1996). Surface water fCO_2 in the eastern equatorial Pacific during the 1992–1993 El Niño, *Journal of Geophysical Research*, 101(C7), 16, 333–16,343.

Wanninkhof, R., Thoning, K. (1993). Measurement of fugacity of CO_2 in surface water using continuous and discrete sampling methods, *Marine Chemistry*, 44, 189–204.

Weiss, R.F. (1981). Determinations of carbon dioxide and methane by dual catalyst flame ionization chromatography and nitrous oxide by electron capture chromatography, *Journal of Chromatographic Science*, 19, 611–616.

Wise, K.D. (1991). The coming opportunities in microsensor systems. Digest technical papers, *International Conference on Solid-State Sensors and Actuators (Transducers 91)*, 2–7.

Wise, K.D. (1990). Integrated microelectromechanical systems: A perspective on MEMS in the 90's. *Proc. IEEE Micro Electro Mechanical Systems Workshop*, 33–38.

Wu, J. and Boyle, E.A. (1997). Lead in the western North Atlantic Ocean: Completed response to leaded gasoline phaseout, *Geochimica et Cosmochimia Acta*, 61(15), 3279–3283.

Yi, Z., Zhuang, G., Brown, P.R. and Duce, R.A. (1992). High-performance liquid chromatographic method for the determination of ultratrace amounts of iron(II) in aerosols, rainwater, and seawater, *Anal. Chem.*, 64, 2826–2830.

Zhang, L., Langmuir, M.E., Bai, M. and Seitz, R.W. (1997). A sensor for pH based on an optical reflective device coupled to swelling of an aminated polystyrene membrane, *Talanta*, 44, 1691–1698.

INDEX